Developments in Mathematics

VOLUME 29

For further volumes:
www.springer.com/series/5834

Zhitao Zhang

Variational, Topological, and Partial Order Methods with Their Applications

 Springer

Zhitao Zhang
Academy of Mathematics & Systems Science
The Chinese Academy of Sciences
Beijing, P.R. China

ISSN 1389-2177   Developments in Mathematics
ISBN 978-3-642-30708-9   ISBN 978-3-642-30709-6 (eBook)
DOI 10.1007/978-3-642-30709-6
Springer Heidelberg New York Dordrecht London

Library of Congress Control Number: 2012949194

Mathematics Subject Classification: 06A06, 34B18, 34B24, 35A15, 35B25, 35B32, 35B33, 35B38, 35B40, 35B45, 35B53, 35B65, 35J20, 35J25, 35J60, 35J70, 35J92, 35J96, 35R35, 45G10, 45G15, 47H10, 47H11, 58E20, 92D25

Preface

Nonlinear functional analysis is an important branch of contemporary mathematics; it has grown from geometry, fluid and elastic mechanics, physics, chemistry, biology, control theory and economics, etc. It is related to many areas of mathematics: topology, ordinary differential equations, partial differential equations, groups, dynamical systems, differential geometry, measure theory, etc.

We mainly present our new results on the three fundamental methods in nonlinear functional analysis: Variational, Topological and Partial Order Methods with their Applications. They have been used extensively to solve questions of the existence of solutions for elliptic equations, wave equations, Schrödinger equations, Hamiltonian systems, etc. Also they have been used to study the existence of multiple solutions and the properties of solutions.

Hilbert posed his famous 23 problems on the occasion of his speech at the centennial assembly of the International Congress 1900 in Paris. Three of these were related to the calculus of variations. Included are minimization methods, minimax methods, Morse theory, category, Ljusternik–Schnirelmann theory, etc. in the calculus of variations. We should mention that Ambrosetti and Rabinowitz's work [11] in the 1970s is the beginning of the minimax method, making it possible for people to deal with functionals that are unbounded from below, which come from the study of nonlinear elliptic equations, Hamiltonian systems, geometry, and mathematical physics. In the 1930s, Morse developed a theory which set up the relationship between critical points of a non-degenerate function and the topology of the underlying compact manifold. In the 1960s Palais [149] and Smale [164] et al. extended Morse theory to infinite-dimensional manifolds by using the Palais–Smale condition.

Topological methods and partial order methods are basic and important tools in nonlinear functional analysis too. The Brouwer degree is a powerful tool in algebraic topology; the Leray–Schauder degree is an extension of the Brouwer degree from finite-dimensional spaces to infinite-dimensional Banach spaces, which has been introduced by Leray and Schauder in the study of nonlinear partial differential equations in the 1930s. Rabinowitz's global bifurcation theorem is based on the computation of the Leray–Schauder degree. In many problems that arise in population biology, economics, and the study of infectious diseases, we need to discuss the existence of nonnegative solutions with some desired qualitative properties, so cones are used to develop partial order methods and fixed point index theory. Then one gets fixed point theorems and applications to many kinds of differential equation, etc.

In Chap. 1, we present preliminaries: some basic concepts, and useful famous theorems and results so that the reader may easily find information if need may be.

In Chap. 2, we introduce three kinds of operator: increasing operators, decreasing operators, and mixed monotone operators. Some fixed point theorems and applications to integral equations and differential equations are included. One equivalent condition of the normal cone is given.

In Chap. 3, we present the minimax methods including the Mountain Pass Theorem, linking methods, local linking methods, and critical groups; next, we treat some applications to elliptic boundary value problems.

In Chap. 4, we use bifurcation and critical point theory together to study the structure of the solutions of elliptic equations; also we have results on three sign-changing solutions.

In Chap. 5, we consider the boundary value problems for a class of Monge–Ampère equations. First we prove that any solution on the ball is radially symmetric by the moving plane argument. Then we show that there exists a critical radius such that, if the radius of a ball is smaller than this critical value, then there exists a solution, and vice versa. Using a comparison between domains we prove that this phenomenon occurs for every domain. By using the Lyapunov–Schmidt reduction method we get the local structure of the solutions near a degenerate point; by Leray–Schauder degree theory, a priori estimates, and using bifurcation theory we get the global structure.

In Chap. 6, on superlinear systems of Hammerstein integral equations and applications, we use the Leray–Schauder degree to obtain new results on the existence of solutions, and apply them to two-point boundary problems of systems of equations. We also are concerned with the existence of (component-wise) positive solutions for a semilinear elliptic system, where the nonlinear term is superlinear in one equation and sublinear in the other equation. By constructing a cone $K_1 \times K_2$, which is the Cartesian product of two cones in the space $C(\overline{\Omega})$, and computing the fixed point index in $K_1 \times K_2$, we establish the existence of positive solutions for the system.

In Chap. 7, we show some results on the Dancer–Fučik spectrum for bounded domains. We are concerned with the Fučik point spectrum for Schrödinger operators, $-\Delta + V$, in $L^2(\mathbb{R}^N)$ for certain types of potential, $V : \mathbb{R}^N \to \mathbb{R}$. We use the Dancer–Fučik spectrum to asymptotically linear elliptic problems to get one-sign solutions.

In Chap. 8, we introduce some results on sign-changing solutions of elliptic and p-Laplacian, including using Nehri manifold, invariant sets of descent flows, Morse theory, etc.

In Chap. 9, we show that if $u_0 \in W_0^{1,p}(\Omega)$ is a local minimizer of J in the C^1-topology, it is still a local minimizer of the functional J in $W_0^{1,p}(\Omega)$. This extends the famous results of Brezis–Nirenberg to $p > 2$. We thus obtain multiple solutions and structures of solutions for p-Laplacian equations. Finally, we also show uniqueness results of various kinds.

In Chap. 10, we obtain nontrivial solutions of a class of nonlocal quasilinear elliptic boundary value problems using the Yang index and critical groups, and we obtain sign-changing solutions of a class of nonlocal quasilinear elliptic boundary value problems using variational methods and invariant sets of descent flows. We also show a uniqueness result.

In Chap. 11, we study free boundary problems, Schrödinger systems from Bose–Einstein condensates, and competing systems with many species. We prove the existence and uniqueness result of the Dirichlet boundary value problem of elliptic competing systems. We show that, for the singular limit, species are spatially segregated; they satisfy a remarkable system of differential inequalities as $\kappa \to +\infty$. We also introduce optimal partition problems related to eigenvalues and nonlinear eigenvalues. Finally, some recent new results on Schrödinger systems from Bose–Einstein condensates are presented.

In preparing this manuscript I have received help and encouragement from several professors and from my students. I wish to thank Professor Shujie Li for his kind suggestions. Special thanks go to my students; to Prof. Xiyou Cheng, Dr. Kelei Wang, Dr. Yimin Sun for useful corrections, and to Dr. Yimin Sun and Liming Sun for wonderful typesetting of parts of Chaps. 1, 2, 3, and 11 of this manuscript.

I dedicate this book to my father Deren Zhang, my wife Jimin Fang and my son Fan Zhang.

Beijing, China   Zhitao Zhang

Contents

Chapter 1
Preliminaries

1.1 Sobolev Spaces and Embedding Theorems

Let Ω denote a bounded domain in $\mathbb{R}^n$ ($n \geq 1$). For $p \geq 1$ we let $L^p(\Omega)$ denote the class Banach space consisting of measurable functions on Ω that are p-integrable. The norm of Banach space $L^p(\Omega)$ is defined by

$$\|u\|_{p;\Omega} = \|u\|_{L^p(\Omega)} = \left(\int_\Omega |u|^p \, dx \right)^{1/p}. \tag{1.1}$$

Hölder's inequality: For real numbers p, q satisfying $\frac{1}{p} + \frac{1}{q} = 1$,

$$\int_\Omega uv \, dx \leq \|u\|_p \|v\|_q \tag{1.2}$$

for functions $u \in L^p(\Omega)$, $v \in L^q(\Omega)$. It is a consequence of Young's inequality:

$$ab \leq \frac{a^p}{p} + \frac{b^q}{q}, \quad \forall a \geq 0, \ b \geq 0.$$

As $p = q = 2$, it is the Schwarz inequality.

Generalization of Hölder's inequality:

Let $u_i \in L^{p_i}(\Omega)$, $i = 1, 2, \ldots, m$, $\frac{1}{p_1} + \frac{1}{p_2} + \cdots + \frac{1}{p_m} = 1$,

$$\int_\Omega u_1 \cdots u_m \, dx \leq \|u_1\|_{p_1} \cdots \|u_m\|_{p_m}. \tag{1.3}$$

Minkowski inequality:

$$\|f + g\|_{L^p} \leq \|f\|_{L^p} + \|g\|_{L^p}, \quad \forall f, g \in L^p(\Omega). \tag{1.4}$$

As $p = 2$, it is Cauchy inequality.

Z. Zhang, *Variational, Topological, and Partial Order Methods with Their Applications*,
Developments in Mathematics 29,
DOI 10.1007/978-3-642-30709-6_1, © Springer-Verlag Berlin Heidelberg 2013

Definition 1.1.1 (Weak derivatives) Let u be locally integrable in Ω and α any multi-index. Then a locally integrable function v is called the αth weak derivative of u if it satisfies

$$\int_\Omega \varphi v \, dx = (-1)^{|\alpha|} \int_\Omega u D^\alpha \varphi \, dx \quad \text{for all } \varphi \in C_0^{|\alpha|}(\Omega).$$

We write $v = D^\alpha u$, and v is uniquely determined up to sets of measure zero. For a non-negative integer vector $\alpha = (\alpha_1, \ldots, \alpha_n)$, we denote

$$D^\alpha = \frac{D^{|\alpha|}}{\partial x_1^{\alpha_1} \cdots \partial x_n^{\alpha_n}}$$

the differential operator, with $|\alpha| = \alpha_1 + \cdots + \alpha_n$.

A function is called weakly differentiable if all its weak derivatives of first order exist and k times weakly differentiable if all its weak derivatives exist for orders up to and including k. We denote the linear space of k times weakly differentiable functions by $W^k(\Omega)$. Clearly $C^k(\Omega) \subset W^k(\Omega)$. For $p \geq 1$ and k a non-negative integer, let

$$W^{k,p}(\Omega) = \left\{ u \in W^k(\Omega); D^\alpha u \in L^p(\Omega) \text{ for all } |\alpha| \leq k \right\}, \tag{1.5}$$

with a norm

$$\|u\|_{W^{k,p}(\Omega)} = \left(\int_\Omega \sum_{|\alpha| \leq k} |D^\alpha u|^p \, dx \right)^{1/p}.$$

Then $W^{k,p}(\Omega)$ is a Banach space. We also have an equivalent norm

$$\|u\|_{W^{k,p}(\Omega)} = \sum_{|\alpha| \leq k} \|D^\alpha u\|_p.$$

$W_0^{k,p}(\Omega)$ is another Banach space by taking the closure of $C_0^k(\Omega)$ in $W^{k,p}(\Omega)$. $W^{k,p}(\Omega)$, $W_0^{k,p}(\Omega)$ are separable for $1 \leq p < \infty$, and reflexive for $1 < p < \infty$.

As $p = 2$, $W^{k,2}(\Omega)$, $W_0^{k,2}(\Omega)$ written as $H^k(\Omega)$, $H_0^k(\Omega)$ are Hilbert spaces under the scalar product

$$(u, v)_k = \int_\Omega \sum_{|\alpha| \leq k} D^\alpha u D^\alpha v \, dx.$$

$W_{\text{loc}}^{k,p}(\Omega)$ are local spaces to be defined to consist of functions belonging to $W^{k,p}(\Omega')$ for all $\Omega' \Subset \Omega$ (i.e., Ω' has compact closure in Ω).

Definition 1.1.2 Assume E_1, E_2 are two normed linear spaces, we call E_1 embedded in E_2, if:

(1) E_1 is a subspace of E_2,

(2) There exists an identity operator $I : E_1 \to E_2$ such that $I(u) = u$, and

$$\left\| I(u) \right\|_{E_2} \le K \|u\|_{E_1}.$$

If $1 \le p_1 \le p_2 \le \infty$, $u \in L^{p_2}(\Omega)$ then $u \in L^{p_1}(\Omega)$. Also we have

$$L^{p_2}(\Omega) \hookrightarrow L^{p_1}(\Omega).$$

Theorem 1.1.1 *The space $C^\infty(\Omega) \cap W^{k,p}(\Omega)$ is dense in $W^{k,p}(\Omega)$.*

Theorem 1.1.2 (Sobolev embedding theorems)

$$W_0^{1,p}(\Omega) \longrightarrow L^\varphi(\Omega), \qquad
\begin{cases}
\nearrow L^{np/(n-p)}(\Omega), & p < n, \\[2mm]
\quad\ \ \varphi = \exp\!\big(|t|^{n/(n-1)}\big) - 1, & p = n, \\[2mm]
\searrow C^\lambda(\bar{\Omega}), \qquad \lambda = 1 - \dfrac{n}{p}, & p > n,
\end{cases}$$

where $L^\varphi(\Omega)$ denotes the Orlicz space.

The Poincaré inequality: For $u \in W_0^{1,p}(\Omega)$, $1 \le p < \infty$

$$\|u\|_p \le \left(\frac{1}{\omega_n}|\Omega|\right)^{1/n} \|Du\|_p \quad \big(\omega_n = \text{volume of unit ball in } \mathbb{R}^n\big).$$

After extension to the spaces $W_0^{k,p}(\Omega)$, we have

$$W_0^{k,p}(\Omega) \qquad
\begin{cases}
\nearrow L^{np/(n-kp)}(\Omega), & kp < n, \\[3mm]
\searrow C^m(\bar{\Omega}), & 0 \le m < k - \dfrac{n}{p}.
\end{cases}$$

For $W^{k,p}(\Omega)$, if Ω satisfies a uniform interior cone condition (i.e., there exists a fixed cone K_Ω such that each $x \in \Omega$ is the vertex of a cone $K_\Omega(x) \subset \bar{\Omega}$ and congruent to K_Ω), then there is an embedding

$$W^{k,p}(\Omega) \qquad
\begin{cases}
\nearrow L^{np/(n-kp)}(\Omega), & kp < n, \\[3mm]
\searrow C_B^m(\Omega), & 0 \le m < k - \dfrac{n}{p},
\end{cases}$$

where $C_B^m(\Omega) = \{u \in C^m(\Omega) \mid D^\alpha u \in L^\infty(\Omega) \text{ for } |\alpha| \le m\}$.

Theorem 1.1.3 (Compactly embedded theorems) *The spaces $W_0^{1,p}(\Omega)$ are compactly embedded* (i) *in the spaces $L^q(\Omega)$ for any $q < np/(n-p)$, if $p < n$, and* (ii) *in $C^0(\bar{\Omega})$, if $p > n$.*

An extension of the above theorem show that the embeddings

$$\nearrow L^{np/(n-p)}(\Omega), \quad \text{for } kp < n,\ q < \frac{np}{n-kp},$$

$$W_0^{k,p}(\Omega)$$

$$\searrow C^m(\bar{\Omega}), \qquad\qquad \text{for } 0 \le m < k - \frac{n}{p}$$

are compact.

Next define the space

$$H^1(\mathbb{R}^n) := \left\{ u \in L^2(\mathbb{R}^n) : \nabla u \in L^2(\mathbb{R}^n) \right\}$$

with the inner product

$$(u, v)_1 := \int_{\mathbb{R}^n} [\nabla u \cdot \nabla v + uv]$$

and the corresponding norm

$$\|u\|_1 := \left(\int_{\mathbb{R}^n} \left[|\nabla u|^2 + u^2 \right] \right)^{1/2}.$$

It is a Hilbert space.

Let $\mathcal{D}(\Omega) := \{ u \in C^\infty(\Omega) : \operatorname{supp} u$ is a compact subset of $\Omega \}$.

Let Ω be an open subset of $\mathbb{R}^n$, the space $H_0^1(\Omega)$ is the closure of $\mathcal{D}(\Omega)$ in $H^1(\mathbb{R}^n)$.

Let $n \ge 3$ and $2^* := 2n/(n-2)$. The space

$$\mathcal{D}^{1,2}(\mathbb{R}^n) := \left\{ u \in L^{2^*}(\mathbb{R}^n) : \nabla u \in L^2(\mathbb{R}^n) \right\}$$

with the inner product $\int_{\mathbb{R}^n} \nabla u \cdot \nabla v$ and the corresponding norm $(\int_{\mathbb{R}^n} |\nabla u|^2)^{1/2}$ is a Hilbert space. The space $\mathcal{D}_0^{1,2}(\Omega)$ is the closure of $\mathcal{D}(\Omega)$ in $\mathcal{D}^{1,2}(\mathbb{R}^n)$. We denote $2^* = \infty$ when $n = 1, 2$.

Theorem 1.1.4 (Sobolev embedding theorem) *The following embeddings are continuous*:

$$H^1(\mathbb{R}^n) \hookrightarrow L^p(\mathbb{R}^n), \qquad 2 \le p < \infty,\ n = 1, 2,$$

$$H^1(\mathbb{R}^n) \hookrightarrow L^p(\mathbb{R}^n), \qquad 2 \le p \le 2^*,\ n \ge 3,$$

$$\mathcal{D}^{1,2}(\mathbb{R}^n) \hookrightarrow L^{2^*}(\mathbb{R}^n), \qquad n \ge 3.$$

In particular, the Sobolev inequality holds:

$$S := \inf_{u \in \mathcal{D}^{1,2}(\mathbb{R}^n),\, |u|_{2^*}=1} |\nabla u|_2^2 > 0. \tag{1.6}$$

Then it is clear that $H_0^1(\Omega) \subset \mathcal{D}_0^{1,2}(\Omega)$. If $|\Omega| < \infty$, Poincaré inequality implies that $H_0^1(\Omega) = \mathcal{D}_0^{1,2}(\Omega)$.

The instanton:

$$U(x) := \frac{[n(n-2)]^{(n-2)/4}}{[1+|x|^2]^{(n-2)/2}} \tag{1.7}$$

is a minimizer for $S, n \geq 3$ (Aubin and Talenti, see [193]). For every open subset Ω of $\mathbb{R}^n$,

$$S(\Omega) := \inf_{u \in \mathcal{D}^{1,2}(\Omega),|u|_{2*}=1} |\nabla u|_2^2 = S, \tag{1.8}$$

and $S(\Omega)$ is never achieved except when $\Omega = \mathbb{R}^n$.

By Theorem 4.7.8 of [50] and [39], $U(x)$ is a minimizer for S (1.6) iff $U(x)$ has the form

$$U(x) := \frac{[n(n-2)\theta]^{(n-2)/4}}{[\theta^2 + |x-y|^2]^{(n-2)/2}}, \quad \forall \theta > 0, \ \forall y \in \mathbb{R}^n. \tag{1.9}$$

Theorem 1.1.5 (Strauss [50]) *Let $H_r^1(\mathbb{R}^n)$ be the subspace of $H^1(\mathbb{R}^n)$ consisting of radial symmetric functions. The embedding $H_r^1(\mathbb{R}^n) \hookrightarrow L^p(\mathbb{R}^n)$ is compact as $2 < p < 2^*, n \geq 2$.*

Remark 1.1.1 About Sobolev spaces and embedding theorems above, please see [95, 193] etc.

1.2 Critical Point

Definition 1.2.1 Let $J : U \to \mathbb{R}$ where U is an open subset of a Banach space E. The functional J has a Gateaux derivative $f \in E^*$ at $u \in U$, if for every $h \in E$,

$$\lim_{t \to 0} \frac{1}{t} \left| J(u+th) - J(u) - \langle f, th \rangle \right| = 0. \tag{1.10}$$

The functional J has a Fréchet derivative $f \in E^*$ at $u \in U$, if

$$\lim_{h \to 0} \frac{1}{\|h\|} \left| J(u+h) - J(u) - \langle f, h \rangle \right| = 0. \tag{1.11}$$

The functional J belongs to $C^1(U, \mathbb{R})$ if the Fréchet derivative of J exists and is continuous on U.

Any Fréchet derivative is a Gateaux derivative. Using the mean value theorem, it is easy to know that if J has a continuous Gateaux derivative on U, then $J \in C^1(U, \mathbb{R})$.

Suppose J is a Fréchet differentiable functional on a Banach space E with normed dual E^* and duality pairing $\langle \cdot, \cdot \rangle : E \times E^* \to \mathbb{R}$, and let $DJ : E \to E^*$ denote the Fréchet-derivative of J. Then the directional (Gateaux-) derivative of J

at u in direction v is given by

$$\left.\frac{d}{dt}J(u+tv)\right|_{t=0} = \langle v, DJ(u)\rangle = DJ(u)v.$$

For such J, we call $u \in E$ a critical point if $J'(u) := DJ(u) = 0$; otherwise u is called a regular point. A number $\alpha \in \mathbb{R}$ is a critical value of J if there exists a critical point u of J with $J(u) = \alpha$. Otherwise, α is called regular.

Let $C^1(E, \mathbb{R})$ denote the set of functionals that are Fréchet differentiable and whose Fréchet derivatives are continuous on E.

Definition 1.2.2 For $J \in C^1(E, \mathbb{R})$, we say J satisfies the *Palais–Smale condition* (henceforth denoted by (PS) condition) if any sequence $\{u_m\} \subset E$ for which $J(u_m)$ is bounded and $J'(u_m) \to 0$ as $m \to \infty$ possesses a convergent subsequence.

Definition 1.2.3 For $J \in C^1(E, \mathbb{R})$, we say J satisfies the $(PS)_c$ condition if any sequence $\{u_m\} \subset E$ for which $J(u_m) \to c$ and $J'(u_m) \to 0$ as $m \to \infty$ possesses a convergent subsequence.

It is clear that if J satisfies the $(PS)_c$ condition, for $\forall c$, then J satisfies the (PS) condition.

The (PS) condition is a kind of compact condition. Indeed observe that the (PS) condition implies that $K_c \equiv \{u \in E | J(u) = c, J'(u) = 0\}$, i.e., the set of critical points having critical value c, is compact for any $c \in R$.

Theorem 1.2.1 (Ekeland variational principle [50]) *Let (X, d) be a complete metric space, and let $f : X \to \mathbb{R} \cup \{+\infty\}$, but $f \not\equiv +\infty$. If f is bounded from below and lower semi-continuous (l.s.c., $\forall \lambda \in \mathbb{R}$, the level set $f_\lambda = \{x \in X | f(x) \le \lambda\}$ is closed), and if $\exists \varepsilon > 0$, $\exists x_\varepsilon \in X$ satisfying $f(x_\varepsilon) < \inf_X f + \varepsilon$. Then $\exists y_\varepsilon \in X$ such that*

1. $f(y_\varepsilon) \le f(x_\varepsilon)$,
2. $d(x_\varepsilon, y_\varepsilon) \le 1$,
3. $f(x) > f(y_\varepsilon) - \varepsilon d(y_\varepsilon, x)$, $\forall x \ne y_\varepsilon$.

Theorem 1.2.2 (Pohozaev identity, 1965) *For the solution of*

$$-\Delta u = f(u), \quad u \in H_0^1(\Omega), \tag{1.12}$$

where $f \in C^1(\mathbb{R}, \mathbb{R})$ and Ω is a smooth bounded domain of $\mathbb{R}^n$, $n \ge 3$. Let $F(u) = \int_0^u f(s)\,ds$.

Let $u \in H_{\text{loc}}^2(\bar{\Omega})$ be a solution of (1.12) such that $F(u) \in L^1(\Omega)$. Then u satisfies the following:

$$\frac{1}{2}\int_{\partial\Omega} |\nabla u|^2 \sigma \cdot v\,d\sigma = n \int_\Omega F(u)\,dx - \frac{n-2}{2}\int_\Omega |\nabla u|^2\,dx,$$

where v denotes the unit outward normal to $\partial\Omega$. (For the proof see [168, 193].)

For P.L. Lions' Concentration-Compactness Principle   On the basis of this principle, for many constrained minimization problems it is possible to state necessary and sufficient conditions for the convergence of all minimizing sequences satisfying the given constraint.

Theorem 1.2.3 (Concentration-Compactness Principle, see [135, 136, 168]) *Suppose that μ_m is a sequence of probability measures on $\mathbb{R}^n : \mu_m \geq 0$, $\int_{\mathbb{R}^n} \mu_m = 1$. Then there is a subsequence (μ_m) satisfying one of the following three possibilities:*

(1) *(Compactness)* $\exists \{x_m\} \subset \mathbb{R}^n$ *such that for any* $\varepsilon > 0$, $\exists R > 0$ *with the property that*

$$\int_{B_R(x_m)} d\mu_m \geq 1 - \varepsilon \quad \text{for all } m.$$

(2) *(Vanishing) For all $R > 0$, there holds*

$$\lim_{m \to \infty} \left(\sup_{x \in \mathbb{R}^n} \int_{B_R(x)} d\mu_m \right) = 0.$$

(3) *(Dichotomy)* $\exists \lambda$, $0 < \lambda < 1$, *such that* $\forall \varepsilon > 0$, $\exists R > 0$ *and* $\exists \{x_m\}$ *with the following property: Given $R' > R$ there are non-negative measures μ_m^1, μ_m^2 such that*

$$0 \leq \mu_m^1 + \mu_m^2 \leq \mu_m, \qquad \mathrm{supp}\big(\mu_m^1\big) \subset B_R(x_m), \qquad \mathrm{supp}\big(\mu_m^2\big) \subset \mathbb{R}^n \setminus B_{R'}(x_m),$$

$$\limsup_{m \to \infty} \left(\left| \lambda - \int_{\mathbb{R}^n} d\mu_m^1 \right| + \left| (1 - \lambda) - \int_{\mathbb{R}^n} d\mu_m^2 \right| \right) \leq \varepsilon.$$

Let X be a topological space. A deformation of X is a continuous map $\eta : X \times [0, 1] \to X$ such that $\eta(\cdot, 0) = \mathrm{id}$.

Definition 1.2.4 For a topological pair $Y \subset X$. A continuous map $r : X \to Y$ is called a deformation retract, if $r \circ i = \mathrm{id}_Y$ and $i \circ r \sim \mathrm{id}_X$, where $i : Y \to X$ is the injection. In this case Y is called a deformation retraction of X.

Definition 1.2.5 A deformation retract r is called a strong deformation retract, if there exists a deformation $\eta : X \times [0, 1] \to X$, such that $\eta(\cdot, t)|_Y = \mathrm{id}_Y, \forall t \in [0, 1]$ and $\eta(\cdot, 1) = i \circ r$. Then Y is called a strong deformation retraction of X.

Definition 1.2.6 Let E be a real Banach space, $U \subset E$, and $I \in C^1(U, \mathbb{R})$. Then $v \in E$ is called a pseudo-gradient vector for I at $u \in U$ if

$$\text{(i)} \quad \|v\| \leq 2\|I'(u)\|,$$

$$\text{(ii)} \quad I'(u)v \geq \|I'(u)\|^2.$$

Note that a pseudo-gradient vector is not unique in general and any convex combination of pseudo-gradient vectors for I at u is also a pseudo-gradient vector for I at u.

Let $I \in C^1(E, \mathbb{R})$ and $K \equiv \{u \in E | I'(u) = 0\}$, $\tilde{E} \equiv E \setminus K \equiv \{u \in E | I'(u) \neq 0\}$. Then $V : \tilde{E} \to E$ is called a pseudo-gradient vector field on $\tilde{E}$ if V is locally Lipschitz continuous and $V(u)$ is a pseudo-gradient vector for I for all $u \in \tilde{E}$.

Theorem 1.2.4 (See [159]) *If $I \in C^1(E, \mathbb{R})$, there exists a pseudo-gradient vector field for I on $\tilde{E}$. If $I(u)$ is even in u, I has a pseudo-gradient vector field on $\tilde{E}$ given by an odd function W.*

Using the pseudo-gradient vector field, one can construct a deformation by modified negative gradient flow for I.

Recall that $I_s \equiv \{u \in E | I(u) \leq s\}$ for $s \in \mathbb{R}$, sometimes also write $I^s := \{u \in E | I(u) \leq s\}$ and we recall the following version of Deformation Theorems.

Theorem 1.2.5 (Noncritical interval theorem, see [50]) *If $I \in C^1(E, \mathbb{R})$ satisfies* $(PS)_c, \forall c \in [a, b]$ *and if $K \cap I^{-1}[a, b] = \emptyset$, then I_a is a strong deformation retraction of I_b.*

Theorem 1.2.6 (Second deformation theorem, see [50]) *If $I \in C^1(E, \mathbb{R})$ satisfies* $(PS)_c$, $\forall c \in [a, b]$, *if $K \cap I^{-1}(a, b] = \emptyset$ and the connected components of $K \cap I^{-1}(a)$ are only isolated points, then I_a is a strong deformation retraction of I_b.*

Theorem 1.2.7 (See [159]) *Let E be a real Banach space and let $I \in C^1(E, \mathbb{R})$ and satisfy* (PS) *condition. If $c \in R$, $\bar{\varepsilon} > 0$, and Θ is any neighborhood of K_c, then there exist an $\varepsilon \in (0, \bar{\varepsilon})$ and $\eta \in C([0, 1] \times E, E)$ such that*

(1) $\eta(0, u) = u$ *for all $u \in E$.*
(2) $\eta(t, u) = u$ *for all $t \in [0, 1]$ if $I(u) \notin [c - \bar{\varepsilon}, c + \bar{\varepsilon}]$.*
(3) $\eta(t, u)$ *is a homeomorphism of E onto E for each $t \in [0, 1]$.*
(4) $\|\eta(t, u) - u\| \leq 1$ *for all $t \in [0, 1]$ and $u \in E$.*
(5) $I(\eta(t, u)) \leq I(u)$ *for all $t \in [0, 1]$ and $u \in E$.*
(6) $\eta(1, I_{c+\varepsilon} \setminus \Theta) \subset I_{c-\varepsilon}$.
(7) *If $K_c = \emptyset$, $\eta(1, I_{c+\varepsilon}) \subset I_{c-\varepsilon}$.*
(8) *If $I(u)$ is even in u, $\eta(t, u)$ is odd in u.*

Definition 1.2.7 The action of a topological group G on a normed space X is a continuous map

$$G \times X \to X : [g, u] \to gu$$

such that $1 \cdot u = u$, $(gh)u = g(hu)$, $u \to gu$ is linear.

The action is isometric if

$$\|gu\| = \|u\|.$$

The space of invariant points is defined by

$$\text{Fix}(G) := \{u \in X : gu = u, \forall g \in G\}.$$

A set $A \subset X$ is invariant if $gA = A$ for every $g \in G$. A function $\varphi : X \to \mathbb{R}$ is invariant if $\varphi \circ g = \varphi$ for every $g \in G$. A map $f : X \to X$ is equivariant if $g \circ f = f \circ g$ for every $g \in G$.

Theorem 1.2.8 (Principle of symmetric criticality, Palais [150]) *Assume that the action of the topological group G on the Hilbert space X is isometric. If $\varphi \in C^1(X, \mathbb{R})$ is invariant and if u is a critical point of φ restricted to $\text{Fix}(G)$, then u is a critical point of φ.*

The following part of this section can be seen in [49, 50].

Definition 1.2.8 Let I be a C^1 function defined on a Banach space E, let p be an isolated critical point of I, and let $c = I(p)$.

$$C_q(I, p) = H_q\big(I_c \cap U, (I_c \setminus \{p\}) \cap U; G\big) \tag{1.13}$$

is called the qth critical group of I at $p, q = 0, 1, 2, \ldots$, where G is the coefficient group, U is a neighborhood of p such that $K \cap (I_c \cap U) = \{p\}$, and $H_*(X, Y; G)$ stands for the singular relative homology groups with the Abelian coefficient group G.

Definition 1.2.9 Let p be a non-degenerate critical point of I, we call the dimension of the negative space corresponding to the spectral decomposition of $I''(p)$, the Morse index of p, and denote it by $\text{ind}(I, p)$.

Example 1.2.1 If p is an isolated minimum point of I, then

$$C_q(I, p) = \delta_{q0} \cdot G.$$

Example 1.2.2 If E is n-dimensional, and p is an isolated local maximum point of I, then

$$C_q(I, p) = \delta_{qn} \cdot G.$$

Example 1.2.3 If $I \in C^2(E, \mathbb{R})$ and p is a non-degenerate critical point of I with Morse index j; then

$$C_q(I, p) = \delta_{qj} \cdot G.$$

Suppose that $f \in C^1(E, \mathbb{R})$ has only isolated critical values, and that each of them corresponds to a finite number of critical points; say

$$\cdots < c_{-2} < c_{-1} < c_0 < c_1 < c_2 < \cdots$$

are critical values with

$$K \cap f^{-1}(c_i) = \{z^i_j\}^{m_i}_{j=1}, \quad i = 0, \pm 1, \pm 2, \ldots.$$

One chooses

$$0 < \varepsilon_i < \max\{c_{i+1} - c_i, c_i - c_{i-1}\}, \quad i = 0, \pm 1, \pm 2, \ldots.$$

Definition 1.2.10 For a pair of regular values $a < b$, we call

$$M_q(a, b) = \sum_{a < c_i < b} \operatorname{rank} H_q(f_{c_i+\varepsilon_i}, f_{c_i-\varepsilon_i}; G)$$

the qth Morse type number of the function f on (a, b), $q = 0, 1, 2, \ldots$.

Theorem 1.2.9 (See [50]) *Assume that $f \in C^1(E, \mathbb{R})$ satisfies the* (PS) *condition, and has an isolated critical value c, with $K \cap f^{-1}(c) = \{z_j\}^m_{j=1}$. Then for sufficiently small $\varepsilon > 0$ we have*

$$H_*(f_{c+\varepsilon}, f_{c-\varepsilon}; G) = \bigoplus_{j=1}^{m} C_*(f, z_j) \quad and \quad M_*(a, b) = \sum_{a < c_i < b} \sum_{j=1}^{m_i} \operatorname{rank} C_*\left(f, z^i_j\right).$$

Define the qth Betti number

$$\beta_q = \beta_q(a, b) = \operatorname{rank} H_q(f_b, f_a; G), \quad q = 0, 1, \ldots.$$

Theorem 1.2.10 (Morse relation [50]) *Suppose that $f \in C^1(E, \mathbb{R})$ satisfies* (PS)$_c$, $\forall c \in [a, b]$, *where a and b are regular values. Assume $(K \cap f^{-1}[a, b])$ is finite. Moreover, if all $M_q(a, b)$ and $\beta_q(a, b)$ are finite, and only finitely many of them are non-zeroes, then*

$$\sum_{q=0}^{\infty} \left(M_q(a, b) - \beta_q(a, b)\right)t^q = (1 + t)Q(t), \tag{1.14}$$

where $Q(t)$ is a formal series with non-negative coefficients. In particular, $\forall p = 0, 1, 2, \ldots$,

$$\sum_{q=0}^{p}(-1)^{p-q} M_q(a, b) \geq \sum_{q=0}^{p}(-1)^{p-q} \beta_q(a, b). \tag{1.15}$$

More specifically,

$$\sum_{q=0}^{\infty}(-1)^q M_q(a, b) = \sum_{q=0}^{\infty}(-1)^q \beta_q(a, b). \tag{1.16}$$

1.3 Cone and Partial Order

Definition 1.3.1 Let E be a real Banach space. A nonempty convex closed set $P \subset E$ is called a cone if it satisfies the following two conditions:

(i) $x \in P$, $\lambda \geq 0$ implies $\lambda x \in P$,
(ii) $x \in P$, $-x \in P$ implies $x = \theta$, where θ denotes the zero element of E.

Every cone P in E defines a partial ordering in E given by $x \leq y$ iff $y - x \in P$. If $x \leq y$ and $x \neq y$, we write $x < y$.

Definition 1.3.2 A cone P is called solid if it contains interior points, i.e., $\text{int}(P) \neq \emptyset$, or denote $\mathring{P} \neq \emptyset$.

Definition 1.3.3 A cone P is called generating if $E = P - P$, i.e., every element $x \in E$ can be represented in the form $x = u - v$, where $u, v \in P$.

Definition 1.3.4 A cone $P \subset E$ is said to be normal if there exists a positive constant δ such that $\|x + y\| \geq \delta$, $\forall x, y \in P$ and $\|x\| = \|y\| = 1$.

If cone P is solid and $y - x \in \mathring{P} \neq \emptyset$, we write $x \ll y$.
Here we list the definitions of different cones:

(a) A cone $P \subset E$ is called regular if every increasing and bounded in order sequence in E has a limit, i.e., if $\{x_n\} \subset E$ and $y \in E$ satisfy

$$x_1 \leq x_2 \leq \cdots \leq x_n \leq \cdots \leq y,$$

then there exists $x^* \in E$ such that $\|x_n - x^*\| \to 0$.
(b) A cone $P \subset E$ is called fully regular if every increasing and bounded in norm sequence in E has a limit, i.e., if $\{x_n\} \subset E$ satisfies

$$x_1 \leq x_2 \leq \cdots \leq x_n \leq \cdots, \qquad M = \sup_n \|x_n\| < \infty,$$

then there exists $x^* \in E$ such that $\|x_n - x^*\| \to 0$.
(c) A cone $P \subset E$ is called minihedral if $\sup\{x, y\}$ exists for any pair x, y, where $\sup D$ is the least upper bound of a set D.
(d) A cone $P \subset E$ is called strongly minihedral if $\sup D$ exists for any bounded above in order set $D \subset E$.

For normal cones, we have

Theorem 1.3.1 *Assume P is a cone of E, the following conclusions are equivalent:*

(a) *P is normal;*
(b) *there exists a constant $\delta > 0$ such that $\|x + y\| \geq \delta \max\{\|x\|, \|y\|\}$ for all x, $y \in P$;*

(c) *there exists a constant $N > 0$ such that $\theta \leq x \leq y$ implies $\|x\| \leq N\|y\|$;*
(d) *$x_n \leq z_n \leq y_n$ $(n = 1, 2, 3, \ldots)$ and $\|x_n - x\| \to 0$, $\|y_n - x\| \to 0$ imply $\|z_x - x\| \to 0$;*
(e) *set $(B + P) \cap (B - P)$ is bounded, where $B = \{x \in E \mid \|x\| \leq 1\}$;*
(f) *every order interval $[x, y] = \{z \in E \mid x \leq z \leq y\}$ is bounded.*

Theorem 1.3.2 *The following assertions hold:*

(i) *If E is reflexive, then P is normal $\Leftrightarrow$ P is regular $\Leftrightarrow$ P is fully regular.*
(ii) *If E is separable and reflexive and the cone $P \subset E$ is normal and minihedral, P is strongly minihedral.*

Zorn's lemma *Suppose a partially ordered set S has the property that every chain (i.e., totally ordered subset) has an upper bound in S. Then the set S contains at least one maximal element.*

Note that the content of this section can be seen in [100, 110].

1.4 Brouwer Degree

Theorem 1.4.1 (Sard, see [165]) *Let U be an open set of $\mathbb{R}^p$ and $f : U \to \mathbb{R}^q$ be a C^s map where $s > \max\{p - q, 0\}$. Then the set of critical values in $\mathbb{R}^q$ has measure zero.*

(This section is included in [81].)

Definition 1.4.1 Let $\Omega \subset \mathbb{R}^n$ be open and bounded, $f \in C^1(\bar{\Omega})$ and $y \in \mathbb{R}^n \setminus f(\partial\Omega \cup S_f)$, where $S_f(\Omega) = \{x \in \Omega : J_f(x) = 0\}$. Then we define

$$d(f, \Omega, y) = \sum_{x \in f^{-1}(y)} \operatorname{sgn} J_f(x) \quad \left(\text{agreement: } \sum_{\emptyset} = 0\right).$$

If y is a regular value of f then $f(x) = y$ has at most finitely many solutions. So Definition 1.4.1 is reasonable. When y is a singular value of f, we have

Definition 1.4.2 Let $\Omega \subset \mathbb{R}^n$ be open and bounded, $f \in C^2(\bar{\Omega})$ and $y \notin f(\partial\Omega)$. Then we define $d(f, \Omega, y) = d(f, \Omega, y^1)$, where y^1 is any regular value of f such that $|y^1 - y| < \varrho(y, f(\partial\Omega))$ and $d(f, \Omega, y^1)$ is given by Definition 1.4.1.

In fact, the smooth assumption of f in Definitions 1.4.1 and 1.4.2 can be relaxed to $C(\bar{\Omega})$.

Definition 1.4.3 Let $f \in C(\bar{\Omega})$ and $y \in \mathbb{R}^n \setminus f(\partial\Omega)$. Then we define $d(f, \Omega, y) := d(g, \Omega, y)$, where $g \in C^2(\bar{\Omega})$ is any map such that $|g - f|_0 < \varrho(y, f(\partial\Omega))$ and $d(g, \Omega, y)$ is given by Definition 1.4.2.

Theorem 1.4.2 *Let $M = \{(f, \Omega, y) : \Omega \subset \mathbb{R}^n \text{ open bounded}, f \in C(\bar{\Omega}) \text{ and } y \notin f(\partial\Omega)\}$ and $d : M \to \mathbb{Z}$ the topological degree defined by Definition 1.4.3. Then d has the following properties:*

(d1) $d(\mathrm{id}, \Omega, y) = 1$ *for* $y \in \Omega$.

(d2) $d(f, \Omega, y) = d(f, \Omega_1, y) + d(f, \Omega_2, y)$ *whenever* Ω_1 *and* Ω_2 *are disjoint open subsets of* Ω *such that* $y \notin \bar{\Omega}\backslash(\Omega_1 \cup \Omega_2)$.

(d3) $d(h(t, \cdot), \Omega, y(t))$ *is independent of* $t \in [0, 1]$ *whenever* $h : [0, 1] \times \bar{\Omega} \to \mathbb{R}^n$ *is continuous,* $y : [0, 1] \to \mathbb{R}^n$ *is continuous and* $y(t) \notin h(t, \cdot)(\partial\Omega)$ *on* $[0, 1]$.

(d4) $d(f, \Omega, y) \neq 0$ *implies* $f^{-1}(y) \neq \emptyset$.

(d5) $d(\cdot, \Omega, y)$ *and* $d(f, \Omega, \cdot)$ *are constant on* $\{g \in C(\bar{\Omega}) : |g - f|_0 < r\}$ *and* $B_r(y) \subset \mathbb{R}^n$, *respectively, where* $r = \varrho(y, f(\partial\Omega))$. *Moreover,* $d(f, \Omega, \cdot)$ *is constant on every connected component of* $\mathbb{R}^n \backslash f(\partial\Omega)$.

(d6) $d(g, \Omega, y) = d(f, \Omega, y)$ *whenever* $g|_{\partial\Omega} = f|_{\partial\Omega}$.

(d7) $d(f, \Omega, y) = d(f, \Omega_1, y)$ *for every open subset* Ω_1 *of* Ω *such that* $y \notin f(\bar{\Omega}\backslash\Omega_1)$.

Theorem 1.4.3 *Let X_n be a real topological vector space of* $\dim X_n = n$, X_m *a subspace with* $\dim X_m = m < n$, $\Omega \subset X_n$ *open bounded,* $f : \bar{\Omega} \to X_m$ *continuous and* $y \in X_m \backslash g(\partial\Omega)$, *where* $g = \mathrm{id} - f$. *Then* $d(g, \Omega, y) = d(g|_{\overline{\Omega \cap X_m}}, \Omega \cap X_m, y)$.

1.5 Compact Map and Leray–Schauder Degree

This section is included in Deimling [81].

1.5.1 Definitions

Consider two Banach spaces X and Y, a subset Ω of X and a map $F : \Omega \to Y$. Then F is said to be compact if it is continuous and such that $F(\Omega)$ is relatively compact. $\mathcal{K}(\Omega, Y)$ will denote the class of compact maps and we shall write $\mathcal{K}(\Omega)$ instead of $\mathcal{K}(\Omega, X)$.

F is said to be completely continuous if it is continuous and maps bounded subsets of Ω into relatively compact sets. F is said to be finite-dimensional if $F(\Omega)$ is contained in a finite-dimensional subspace of Y. The class of all finite-dimensional compact maps will be denoted by $\mathcal{F}(\Omega, Y)$ and we shall write $\mathcal{F}(\Omega)$ instead of $\mathcal{F}(\Omega, X)$. Instead of "maps" we shall also speak of "operators".

If $F : X \to Y$ is linear and maps bounded sets into relatively compact sets then it is automatically continuous, and if it is linear and finite-dimensional then it is automatically compact.

Finally, let $\Omega \subset X$ be closed and bounded. Then $F : \Omega \to Y$ is said to be proper if $F^{-1}(K)$ is compact whenever K is compact. Let us note that a continuous proper map is closed, that is, $F(A)$ is closed whenever $A \subset \Omega$ is closed. In fact, if $(x_n) \subset A$ and $Fx_n \to y$ then $(x_n) \subset F^{-1}(\{Fx_n : n \geq 1\} \cup \{y\})$ and therefore (x_n) has a cluster point $x_0 \in A$, and $y = Fx_0 \in F(A)$. Next, we introduce some useful properties.

1.5.2 Properties of Compact Maps

Definition 1.5.1 Let X be a Banach space and $\mathbb{B}$ its bounded sets. Then $\alpha : \mathbb{B} \to \mathbb{R}^+$, defined by

$$\alpha(B) = \inf\{d > 0 : B \text{ admits a finite cover by sets of diameter } \leq d\},$$

is called the (Kuratowski-) measure of noncompactness, the α-MNC for short, and $\beta : \mathbb{B} \to \mathbb{R}^+$ defined by

$$\beta(B) = \inf\{r > 0 : B \text{ can be covered by finitely many balls of radius } r\},$$

is called the ball measure of noncompactness. (Here $\operatorname{diam} B = \sup\{|x - y| : x \in B, y \in B\}$.)

Proposition 1.5.1 *Let X be a Banach space with $\dim X = \infty$, $\mathbb{B}$ the family of all bounded sets of X, and $\gamma : \mathbb{B} \to \mathbb{R}^+$ be either α or β. Then*

(a) $\gamma(B) = 0$ *iff $\bar{B}$ is compact.*
(b) γ *is a seminorm, i.e.,* $\gamma(\lambda B) = |\lambda|\gamma(B)$ *and* $\gamma(B_1 + B_2) \leq \gamma(B_1) + \gamma(B_2)$.
(c) $B_1 \subset B_2$ *implies* $\gamma(B_1) \leq \gamma(B_2)$; $\gamma(B_1 \cup B_2) = \max\{\gamma(B_1), \gamma(B_2)\}$.
(d) $\gamma(\operatorname{conv} B) = \gamma(B)$.
(e) γ *is continuous with respect to the Hausdorff distance ϱ_H, defined by*

$$\varrho_H(B_1, B_2) = \max\left\{\sup_{B_1} \varrho(x, B_2), \ \sup_{B_2} \varrho(x, B_1)\right\};$$

in particular $\gamma(\bar{B}) = \gamma(B)$.

Together with the degree for finite-dimensional spaces the following proposition will be essential to obtain a degree for compact perturbations of the identity.

Proposition 1.5.2 *Let X and Y be Banach spaces, and $B \subset X$ closed bounded. Then*

(a) $\mathcal{F}(B, Y)$ *is dense in $\mathcal{K}(B, Y)$ with respect to the sup norm, i.e. for $F \in \mathcal{K}(B, Y)$ and $\varepsilon > 0$ there exists $F_\varepsilon \in \mathcal{F}(B, Y)$ such that $\sup_B |Fx - F_\varepsilon x| \leq \varepsilon$.*
(b) *If $F \in \mathcal{K}(B)$ then $I - F$ is proper.*

Proof To prove (a), let $F \in \mathcal{K}(B, Y)$, $\varepsilon > 0$ and $y_1, y_2, \ldots, y_p$ such that $\overline{F(B)} \subset \bigcup_{i=1}^p B_i(y_i)$. Let $\varphi_i(y) = \max\{0, \ \varepsilon - |y - y_i|\}$ and $\psi_i(y) = \varphi_i(y)/\sum_{j=1}^p \varphi_j(y)$ for $y \in \overline{F(B)}$, and define $F_\varepsilon(x) = \sum_{i=1}^p \psi_i(Fx)y_i$ for $x \in B$. Then F_ε is continuous, $F_\varepsilon(B) \subset \{y_1, \ldots, y_p\}$, $F_\varepsilon(B)$ is relatively compact and $\sup_B |F_\varepsilon x - Fx| \leq \varepsilon$.

To prove (b), let $A = (I - F)^{-1}(K)$ and K compact. Then $\alpha(A) \leq \alpha(F(A)) + \alpha(K) = 0$ and A is closed, and therefore compact. $\qquad\square$

For differentiable compact maps we have

Proposition 1.5.3 *Let X, Y be Banach space, $\Omega \subset X$ be open, $F \in \mathcal{K}(\Omega, Y)$ and F is differentiable at $x_0 \in \Omega$. Then $F'(x_0)$ is completely continuous.*

Proof Since $F'(x_0) \in L(X, Y)$, it is sufficient to prove that $F'(x_0)(B_1(0))$ is relatively compact. Recall that $F(x_0 + h) = Fx_0 + F'(x_0)h + \omega(x_0; h)$ with $|\omega(x_0; h)| \le \varepsilon\delta$ for $|h| \le \delta = \delta(x_0, \varepsilon)$. Therefore,

$$\delta F'(x_0)(B_1(0)) = F'(x_0)(B_\delta(0)) \subset -Fx_0 + F(B_\delta(0)) + \delta B_\varepsilon(0),$$

and this implies $\delta \cdot \alpha(F'(x_0)(B_1(0))) \le 2\varepsilon\delta$, i.e., $\alpha(F'(x_0)(B_1(0))) = 0$ since $\varepsilon > 0$ has been arbitrary. $\qquad\square$

Proposition 1.5.4 *Let X, Y be Banach spaces, $A \subset X$ closed bounded and $F \in \mathcal{K}(A, Y)$. Then F has an extension $\tilde{F} \in \mathcal{K}(X, Y)$ such that $\tilde{F}(X) \subset \operatorname{conv} F(A)$.*

1.5.3 The Leray–Schauder Degree

Let X be a real Banach space, $\Omega \subset X$ open bounded, $F \in \mathcal{K}(\bar\Omega)$ and $y \notin (I - F)(\partial\Omega)$. On these admissible triplets $(I - F, \Omega, y)$ we want to define a $\mathbb{Z}$-valued function D that satisfies the three basic conditions corresponding to (D1)–(D3) of the Brouwer degree, namely

(D1) $D(I, \Omega, y) = 1$ for $y \in \Omega$;
(D2) $D(I - F, \Omega, y) = D(I - F, \Omega_1, y) + D(I - F, \Omega_2, y)$ whenever Ω_1 and Ω_2 are disjoint open subsets of Ω such that $y \notin (I - F)(\bar\Omega \backslash (\Omega_1 \cup \Omega_2))$;
(D3) $D(I - H(t, \cdot), \Omega, y(t))$ is independent of $t \in [0, 1]$ whenever $H : [0, 1] \times \bar\Omega \to X$ is compact, $y : [0, 1] \to X$ is continuous and $y(t) \notin (I - H(t, \cdot))(\partial\Omega)$ on $[0, 1]$.

Since $G = I - F$ is proper and $y \notin G(\partial\Omega)$, we have $\varrho = \varrho(y, G(\partial\Omega)) > 0$, and if we choose $F_1 \in \mathcal{F}(\bar\Omega)$ such that $\sup\{|F_1 x - Fx| : x \in \bar\Omega\} < \varrho$, then $H(t, x) = Fx + t(F_1 x - Fx)$ satisfies (D3) with $y(t) \equiv y$, and therefore $D(I - F, \Omega, y) = D(I - F_1, \Omega, y)$.

Next, since $F_1(\bar\Omega)$ is contained in a finite-dimensional subspace, we may choose a subspace X_1 with $\dim X_1 < \infty$ such that $y \in X_1$ and $F(\bar\Omega) \subset X_1$.

Then $x - F_1 x = y$ for some $x \in \Omega$ implies that x is already in $\Omega \cap X_1$ and this suggests that $D(I - F_1, \Omega, y)$ should already be determined by the Brouwer degree of $(I - F_1)|_{\overline{\Omega \cap X_1}}$ with respect to $\Omega \cap X_1$ and y. Notice, in particular, that $\Omega \cap X_1 = \emptyset$ implies $0 = D(I - F_1, \Omega, y) = D(I - F, \Omega, y)$, by (D2).

To make this precise, notice first that there exists a continuous projection P_1 from X onto X_1. Then $X = X_1 \oplus X_2$, where $X_2 = P_2(x)$, $P_2 = I - P_1$, and X_2 is closed since P_2 is continuous. Let $\Omega_1 = \Omega \cap X_1 \ne \emptyset$ and $\tilde{F}_1 : X_1 \to X_1$ be any continuous extension of $F_1|_{\bar\Omega_1}$. Then we obtain $D(I - F_1, \Omega; y) = D(I - \tilde{F}_1 P_1, \Omega, y)$, by means of (D3) applied to $H(t, x) = t F_1 x + (1 - t)\tilde{F}_1 P_1 x$ and $y(t) \equiv y$. But all solutions in Ω of $x - \tilde{F}_1 P_1 x = y$ belong to Ω_1 and therefore (D2) tells us that we may replace Ω by any bounded open set which contains Ω_1, for example by $\Omega_1 + B_1(0)$, where $B_1(0)$ is the unit ball of X_2. Hence, we have

$$D(I - F, \Omega, y) = D(I - F_1, \Omega, y) = D\big(I - \tilde{F}_1 P_1, \Omega_1 + B_1(0), y\big)$$
$$= D\big(I - F_1 P_1, \Omega_1 + B_1(0), y\big).$$

Now, you will guess how we have to proceed. Given any open bounded set $\Omega_1 \subset X_1$, $f \in \bar{\Omega}_1 \to X_1$ continuous and $y \in X_1 \setminus f(\partial\Omega_1)$, we define

$$d_0(f, \Omega_1, y) = D\big(I - (I - f)P_1, \Omega_1 + B_1(0), y\big)$$
$$= D\big(f P_1 + P_2, \Omega_1 + B_1(0), y\big).$$

Then (D1)–(D3) imply that d_0 satisfies (d1)–(d3), and therefore d_0 is the Brouwer degree for X_1. In particular, choosing $f = (I - F)|_{\overline{\Omega \cap X_1}}$, we obtain

$$D(I - F_1, \Omega, y) = d_0\big((I - F)|_{\overline{\Omega \cap X_1}}, \overline{\Omega \cap X_1}, y\big).$$

Thus, there is at most on function D. But the construction of D is now a simple exercise in using Theorem 1.4.3. In fact, if F_2 and X_2 satisfy the same conditions as F_1 and X_1, we let X_0 be the span of X_1 and X_2 and $\Omega_0 = \Omega \cap X_0$. Then Theorem 1.4.3 implies

$$d\big((I - F_i)|_{\bar{\Omega}_0}, \Omega_0, y\big) = d\big((I - F_i)|_{\bar{\Omega}_i}, \Omega_i, y\big) \quad \text{for } i = 1, 2$$

and since $x - h(t, x) \neq y$ on $[0, 1] \times \partial\Omega_0$ for $h(t, x) = t F_1 x + (1 - t)F_2 x$, (d3) implies $d((I - F_1)|_{\bar{\Omega}_0}, \Omega_0, y) = d((I - F_2)|_{\bar{\Omega}_0}, \Omega_0, y)$. Therefore, we define $D(I - F, \Omega, y)$ by $d((I - F_1)|_{\bar{\Omega}_1}, \Omega_1, y)$ for any pair F_1 and X_1 of the type mentioned above. Let us write down this result as

Theorem 1.5.1 *Let X be a real Banach space and*

$$M = \big\{(I - F, \Omega, y) : \Omega \subset X \text{ open bounded, } F \in \mathcal{K}(\bar{\Omega}) \text{ and } y \notin (I - F)(\partial\Omega)\big\}.$$

Then there exists exactly a function $D : M \to \mathbb{Z}$, the Leray–Schauder degree, satisfying (D1)–(D3). *The integer $D(I - F, \Omega, y)$ is given by $d((I - F_1)|_{\Omega_1}, \Omega_1, y)$, where F_1 is any map in $\mathcal{F}(\bar{\Omega})$ such that $\sup_{\bar{\Omega}} |F_1 x - F x| \leq \varrho(y, (I - F)(\partial\Omega))$, $\Omega_1 = \Omega \cap X_1$, and X_1 is any subsequence of X such that $\dim X_1 < \infty$, $y \in X_1$ and $F_1(\bar{\Omega}) \subset X_1$, and d is the Brouwer degree of X_1.*

Further properties of the Leray–Schauder degree

Theorem 1.5.2 *Besides* (D1)–(D3), *the Leray–Schauder degree has the following properties*:

(D4) $D(I - F, \Omega, y) \neq 0$ *implies* $(I - F)^{-1}(y) \neq \emptyset$;

(D5) $D(I - G, \Omega, y) = D(I - F, \Omega, y)$ *for* $G \in \mathcal{K}(\bar{\Omega}) \cap B_r(F)$ *and* $D(I - F, \Omega, \cdot)$
   is constant on $B_r(y)$, *where* $r = \varrho(y, (I - F)(\partial\Omega))$. *Even more:* $D(I - F, \Omega, \cdot)$ *is constant on every connected component of* $X \backslash (I - F)(\partial\Omega)$;
(D6) $D(I - G, \Omega, y) = D(I - F, \Omega, y)$ *whenever* $G|_{\partial\Omega} = F|_{\partial\Omega}$;
(D7) $D(I - F, \Omega, y) = D(I - F, \Omega_1, y)$ *for every open subset* Ω_1 *of* Ω *such that* $y \notin (I - F)(\bar{\Omega} \backslash \Omega_1)$.

We have a product formula

Theorem 1.5.3 *Let* $\Omega \subset X$ *be open bounded,* $F_0 \in \mathcal{K}(\bar{\Omega})$ *and* $F = I - F_0$, $G_0 : X \to X$ *completely continuous and* $G = I - G_0$, $y \notin GF(\partial\Omega)$ *and* $(K_\lambda)_{\lambda \in \Lambda}$ *the connected components of* $X \backslash F(\partial\Omega)$. *Then*

$$D(GF, \Omega, y) = \sum_{\lambda \in \Lambda} D(F, \Omega, K_\lambda) D(G, K_\lambda, y)$$

where only finitely many terms are non-zero and $D(F, \Omega, K_\lambda)$ *is* $D(F, \Omega, z)$ *for any* $z \in K_\lambda$.

1.6 Fredholm Operators

Definition 1.6.1 Suppose that X, Y are Banach spaces, $L \in \mathcal{L}(X, Y)$ (linear bounded maps) is called a Fredholm operator, if

(1) Range L is closed;
(2) dim Ker $L < \infty$;
(3) Coker $L = Y / \text{Range } L$ has finite dimension.

We denote $\mathcal{F}(X, Y)$ all Fredholm operators from X to Y. Especially as $Y = X$, we denote $\mathcal{F}(X)$.

Definition 1.6.2 Assume $L \in \mathcal{F}(X, Y)$, let

$$\text{ind}(L) \triangleq q \dim \text{Ker } L - \dim \text{Coker } L,$$

it is called the index of L.

Example 1.6.1 If $F : X \to X$ is linear compact, then $T = I - F \in \mathcal{F}(X)$, and $\text{ind}(T) = 0$.

For the Leray–Shauder degree theory extending to Fredholm operators of index 0, please see [50].

Theorem 1.6.1 (Gohberg and Krein, see [165]) *The set* $\mathcal{F}(X, Y)$ *of Fredholm operators is open in the space of all bounded operators* $\mathcal{L}(X, Y)$ *in the norm topology. Furthermore the index is continuous on* $\mathcal{F}(X, Y)$.

Definition 1.6.3 Suppose that X, Y are C^1 Banach manifolds and $f : X \to Y$ is a C^1 map. A point $x \in X$ is called a regular point of f if $Df(x) : T_x(X) \to T_{f(x)}(Y)$ is surjective, and is singular if not regular. The images of the singular points under f are called the singular values or critical values, their complement the regular values.

Note that if $y \in Y$ is not in the image of f it is automatically a regular value.

Definition 1.6.4 Assume $U \subset X$, a map $f \in C^1(U, Y)$ is called Fredholm map if for each $x \in U$, the derivative $Df(x) : T_x(U) \to T_{f(x)}(Y)$ is a Fredholm operator. The index of f is defined to be the index of $Df(x)$ for some x. By Theorem 1.6.1, if U is connected, then ind $f'(x)$ (or $Df(x)$) does not depend on x, it is denoted by $\mathrm{ind}(f)$.

Theorem 1.6.2 (Sard–Smale, see [165]) *Assume X is a separable Banach space and Y is a Banach space. Let $f : X \to Y$ be a C^q Fredholm map with $q > \max\{\mathrm{ind}(f), 0\}$. Then the regular values of f are almost all of Y (or the set of critical values is of first category).*

Corollary 1.6.1 (See [165]) *X is a separable Banach space and Y is a Banach space. Let $f : X \to Y$ be a C^1 Fredholm map of negative index, its image contains no interior points.*

Corollary 1.6.2 (See [165]) *X is a separable Banach space and Y is a Banach space. Let $f : X \to Y$ be a C^q Fredholm map with $q > \max\{\mathrm{ind}(f), 0\}$, then for almost all $y \in Y$, $f^{-1}(y)$ is a submanifold of X whose dimension is equal to index of f or is empty.*

Definition 1.6.5 A map is proper if the inverse image of a compact set is compact.

Theorem 1.6.3 (See [165]) *A Fredholm map is locally proper. In other words, if $f : X \to Y$ is Fredholm and $x \in X$, there exists a neighborhood N of x such that f restricted to N is proper.*

1.7 Fixed Point Index

Remember that a subset $K \neq \emptyset$ of X is called a retract of X if there is a continuous map $R : X \to K$, a retraction, such that $Rx = x$ on K. Recall also that every closed convex subset is a retract and that every retract is closed but not necessarily convex; remember that $\partial B_1(0)$ is a retract of X if $\dim X = \infty$.

Whenever we are concerned with subsets of a retract K, it is understood that all topological notions are understood with respect to the topology induced by $|\cdot|$ on K.

Now, let $\Omega \subset K$ be open and $F : \bar{\Omega} \to K$ compact and such that $\mathrm{Fix}(F) \cap \partial\Omega = \emptyset$, where $\mathrm{Fix}(F) = \{x \in \bar{\Omega} \mid F(x) = x\}$. If $R : X \to K$ is retraction, then

$D(I - FR, R^{-1}(\Omega, 0))$ is defined, and it follows immediately from the homotopy invariance and the excision property (D3) and (D7) that this integer is the same for all retractions of X onto K. Conventionally, this number is called the fixed point index over Ω with respect to K for the compact F, $i(F, \Omega, K)$ for short. The map $i : M \to \mathbb{Z}$ with

$$M = \{(F, \Omega, K) : K \subset X \text{ retract, } \Omega \subset K \text{ open, } F : \bar{\Omega} \to K \text{ compact,}$$
$$\mathrm{Fix}(F) \cap \partial\Omega = \emptyset\},$$

inherits the properties of Leray–Schauder degree D.

Let X be a Banach space, $K \subset X$ a cone and $F : K \to X$. Since one often knows $F(0) = 0$ but fixed points in $K \setminus \{0\}$ are of interest, the simplest abstract approach is to consider a shell $\{x \in K : 0 < \varrho \le \|x\| \le r\}$ and to impose conditions at the lower and upper boundary sufficient for F to have a fixed point in the shell. In the sequel, we let $K_r = K \cap B_r(0)$ and we shall write $i(F, \Omega)$ for $i(F, \Omega, K)$ whenever the index is defined. Let us start with

Theorem 1.7.1 *Let X be a Banach space, $K \subset X$ a cone and $F : \bar{K}_r \to K$ is γ-condensing. Suppose that*

(a) *$Fx \ne \lambda x$ for $\|x\| = r$ and $\lambda > 1$;*
(b) *there exist a smaller radius $\varrho \in (0, r)$ and an $e \in K \setminus \{0\}$ such that $x - Fx \ne \lambda e$ for $\|x\| = \varrho$ and $\lambda > 0$.*

Then F has a fixed point in $\{x \in K : \varrho \le \|x\| \le r\}$.

As a trivial consequence we have the following corollary on 'compression of conical shells':

Corollary 1.7.1 *Suppose that $F : \bar{K}_r \to K$ is γ-condensing and such that*

(a) *$Fx \not\ge x$ on $\|x\| = r$.*
(b) *$Fx \not\le x$ on $\|x\| = \varrho$, for some $\varrho \in (0, r)$.*

Then F has a fixed point in $\{x \in K : \varrho < \|x\| < r\}$.

Theorem 1.7.2 *Let $0 < \varrho < r$, $F : \bar{K}_r \to K$ compact and such that*

(a) *$Fx \ne \lambda x$ on $\|x\| = r$ and $\lambda > 1$.*
(b) *$Fx \ne \lambda x$ on $\|x\| = \varrho$ and $\lambda < 1$.*
(c) *$\inf\{|Fx| : \|x\| = \varrho\} > 0$.*

Then F has a fixed point in $\bar{K}_r \setminus K_\varrho$.

Remark 1.7.1 Note that this section is included in Deimling [81].

1.8 Banach's Contract Theorem, Implicit Functions Theorem

Theorem 1.8.1 *Let X be a Banach space, $D \subset X$ closed and $F : D \to D$ a strict contraction, i.e., $\|F(x) - F(y)\| \leq k\|x - y\|$ for some $0 < k < 1$ and all $x, y \in D$. Then F has a unique fixed point x^*. For any $x_0 \in D$, let $x_{n+1} = F(x_n) = F^n x_0$, then $x_n \to x^*$, and $\|x_n - x^*\| \leq (1 - k)^{-1} k^n \|F(x_0) - x_0\|$.*

Theorem 1.8.2 (Implicit function theorem) *Let X, Y, Z be Banach spaces, $U \subset X$ and $V \subset Y$ neighborhoods of x_0 and y_0, respectively, $F : U \times V \to Z$ continuous and continuously differentiable with respect to y. Suppose also that $F(x_0, y_0) = 0$ and $F_y^{-1}(x_0, y_0) \in L(Z, Y)$. Then there exist balls $\bar{B}_r(x_0) \subset U$, $\bar{B}_\delta(y_0) \subset V$ and exactly one map $T : B_r(x_0) \to B_\delta(y_0)$ such that $T x_0 = y_0$ and $F(x, Tx) = 0$ on $B_r(x_0)$. This map T is continuous.*

Moreover, T is as smooth as F, possibly on a smaller ball $B_\varrho(x_0) \subset B_r(x_0)$, i.e., $F \in C^m(U \times V)$ implies that $T \in C^m(B_\varrho(x_0))$.

Theorem 1.8.3 (Inverse function theorem) *Let X, Y be Banach space, U_0 a neighborhood of x_0, $G : U_0 \to Y$ continuously differentiable and $G'(x_0)^{-1} \in L(Y, X)$. Then G is a local homeomorphism, i.e., there is a neighborhood $U \subset U_0$ of x_0 such that $G|_U$ is a homeomorphism onto the neighborhood $G(U)$ of $y_0 = G x_0$. Furthermore, there is a possibly small neighborhood $V \subset U$ such that $G|_U^{-1} \in C^1(G(V))$ and*

$$\left(G|_U^{-1}\right)'(G_X) = G'(x)^{-1} \quad on\ V.$$

Actually $G|_U^{-1}$ is as smooth as G, i.e., $G|_U^{-1} \in C^m(G(V))$ if $G \in C^m(U_0)$, also for $m = \infty$.

1.9 Krein–Rutman Theorem

Let E be a real Banach space. We denote by $L(E) := L(E, E)$ the Banach space of all continuous linear operators in E. Let $P^* = \{f \in E^* | f(x) \geq 0, \forall x \in P\}$, if $\overline{P - P} = E$ (i.e., P is total), then $P^* \subset E^*$ is a cone. Then for every $T \in L(E)$, the limit

$$r(T) := \lim_{n \to \infty} \|T^k\|^{1/k}$$

exists and is called the spectral radius of T.

Recall that a linear operator $T \in L(E)$ is called compact if the image of the unit ball is relatively compact in E. An eigenvalue λ of a linear operator T is called simple if $\dim(\bigcup_{k=1}^{\infty} \ker(\lambda - T)^k) = 1$.

Theorem 1.9.1 (Krein and Rutman, see [8]) *Let (E, P) be an OBS with total positive cone. Suppose that $T \in L(E)$ is compact and has a positive spectral radius $r(T)$. Then $r(T)$ is an eigenvalue of T and of the dual operator T^* with eigenvector in P and in P^*, respectively.*

Theorem 1.9.2 (Theorem 3.2, see [8]) *Let (E, P) be an OBS whose positive cone has nonempty interior $\mathring{P}$. Let T be a strongly positive (i.e., $T : \dot{P} \to \mathring{P}$, where $\dot{P} = P \setminus \{\theta\}$) compact endomorphism of E. Then the following is true:*

 (i) *The spectral radius $r(T)$ is positive;*
 (ii) *$r(T)$ is a simple eigenvalue of T having a positive eigenvector and there is no other eigenvalue with a positive eigenvector;*
(iii) *$r(T)$ is a simple eigenvalue of T^* having a strictly positive eigenvector;*
(iv) *For every $y \in \dot{P}$ the equation*

$$\lambda x - T x = y$$

 has exactly one positive solution if $\lambda > r(T)$, and no positive solution for $\lambda \leq r(T)$. The equation $r(T)x - Tx = -y$ has no positive solution.
 (v) *For every $S \in L(E)$ satisfying $S \geq T$ (i.e., $S(x) \geq T(x)$ for all $x \in P$), $r(S) \geq r(T)$. If $S - T$ is strongly positive, then $r(S) > r(T)$.*

1.10 Bifurcation Theory

Definition 1.10.1 Let X, Y be Banach spaces. Suppose that $F : X \times \mathbb{R} \to Y$ is a continuous map. $\forall \lambda \in \mathbb{R}$, let

$$S_\lambda = \left\{ x \in X \,|\, F(x, \lambda) = \theta \right\}$$

be the solution set of the equation $F(x, \lambda) = \theta$. Assume $\theta \in S_\lambda$, $\forall \lambda \in \mathbb{R}$. We call (θ, λ_0) is a bifurcation point, if for any neighborhood U of $(0, \lambda_0)$, there exists $(x, \lambda) \in U$ with $x \in S_\lambda \setminus \{\theta\}$.

The following is a local bifurcation result obtained by Crandall and Rabinowitz.

Theorem 1.10.1 (Crandall and Rabinowitz, see [60]) *Let X, Y be Banach spaces, V a neighborhood of θ in X and*

$$F : V \times (-1, 1) \to Y$$

have the properties

(a) *$F(\theta, t) = \theta$ for $|t| < 1$,*
(b) *The partial derivatives F_t, F_x and F_{xt} exist and are continuous,*
(c) *$N(F_x(\theta, 0))$ and $Y/R(F_x(\theta, 0))$ are one dimensional,*
(d) *$F_{xt}(\theta, 0)x_0 \notin R(F_x(\theta, 0))$, where*

$$N\big(F_x(\theta, 0)\big) = \mathrm{span}\{x_0\}.$$

If Z is any complement of $N(F_x(\theta,0))$ in X, then there is a neighborhood U of $(\theta,0)$ in $X \times \mathbb{R}$, an interval $(-a,a)$, and continuous functions $\varphi : (-a,a) \to \mathbb{R}$, $\psi : (-a,a) \to Z$ such that $\varphi(0) = 0$, $\psi(0) = \theta$ and

$$F^{-1}(\theta) \cap U = \big\{ \big(\alpha x_0 + \alpha\psi(\alpha), \varphi(\alpha)\big) : |\alpha| < a \big\} \cup \big\{ (\theta,t) : (\theta,t) \in U \big\}.$$

If F_{xx} is also continuous, the functions φ and ψ are once continuously differentiable.

Consider $F : X \times \mathbb{R} \to X$ in the following form:

$$F(x,\lambda) = Lx - \lambda x + N(x,\lambda),$$

where X is a real Banach space, $L \in \mathcal{L}(X,X)$, $\lambda \in \mathbb{R}$, and $\|N(x,\lambda)\| = o(\|x\|)$ as $x \to 0$ uniformly in any finite interval of λ.

Theorem 1.10.2 (Krasnosel'skii, see [122]) *Suppose that L is a linear compact operator on X, and that $\lambda_0 \neq 0$ is an eigenvalue of L with odd multiplicity, i.e., $\beta = \dim \bigcup_{k=1}^{\infty} \ker(L - \lambda_0 I)^k$ is odd. If $\forall \lambda$, $N(\cdot,\lambda)$ is compact, and N is continuous in x and λ, and satisfies*

$$\big\| N(x,\lambda) \big\| = o\big(\|x\|\big)$$

uniformly in any finite interval of λ, then (θ,λ_0) is a bifurcation point of the equation $F(x,\lambda) = \theta$.

Theorem 1.10.3 (Leray–Schauder) *Let X be a real Banach space, $T : X \times \mathbb{R} \to X$ be a compact map satisfying $T(x,\theta) = \theta$ and $f(x,\lambda) = x - T(x,\lambda)$. Let*

$$S = \big\{ (x,\lambda) \in X \times \mathbb{R} \,|\, f(x,\lambda) = \theta \big\},$$

and let ζ be the component of S passing through $(\theta,0)$. If

$$\zeta^{\pm} = \zeta \cap (X \times \mathbb{R}_{\pm}),$$

then both ζ^+ and ζ^- are unbounded.

Theorem 1.10.4 (Rabinowitz, see [156]) *Let X be a real Banach space and $F(x,\lambda) = x - \lambda Lx - N(x,\lambda)$, where $L \in \mathcal{L}(X,X)$ and $N : X \times \mathbb{R} \to X$ are compact. Let S be the solution set of $F(x,\lambda) = \theta$, $S_+ = \overline{S \setminus (\{\theta\} \times \mathbb{R})}$, and let ζ be the component of S_+, containing (θ,λ_1). Assume that $N(x,\lambda) = o(\|x\|)$ uniformly on any finite interval in λ and that $\lambda_1^{-1} \in \sigma(L)$ is an eigenvalue of odd multiplicity. Then the following alternatives hold: Either*

1. *ζ is unbounded; or*
2. *there are only finite number of points $\{(\theta,\lambda_i)|i = 1,\ldots,l\}$ lying on ζ, where $\lambda_i^{-1} \in \sigma(L)$, $i = 1,2,\ldots,l$. Furthermore, if β_i is the algebraic multiplicity of λ_i^{-1}, then $\sum_{i=1}^{l} \beta_i$ is even.*

Theorem 1.10.5 (Rabinowitz, see [158, 159]) *Suppose E is a real Hilbert space and $\Phi \in C^2(E, \mathbb{R})$ with $\Phi'(u) = Lu + H(u)$, where $L \in \mathcal{L}(E, E)$ is symmetric and $H(u) = o(\|u\|)$ as $u \to \theta$. If $\mu \in \sigma(L)$ is an isolated eigenvalue of finite multiplicity, then (θ, μ) is a bifurcation point for*

$$\mathcal{F}(u, \lambda) \equiv \Phi'(u) - \lambda u = Lu + H(u) - \lambda u.$$

Moreover, the following alternatives hold: Either

(i) *(θ, μ) is not an isolated solution of $\mathcal{F}(x, \lambda) = \theta$, or*
(ii) *there is a one sided neighborhood Λ of μ such that for all $\lambda \in \Lambda \setminus \{\mu\}$, $\mathcal{F}(x, \lambda) = \theta$ possesses at least two distinct non-trivial solutions, or*
(iii) *there is a neighborhood Λ of μ such that for all $\lambda \in \Lambda \setminus \{\mu\}$, the equation $\mathcal{F}(x, \lambda) = \theta$ possesses at least one non-trivial solution.*

1.11 Rearrangements of Sets and Functions

Let $A \subset \mathbb{R}^n$ is a Borel set of finite Lebesgue measure, we define A^*, the symmetric rearrangement of the set A, to be the open ball centered at the origin whose volume is that of A. Thus

$$A^* = \{x : |x| < r\} \quad \text{with } \left(\left|\mathbb{S}^{n-1}\right|/n\right)r^n = \text{meas}(A),$$

where meas denotes Lebesgue measure, $\left|\mathbb{S}^{n-1}\right|$ is the surface area of $\mathbb{S}^{n-1}$. The symmetric-decreasing rearrangement of a characteristic function of a set is obvious, namely $\chi_A^* := \chi_{A^*}$. For a Borel measure function $f : \mathbb{R}^n \to \mathbb{C}$ vanishing at infinity (i.e., $\text{meas}\{x : |f(x)| > t\}$ is finite for all $t > 0$), we define the symmetric-decreasing rearrangement,

$$f^*(x) = \int_0^\infty \chi^*_{\{|f|>t\}}(x)\, dt,$$

which is to be compared with

$$|f(x)| = \int_0^\infty \chi_{\{|f|>t\}}(x)\, dt.$$

$f^*(x)$ has some obvious properties:

(i) $f^*(x)$ is non-negative.
(ii) $f^*(x)$ is radially symmetric and non-increasing, i.e., $f^*(x) = f^*(y)$ if $|x| = |y|$ and $f^*(x) \geq f^*(y)$ if $|x| \leq |y|$.
(iii) $f^*(x)$ is lower semi-continuous and measurable.
(iv) $\{x : f^*(x) > t\} = \{x : |f(x)| > t\}^*$, and consequently $\text{meas}\{x : f^*(x) > t\} = \text{meas}\{x : |f(x)| > t\}$ for every $t > 0$. With the layer cake representation,

$\int_{\mathbb{R}^n} \phi(|f(x)|)\,dx = \int_{\mathbb{R}^n} \phi(|f^*(x)|)\,dx$ for any function ϕ that is the difference of two monotone functions ϕ_1, ϕ_2 such that either $\int_{\mathbb{R}^n} \phi_1(|f(x)|)\,dx$ or $\int_{\mathbb{R}^n} \phi_2(|f(x)|)\,dx$ is finite.

In particular, for $f \in L^p(\mathbb{R}^n)$, $\|f\|_p = \|f^*\|_p$ for all $1 \le p \le \infty$.

(v) If $\Phi : \mathbb{R}^+ \to \mathbb{R}^+$ is non-decreasing, then $(\Phi \circ |f|)^* = \Phi \circ f^*$.

(vi) If $f(x) \le g(x)$ for all $x \in \mathbb{R}^n$, then $f^*(x) \le g^*(x)$ for all $x \in \mathbb{R}^n$.

Theorem 1.11.1 (The simplest rearrangement inequality) *Let f and g be non-negative functions on $\mathbb{R}^n$, vanishing at infinity, and let f^* and g^* be their symmetric-decreasing rearrangements. Then*

$$\int_{\mathbb{R}^n} f(x)g(x)\,dx \le \int_{\mathbb{R}^n} f^*(x)g^*(x)\,dx, \tag{1.17}$$

with the understanding that the left side is infinity so is the right side.

If f is strictly symmetric-decreasing, then there is equality in (1.17) if and only if $g = g^*$.

Lemma 1.11.1 *Let $f : \mathbb{R}^n \to \mathbb{R}$ be a non-negative measurable function that vanishes at infinity. Assume that ∇f, in the sense of distribution, is a function such that $\|\nabla f\|_2 < \infty$. Then ∇f^* has the same property and $\|\nabla f^*\|_2 \le \|\nabla f\|_2$.*

Remark 1.11.1 The rearrangement above is introduced in [131] (see pp. 71, 174).

We also can give another definition of the above rearrangement (see [23]). The Schwarz symmetrization: Let u be a real valued function in a bounded domain $D \subset \mathbb{R}^n$.

Definition 1.11.1 If $D(\mu) = \{x \in D : u(x) \ge \mu\}$, then for any given $x \in D^*$, define $u^*(x) = \sup\{\mu : x \in D(\mu)^*\}$.

$u^*(x)$ has the following simple properties:

(A) $u^*(x)$ is a radially symmetric function; that is $u^*(x) = u^*(|x|)$.

(B) $u^*(|x|)$ is a non-increasing function of $|x|$.

(C) $\inf_{x \in D} u(x) = \inf_{x \in D^*} u^*(x)$ and $\sup_{x \in D} u(x) = \sup_{x \in D^*} u^*(x)$.

(D) If $u_1(x) \le u_2(x)$ in D, then $u_1^*(x) \le u_2^*(x)$ in D^*.

(E) If $u(x)$ is continuous in D, then $u^*(x)$ is also continuous in D^*.

(F) $\text{meas}\{x \in D : u(x) \ge \mu\} = \text{meas}\{x \in D^* : u^*(x) \ge \mu\}$.

Moreover, for non-negative functions u, v with compact support in $\bar{\Omega}$, in [120], p. 23, we have

$$\int_{\bar{D}} u(x)v(x)\,dx \le \int_{\bar{D}^*} u^*(x)v^*(x)\,dx.$$

Definition 1.11.2 $\mathcal{P}_0(D) = \{u(x) \mid u : D \to \mathbb{R}^+, u(x) = 0 \text{ on } \partial D\}$.

Lemma 1.11.2 *If $u(x) \in \mathcal{P}_0(D)$ satisfies the Lipschitz condition $|u(x) - u(y)| \leq K|x - y|$ for all x, y in $\bar{D}$, then $u^*(x)$ satisfies the same condition in $\bar{D}^*$.*

The property (F) implies that if $\psi(t)$ is a continuous function, then

$$\int_D \psi\big[u(x)\big]\,dx = \int_{D^*} \psi\big[u^*(x)\big]\,dx.$$

So we have

$$\|u\|_{L^p(D)} = \|u^*\|_{L^p(D^*)}. \tag{1.18}$$

Theorem 1.11.2 (See [23]) *Let the function $g(t)$ be continuous and positive and let $F(t)$ be a convex non-decreasing function for $t \geq 0$. Suppose $u(x) \geq 0$ is analytic and belongs to $\mathcal{P}_0(D)$,*

$$\int_{D(m)} g\big[u(x)\big] F\big(|\operatorname{grad} u(x)|\big)\,dx \geq \int_{D(m)^*} g\big[u^*(x)\big] F\big(|\operatorname{grad} u^*(x)|\big)\,dx.$$

And the Schwarz symmetrization diminishes the integral $\int_D g[u]F(|\operatorname{grad} u|)$. Here $D(m) := \{x \in D : u(x) \geq m\}$.

So for $u \in W_0^{1,p}(D)$ $(1 \leq p < \infty)$, $u \geq 0$, then

$$\int_{D^*} |\nabla u^*|^p\,dx \leq \int_D |\nabla u|^p\,dx. \tag{1.19}$$

Example 1.11.1 If $u(x)$ is a minimizer for S (1.6), we can assume that $u \geq 0$ since $\|u\|_{\mathcal{D}^{1,2}(\mathbb{R}^n)} \leq \|u\|_{\mathcal{D}^{1,2}(\mathbb{R}^n)}$, $\|u\|_{L^p(\mathbb{R}^n)} = \|u\|_{L^p(\mathbb{R}^n)}$. Then $\|u^*\|_{L^{2^*}} = \|u\|_{L^{2^*}}$, and by Lemma 1.11.1 we have $\|\nabla u^*\|_2 \leq \|\nabla u\|_2$, so S has a radially symmetric and non-increasing minimizer u^*. This fact is useful to prove that the minimizer is achieved.

1.12 Genus and Category

Definition 1.12.1 (See [11, 159]) Suppose E is a real Banach space, let

$$\Sigma(E) = \big\{ A \,|\, A \subset E, \text{ symmetric with respect to } \theta, A \subset E \setminus \{\theta\} \big\}.$$

Assume $A \in \Sigma(E)$. If $A = \emptyset$ we say the genus of A is 0, denote $\gamma(A) = 0$. If $A \neq \emptyset$, and $\exists n \in \mathbb{N}_* := \{1, 2, \ldots\}$ such that there exists a continuous odd map $\varphi : A \to \mathbb{R}^n \setminus \{\theta\}$, the minimum n_0 is called the genus of A, we denote $\gamma(A) = n_0$; if there exists no such n, we say that the genus of A is $+\infty$, denoted $\gamma(A) = \infty$.

Proposition 1.12.1 (Some properties, see [11]) *Assume* $A, B \in \Sigma(E)$.

(i) *If there exists a continuous odd map* $f : A \to B$, *then* $\gamma(A) \le \gamma(B)$;
(ii) *if* $A \subset B$, *then* $\gamma(A) \le \gamma(B)$;
(iii) *if there exists an odd homeomorphism between* A *and* B, *then* $\gamma(A) = \gamma(B)$;
(iv) $\gamma(A \cup B) \le \gamma(A) + \gamma(B)$;
(v) *if* $\gamma(B) < +\infty$, *then* $\gamma(\overline{A \setminus B}) \ge \gamma(A) - \gamma(B)$;
(vi) *if* A *is compact, then* $\gamma(A) < +\infty$, *and* $\exists \delta > 0$ *such that* $\gamma(\overline{N_\delta(A)}) = \gamma(A)$, *where* $N_\delta(A) = \{x \in E \,|\, d(x, A) = \inf_{z \in A} \|x - z\| < \delta\}$;
(vii) $\gamma(A) \le \dim E$, $\gamma(\partial\Omega) = \dim E$, *where* $\Omega \subset E$ *is a symmetric open bounded set,* $\theta \in \Omega$ *(especially,* Ω *is a ball centered at* θ*).*

The genus can be used to get infinitely many distinct pairs of critical points for even functionals:

Theorem 1.12.1 (See Theorem 8.10 of [159]) *Let* E *be an infinite-dimensional Hilbert space and let* $I \in C^1(E, \mathbb{R})$ *be even. Suppose* $r > 0$, $I|_{\partial B_r}$ *satisfies* (PS) *condition, and* $I|_{\partial B_r}$ *is bounded from below. Then* $I|_{\partial B_r}$ *possesses infinitely many distinct pairs of critical points.*

Definition 1.12.2 Let M be a topological space, $A \subset M$ be a closed subset. Set

$$\mathrm{cat}_M(A) = \inf\left\{ m \in \mathbb{N} \cup \{+\infty\} \,\middle|\, \exists m \text{ contractible closed subsets of} \right.$$
$$\left. M : F_1, F_2, \ldots, F_m \text{ such that } A \subset \bigcup_{i=1}^{m} F_i \right\}.$$

A set F is called contractible (in M) if $\exists \eta : [0, 1] \times M \to M$ such that $\eta(0, \cdot) = \mathrm{id}_M$ and $\eta(1, F) = x_0$ (for some $x_0 \in M$).

Proposition 1.12.2 (Some properties, see [50, 193])

(i) $\mathrm{cat}_M(A) = 0 \Leftrightarrow A = \emptyset$,
(ii) $A \subset B \Rightarrow \mathrm{cat}_M(A) \le \mathrm{cat}_M(B)$,
(iii) $\mathrm{cat}_M(A \cup B) \le \mathrm{cat}_M(A) + \mathrm{cat}_M(B)$,
(iv) *if* $\eta : [0, 1] \times M \to M$ *is continuous such that* $\eta(0, \cdot) = \mathrm{id}_M$, *then* $\mathrm{cat}_M(A) \le \mathrm{cat}_M(\eta(1, A))$,
(v) *if* A *is compact, then there is a closed neighborhood* N *of* A *such that* $A \subset \mathrm{int}(N)$ *and* $\mathrm{cat}_M(A) = \mathrm{cat}_M(N)$,
(vi) $\mathrm{cat}_M(\{p\}) = 1$, $\forall p \in M$.

Example 1.12.1

(i) S^∞ is contractible, so $\mathrm{cat}_{S^\infty}(S^\infty) = 1$.
(ii) A cone P in a Banach space E, then $\mathrm{cat}_E(P) = 1$.

The number of critical points of a C^1 functional φ defined on a compact manifold X is greater or equal to $\mathrm{cat}_X(X)$. The corresponding critical values are given by

$$c_k := \inf_{A \in A_k} \sup_{u \in A} \varphi(u),$$

$$A_k := \{A \subset X : A \text{ is closed, } \mathrm{cat}_X(A) \geq k\}.$$

1.13 Maximum Principles and Symmetry of Solution

Let Ω be a bounded and connected domain in $\mathbb{R}^n$, $n \geq 2$. Consider the operator L in Ω

$$Lu \equiv a_{ij}(x)D_{ij}u + b_i(x)D_iu + c(x)u$$

for $u \in C^2(\Omega) \cap C(\bar{\Omega})$, where $a_{ij} = a_{ji}$, a_{ij}, b_i and c are continuous, hence they are bounded, we assume L is uniformly elliptic in Ω in the following sense:

$$a_{ij}(x)\xi_i\xi_j \geq \lambda|\xi|^2 \quad \text{for any } x \in \Omega \text{ and any } \xi \in \mathbb{R}^n$$

for some positive constant λ.

Lemma 1.13.1 *Suppose $u \in C^2(\Omega) \cap C(\bar{\Omega})$ satisfies $Lu > 0$ in Ω with $c(x) \leq 0$ in Ω. If u has a non-negative maximum in $\bar{\Omega}$, then u cannot attain this maximum in Ω.*

Theorem 1.13.1 (Weak Maximum Principle, see [117]) *Suppose that $u \in C^2(\Omega) \cap C(\bar{\Omega})$ satisfies $Lu \geq 0$ in Ω with $c(x) \leq 0$ in Ω. Then u attains on $\partial\Omega$ its non-negative maximum in $\bar{\Omega}$ if u has a non-negative maximum in $\bar{\Omega}$.*

Remark 1.13.1 If $c(x) \equiv 0$, then the requirement for non-negativeness can be removed. The conclusion of Theorem 1.13.1 is: u attains on $\partial\Omega$ its maximum in $\bar{\Omega}$ (see [95]).

Theorem 1.13.2 (Strong Maximum Principle, see [117]) *Let $u \in C^2(\Omega) \cap C(\bar{\Omega})$ satisfy $Lu \geq 0$ in Ω with $c(x) \leq 0$ in Ω. Then the non-negative maximum of u in $\bar{\Omega}$ can be assumed only on $\partial\Omega$ unless u is a constant.*

Corollary 1.13.1 (See [117]) *Suppose that $u \in C^2(\Omega) \cap C(\bar{\Omega})$ satisfies $Lu \geq 0$ in Ω with $c(x) \leq 0$ in Ω. If $u \leq 0$ on $\partial\Omega$, then $u \leq 0$ in Ω. In fact, either $u < 0$ in Ω or $u \equiv 0$ in Ω.*

Corollary 1.13.2 (See [117]) *Suppose Ω has the interior sphere property and that $u \in C^2(\Omega) \cap C(\bar{\Omega})$ satisfy $Lu \geq 0$ in Ω with $c(x) \leq 0$ in Ω. Assume u attains its* *non-negative maximum at $x_0 \in \bar{\Omega}$. Then $x_0 \in \partial\Omega$ and for any outward direction ν at x_0 to $\partial\Omega$*

$$\frac{\partial u}{\partial \nu}(x_0) > 0$$

unless u is constant in $\bar{\Omega}$.

Serrin generalizes the comparison principle without restriction on $c(x)$.

Theorem 1.13.3 (See [117]) *Suppose that $u \in C^2(\Omega) \cap C(\bar{\Omega})$ satisfies $Lu \geq 0$ in Ω. If $u \leq 0$ in Ω, then either $u < 0$ in Ω or $u \equiv 0$ in Ω.*

A Strong Maximum Principle for some quasilinear elliptic equations was given by J.L. Vázquez [184]:

Theorem 1.13.4 (See [184]) *Let $u \in C^1(\Omega)$ be such that $\Delta_p u \in L^2_{\mathrm{loc}}(\Omega)$, $u \geq 0$ a.e. in Ω, $\Delta_p u \leq \beta(u)$ a.e. in Ω with $\beta : [0, \infty) \to \mathbb{R}$ continuous, non-decreasing, $\beta(0) = 0$ and either $\beta(s) = 0$ for some $s > 0$ or $\beta(s) > 0$ for all $s > 0$ but $\int_0^1 (\beta(s)s)^{-1/p}\, ds = \infty$.*

Then if u does not vanish identically on Ω it is positive everywhere in Ω.

Moreover, if $u \in C^1(\Omega \cup \{x_0\})$ for any $x_0 \in \partial\Omega$ that satisfies an interior sphere condition and $u(x_0) = 0$ then

$$\frac{\partial u}{\partial \nu}(x_0) > 0, \tag{1.20}$$

where ν is an interior normal at x_0, and

$$\Delta_p u = \mathrm{div}\big(|\nabla u|^{p-2}\nabla u\big), \quad 1 < p < \infty.$$

Theorem 1.13.5 (See [117]) *Suppose $u \in C(\bar{B}_1) \cap C^2(B_1)$ is a positive solution of*

$$\Delta u + f(u) = 0 \quad \text{in } B_1$$

$$u = 0 \qquad\qquad \text{on } \partial B_1$$

where f locally Lipschitz in $\mathbb{R}$. Then u is radially symmetric in $B_1 := \{x \in \mathbb{R}^n \mid |x| < 1\}$ and $\frac{\partial u}{\partial r}(x) < 0$ for $x \neq 0$.

1.14 Comparison Theorems

Next we consider the operator $-\mathrm{div}\, A(x, Du)$ in an open set $\Omega \subset \mathbb{R}^N$, $N \geq 2$, and we make the following assumptions on A:

(A_1) $A \in C^0(\bar{\Omega} \times \mathbb{R}^N; \mathbb{R}^N) \cap C^1(\bar{\Omega} \times \mathbb{R}^N \setminus \{0\}; \mathbb{R}^N)$,
(A_2) $A(x, 0) = 0$, $\forall x \in \Omega$,

(A_3) $\sum_{i,j=1}^{N} |\frac{\partial A_j}{\partial \eta_i}(x,\eta)| \leq \Gamma |\eta|^{p-2}$, $\forall x \in \Omega$, $\eta \in \mathbb{R}^N \setminus \{0\}$,

(A_4) $\sum_{i,j=1}^{N} \frac{\partial A_j}{\partial \eta_i}(x,\eta)\xi_i \xi_j \geq \gamma |\eta|^{p-2}|\xi|^2$, $\forall x \in \Omega$, $\eta \in \mathbb{R}^N \setminus \{0\}$, $\xi \in \mathbb{R}^N$

with $1 < p < \infty$ and for suitable constants $\gamma, \Gamma \geq 0$.

In case of the p-Laplacian operator $A = A(\eta) = |\eta|^{p-2}\eta$. (All inequalities in the following of this section are meant to be satisfied in a weak sense.)

Theorem 1.14.1 (Weak maximum Principle, L. Damascelli [63]) *Suppose Ω is bounded and $u \in W^{1,p}(\Omega) \cap L^\infty(\Omega)$, $1 < p < \infty$, satisfies*

$$-\operatorname{div} A(x, Du) + g(x, u) - \Lambda |u|^{p-2}u \leq 0 \, [\geq 0] \quad in \; \Omega \qquad (1.21)$$

where $\Lambda \geq 0$ and $g \in C(\bar{\Omega} \times \mathbb{R})$ satisfies $g(x,s) \geq 0$ if $s \geq 0$ $[g(x,u) \leq 0$ if $s \leq 0]$. Let $\Omega' \subseteq \Omega$ be open and suppose $u \leq 0$ $[\geq 0]$ on $\partial \Omega'$.

Then there exists a constant $c > 0$ depending on p and on γ, Γ in $(A_3), (A_4)$, such that if $\Lambda(\frac{|\Omega'|}{\omega_N})^{p/N} < c$ then $u \leq 0$ $[\geq 0]$ in Ω' (where $|\cdot|$ stands for the Lebesgue measure and ω_N is the measure of the unit ball in $\mathbb{R}^N$). In particular if $\Lambda = 0$ then Ω' can be an arbitrary open subset of Ω.

Let us put, if u, v are functions in $W^{1,\infty}(\Omega)$ and $\Omega_1 \subseteq \Omega$

$$M_{\Omega_1} = M_{\Omega_1}(u, v) = \sup_{\Omega_1}\big(|Du| + |Dv|\big),$$

$$m_{\Omega_1} = m_{\Omega_1}(u, v) = \inf_{\Omega_1}\big(|Du| + |Dv|\big).$$

Theorem 1.14.2 (Weak Comparison Principle, L. Damascelli [63]) *Let Ω be bounded and $u, v \in W^{1,\infty}(\Omega)$ satisfy*

$$-\operatorname{div} A(x, Du) + g(x, u) - \Lambda u \leq -\operatorname{div} A(x, Dv) + g(x, v) - \Lambda v \quad in \; \Omega \quad (1.22)$$

where $\Lambda \geq 0$ and $g \in C(\bar{\Omega} \times \mathbb{R})$ is such that for each $x \in \Omega$ $g(x,s)$ is nondecreasing in s for $|s| \leq \max\{\|u\|_{L^\infty}, \|v\|_{L^\infty}\}$. Let $\Omega' \subseteq \Omega$ be open and suppose $u \leq v$ on $\partial \Omega'$.

(a) *if $\Lambda = 0$ then $u \leq v$ in Ω', $\forall p > 1$.*

(b) *if $p = 2$ there exists $\delta > 0$, depending on Λ and γ, Γ in $(A_3), (A_4)$, such that if $|\Omega'| < \delta$ then $u \leq v$ in Ω'.*

(c) *if $1 < p < 2$ there exist $\delta, M > 0$, depending on $p, \Lambda, \gamma, \Gamma, |\Omega|$ and M_Ω, such that the following holds: if $\Omega' = \Omega_1 \cup \Omega_2$ with $|\Omega_1 \cap \Omega_2| = 0$, $|\Omega_1| < \delta$ and $M_{\Omega_2} < M$ then $u \leq v$ in Ω'.*

(d) *if $p > 2$ and $m_\Omega > 0$, there exist $\delta, m > 0$, depending on $p, \Lambda, \gamma, \Gamma, |\Omega|$ and m_Ω, such that the following holds: if $\Omega' = \Omega_1 \cup \Omega_2$ with $|\Omega_1 \cap \Omega_2| = 0$, $|\Omega_1| < \delta$ and $m_{\Omega_2} > m$ then $u \leq v$ in Ω'.*

Next we deal with a form of the strong comparison principle. The strong maximum principle is well known for the kind of operators we are talking about and can be obtained via Hopf Lemma or as a consequence of a Harnack type inequality. We shall follow the second approach to derive a strong comparison theorem. First we give the following Harnack type comparison inequality.

Theorem 1.14.3 (Harnack type comparison inequality, L. Damascelli [63]) *Suppose u, v satisfy*

$$-\operatorname{div} A(x, Du) + \Lambda u \le -\operatorname{div} A(x, Dv) + \Lambda v, \quad u \le v \quad in \ \Omega, \tag{1.23}$$

where $\Lambda \in \mathbb{R}$ and $u, v \in W^{1,\infty}_{loc}(\Omega)$ if $p \ne 2$; $u, v \in W^{1,2}_{loc}(\Omega) \cap L^{\infty}_{loc}(\Omega)$ if $p = 2$. Suppose $\overline{B(x_0, 5\delta)} \subseteq \Omega$ and, if $p \ne 2$, $\inf_{B(x_0, 5\delta)}(|Du| + |Dv|) > 0$. Then for any positive $s < \frac{N}{N-2}$ we have

$$\|v - u\|_{L^s(B(x_0, 2\delta))} \le c\delta^{\frac{N}{s}} \inf_{B(x_0,\delta)} (v - u) \tag{1.24}$$

where c is a constant depending on N, p, s, Λ, δ, then constants γ, Γ in $(A_3), (A_4)$, and if $p \ne 2$ also depending on m and M, where $m = \inf_{B(x_0, 5\delta)}(|Du| + |Dv|)$, $M = \sup_{B(x_0, 5\delta)}(|Du| + |Dv|)$.

Theorem 1.14.3 implies the following strong comparison principle.

Theorem 1.14.4 (Strong Comparison Principle, L. Damascelli [63]) *Let $u, v \in C^1(\Omega)$ satisfy (1.23) and define $Z = \{x \in \Omega : |Du(x)| + |Dv(x)| = 0\}$ if $p \ne 2$, $Z = \emptyset$ if $p = 2$.*

If $x_0 \in \Omega \setminus Z$ and $u(x_0) = v(x_0)$ then $u = v$ in the connected component of $\Omega \setminus Z$ containing x_0.

If in (1.23) $\Lambda = 0$ we can get further results, as the following corollaries show. The first one is a corollary to Theorem 1.14.2(a) and, in the case when the set S defined below is compact, it has been proved in [98] by another method. The second one is a corollary to Theorem 1.14.4 (and Corollary 1.14.1).

Corollary 1.14.1 (L. Damascelli [63]) *Suppose $u, v \in C^1(\Omega)$ satisfy*

$$-\operatorname{div} A(x, Du) \le -\operatorname{div} A(x, Dv), \quad u \le v \quad in \ \Omega \tag{1.25}$$

Let us define $S = \{x \in \Omega : u(x) = u(x)\}$. If S is either discrete or compact in Ω then it is empty.

Corollary 1.14.2 (L. Damascelli [63]) *Let $u, v \in C^1(\Omega)$ satisfy (1.25). Let us define $Z = \{x \in \Omega : Du(x) = Dv(x)\}$ and suppose that either*

(a) *Ω is connected and Z is discrete or*
(b) *Z is compact and $\Omega \setminus Z$ is connected.*

Then $u < v$ unless $u \equiv v$.

We apply the previous comparison theorems to the study of symmetry and monotonicity properties of solutions to quasi-linear elliptic equations. For simplicity we consider here the case of the p-Laplacian operator that we denote by Δ_p, so that $-\Delta_p u$ stands for $-\operatorname{div}(|Du|^{p-2}Du)$. Let Ω be a bounded domain in $\mathbb{R}^N$, $N > 2$, which is convex and symmetric in the x_1-direction and consider the problem

$$\begin{cases} -\Delta_p u = f(u) & \text{in } \Omega, \\ u > 0 & \text{in } \Omega, \\ u = 0 & \text{on } \Omega. \end{cases} \tag{1.26}$$

We make the following notations.

Let Ω be a bounded domain in $\mathbb{R}^N$, $N > 2$, convex and symmetric in the x_1-direction (i.e., for each $x' \in \mathbb{R}^{N-1}$ the set $\{x_1 \in \mathbb{R} : (x_1, x') \in \Omega\}$ is either empty or an open interval symmetric with respect to 0). For such a domain we set $-a = \inf_{x \in \Omega} x_1$ and for $-a < \lambda < a$ we define

$$T_\lambda = \{x \in \mathbb{R}^N : x_1 = \lambda\}, \qquad \Omega_\lambda = \{x \in \Omega : x_1 < \lambda\},$$

$$\Omega^\lambda = \{x \in \Omega : x_1 > \lambda\}.$$

If $x = (x_1, x')$ let $x_\lambda = (2\lambda - x_1, x')$ be the point corresponding to x in the reflection through T_λ and if u is a real function in Ω let us put $u_\lambda(x) = u(x_\lambda)$ whenever $x, x_\lambda \in \Omega$. Finally if $u \in C^1(\Omega)$ we put

$$Z = \{x \in \Omega : Du(x) = 0\}$$

and

$$Z_\lambda = \{x \in \Omega_\lambda : Du(x) = Du_\lambda(x) = 0\} \quad \text{for } -a < \lambda \leq 0,$$

$$Z^\lambda = \{x \in \Omega^\lambda : Du(x) = Du_\lambda(x) = 0\} \quad \text{for } 0 \leq \lambda < a.$$

Theorem 1.14.5 (L. Damascelli [63]) *Let $1 < p < 2$ and $u \in C^1(\Omega)$ a weak solution of* (1.26) *with f locally Lipschitz continuous. Suppose that the following condition holds*:

- *if $\lambda < 0$ and C_λ is a connected component of Ω_λ then $C_\lambda \setminus Z_\lambda$ is connected, with the analogous condition satisfied by $C^\lambda \setminus Z^\lambda$ for $\lambda > 0$.*

Then u is symmetric with respect to the hyperplane $T_0 = \{x \in \mathbb{R}^N : x_1 = 0\}$ (i.e., $u(x_1, x') = u(-x_1, x')$ if $(x_1, x') \in \Omega$) and $u(x_1, x')$ is non-decreasing in x_1 for $x_1 < 0$ (and $(x_1, x') \in \Omega$).

The condition in the above theorem is in particular satisfied if the set Z is discrete. In this case the solution is strictly monotone:

Corollary 1.14.3 (L. Damascelli [63]) *Suppose that Z is discrete (and $1 < p < 2$). Then $u(x_1, x')$ is strictly increasing in x_1 for $x_1 < 0$ and if $\Omega = B(0, R)$ then u is radial and radially strictly decreasing.*

Theorem 1.14.6 (L. Damascelli [63]) *Let $u \in C^1(\bar{\Omega})$ be a weak solution of problem (1.26), where $p > 2$ and f is locally Lipschitz continuous. Suppose that the set where the gradient of u vanishes is contained in the hyperplane $T_0 = \{x \in \mathbb{R}^N : x_1 = 0\}$. Then u is symmetric with respect to T_0 and $u(x_1, x')$ is strictly increasing in x_1 for $x_1 < 0$.*

Corollary 1.14.4 (L. Damascelli [63]) *Let Ω be a ball $B(0, R)$ in $\mathbb{R}^N$, $N \geq 2$ and suppose f is locally Lipschitz and $u \in C^1(\Omega)$ is a weak solution of (1.26) whose gradient vanishes only at the origin. Then u is radial and radially strictly decreasing.*

Lemma 1.14.1 (L. Damascelli [63]) *There exist constants c_1, c_2, depending on p and on the constants γ, Γ in (A_3), (A_4) such that $\forall \eta, \eta' \in \mathbb{R}^N$ with $|\eta| + |\eta'| > 0$, $\forall x \in \Omega$:*

$$\left| A(x, \eta) - A\left(x, \eta'\right) \right| \leq c_1 \left(|\eta| + \left|\eta'\right| \right)^{p-2} \left| \eta - \eta' \right|, \qquad (1.27)$$

$$\left[A(x, \eta) - A\left(x, \eta'\right) \right] \cdot \left[\eta - \eta' \right] \geq c_2 \left(|\eta| + \left|\eta'\right| \right)^{p-2} \left| \eta - \eta' \right|^2, \qquad (1.28)$$

where the dot stands for the scalar product in $\mathbb{R}^N$. In particular, since (A_2) holds, we have for any $x \in \Omega$, $\eta \in \mathbb{R}^N$:

$$\left| A(x, \eta) \right| \leq c_1 |\eta|^{p-1}, \qquad (1.29)$$

$$A(x, \eta) \cdot \eta \geq c_2 |\eta|^p. \qquad (1.30)$$

Moreover, for each $x \in \Omega$, $\eta, \eta' \in \mathbb{R}^N$, we have

$$\left| A(x, \eta) - A\left(x, \eta'\right) \right| \leq c_1 \left| \eta - \eta' \right|^{p-1} \qquad \textit{if } 1 < p \leq 2, \qquad (1.31)$$

$$\left[A(x, \eta) - A\left(x, \eta'\right) \right] \cdot \left[\eta - \eta' \right] \geq c_2 \left| \eta - \eta' \right|^p \quad \textit{if } p \geq 2. \qquad (1.32)$$

If $u, v \in W_0^{1,p}(\Omega) \cap L^\infty_{\text{loc}}(\Omega)$ and $\beta \in C^0(\bar{\Omega} \times \mathbb{R})$ we say that (in a weak sense)

$$-\operatorname{div} A(x, Du) + \beta(x, u) \leq \begin{cases} -\operatorname{div} A(x, Dv) + \beta(x, v) \\ 0 \end{cases} \quad \text{in } \Omega \qquad (1.33)$$

if for each non-negative $\varphi \in C_c^\infty(\Omega)$ we have

$$\int_\Omega \left[A(x, Du) \cdot D\varphi + \beta(x, u)\varphi \right] dx$$

$$\leq \begin{cases} \int_\Omega [A(x, Dv) \cdot D\varphi + \beta(x, v)\varphi] dx, \\ 0. \end{cases} \qquad (1.34)$$

If Ω is bounded and $u, v \in W_0^{1,p}(\Omega) \cap L^\infty(\Omega)$ since β is continuous and (1.29) holds, by a density argument (1.34) holds for any non-negative $\varphi \in W_0^{1,p}(\Omega)$.

Similarly by $u \leq v$ on $\partial\Omega$ (in the weak sense) we mean $(u - v)^+ \in W_0^{1,p}(\Omega)$. Of course if u and v are continuous in $\bar{\Omega}$ and satisfy $u \leq v$ point-wisely on $\partial\Omega$ then they satisfy the inequality also weakly.

Lemma 1.14.2 (Poincaré's inequality, L. Damascelli [63]) *Let Ω be a bounded open set and suppose $\Omega = A \cup B$, with A, B measurable subset of Ω. If $u \in W_0^{1,p}(\Omega)$, $1 < p < \infty$, then*

$$\|u\|_{L^p(\Omega)} \leq \omega_N^{-\frac{1}{N}} |\Omega|^{\frac{1}{Np}} \left[|A|^{\frac{1}{Np'}} \|Du\|_{L^p(A)} + |B|^{\frac{1}{Np'}} \|Du\|_{L^p(B)} \right] \qquad (1.35)$$

where $p' = \frac{p}{p-1}$.

Theorem 1.14.7 (Harnack type inequality, L. Damascelli [63]) *Suppose that $v \in W_{\mathrm{loc}}^{1,p}(\Omega) \cap L_{\mathrm{loc}}^\infty(\Omega)$ satisfies*

$$-\mathrm{div}\, A(x, Dv) + \Lambda |v|^{p-2} v \geq 0, \quad v \geq 0 \quad in \ \Omega \qquad (1.36)$$

for a constant $\Lambda \in \mathbb{R}$. Let $x_0 \in \Omega$, $\delta > 0$ with $\overline{B(x_0, 5\delta)} \subseteq \Omega$ and $s > 0$ with $s < \frac{N(p-1)}{N-p}$ if $p \leq N$, $s \leq \infty$ if $p > N$.

Then there exists a constant $c > 0$ depending on N, p, s, Λ, δ and on the constants γ, Γ in $(A_3), (A_4)$ such that

$$\|v\|_{L^s(B(x_0,2\delta))} \leq c\delta^{\frac{N}{s}} \inf_{B(x_0,\delta)} v. \qquad (1.37)$$

The following strong maximum principle follows at once from the Harnack inequality.

Theorem 1.14.8 (Strong maximum principle, L. Damascelli [63]) *Suppose that Ω is connected and $v \in W_{\mathrm{loc}}^{1,p}(\Omega) \cap C^0(\Omega)$ satisfies (1.36). Then either $v \equiv 0$ in Ω or $v > 0$ in Ω.*

Chapter 2
Cone and Partial Order Methods

2.1 Increasing Operators

Suppose that E is a real Banach space, θ is the zero element of E, $P \subset E$ is a cone. $A : D \to E$ is called an increasing operator, if $\forall x_1, x_2 \in D \subset E$, $x_1 \leq x_2 \Rightarrow Ax_1 \leq Ax_2$.

Theorem 2.1.1 (Jingxian Sun, see [170, 171]) *Suppose $u_0, v_0 \in E$, $u_0 < v_0$, $A : [u_0, v_0] \to E$ is an increasing operator satisfying*

$$u_0 \leq Au_0, \qquad Av_0 \leq v_0. \tag{2.1}$$

And if $A([u_0, v_0])$ is relatively compact (sequentially compact), then A must have minimal fixed point x_* and maximal fixed point x^*, which satisfy

$$u_0 \leq u_1 \leq \cdots \leq u_n \leq \cdots \leq x_* \leq \cdots \leq x^* \leq \cdots \leq v_n \leq \cdots \leq v_1 \leq v_0. \tag{2.2}$$

Proof (a) Firstly we prove A has a fixed point in $[u_0, v_0]$. Let $D = [u_0, v_0]$. By (2.1) and A is increasing, we know

$$A(D) \subset D. \tag{2.3}$$

Define $F = \{x \in D : Ax \geq x\}$. Because $u_0 \in F$, we have $F \neq \emptyset$. Suppose H is a totally ordered subset of F. Since A is increasing, $A(H)$ is also a totally ordered set. Noticing $A(D)$ is relatively compact thus separable and $A(H) \subset A(D)$, we conclude $A(H)$ is separable. So there exist a countable set $V = \{y_1, y_2, \ldots, y_n, \ldots\} \subset A(H)$ which is dense in $A(H)$. Since V is totally ordered, $z_n = \max\{y_1, y_2, \ldots, y_n\}$ exists (z_n is one of $y_1, y_2, \ldots, y_n$), and

$$z_n \in V \subset A(H) \quad (n = 1, 2, 3, \ldots), \tag{2.4}$$

$$z_1 \leq z_2 \leq \cdots \leq z_n \leq \cdots . \tag{2.5}$$

Since $A(D)$ is relatively compact, we have a subsequence $\{z_{n_i}\} \subset \{z_n\}$, such that

$$z_{n_i} \to z^* \in E \quad (i \to \infty). \tag{2.6}$$

Z. Zhang, *Variational, Topological, and Partial Order Methods with Their Applications,*
Developments in Mathematics 29,
DOI 10.1007/978-3-642-30709-6_2, © Springer-Verlag Berlin Heidelberg 2013

Through (2.3), (2.4), and (2.6), we know

$$z^* \in D. \tag{2.7}$$

Combining (2.5) and (2.6),

$$y_n \leq z_n \leq z^* \quad (n = 1, 2, 3, \ldots). \tag{2.8}$$

Also because V is dense in $A(H)$ we have $z \leq z^*$, $\forall z \in A(H)$. Thus

$$x \leq Ax \leq z^*, \quad \forall x \in H, \tag{2.9}$$

then

$$Ax \leq Az^* \quad \forall x \in H, \tag{2.10}$$

and

$$z_n \leq Az^* \quad (n = 1, 2, 3, \ldots). \tag{2.11}$$

By (2.6) and (2.11), we have

$$z^* \leq Az^*. \tag{2.12}$$

So $z^* \in F$. Also (2.9) implies z^* is an upper bound for H in F. So we can conclude F has a maximal element v^* by Zorn's Lemma. Since $v^* \in F$, $A(v^*) \geq v^*$. Also we have $A(A(v^*)) \geq A(v^*)$, so $A(v^*) \in F$. Because v^* is maximal, we have $A(v^*) = v^*$, which means v^* is a fixed point of A.

Similarly consider $G = \{s \in D : Ax \leq x\}$. We can obtain G has a minimal element u^* and $A(u^*) = u^*$.

(b) Secondly we prove A has minimal fixed point and maximal fixed point in D. Let $\mathrm{Fix}(A) = \{x \in D : Ax = x\}$. Then from the above proof, $\mathrm{Fix}(A) \neq \emptyset$. Let $S = \{I = [u, v] : u, v \in D, u \leq v, u \leq Au, Av \leq v, \mathrm{Fix}(A) \subset [u, v]\}$. Since $D \in S$, then $S \neq \emptyset$. S is a partial ordered set if we define $I_1 \leq I_2$ for $\forall I_1, I_2 \in S$ and $I_1 \subset I_2$. Suppose $S_1 = \{I_r : r \in \Lambda\}$ is a totally ordered subset of S, where $I_r = [u_r, v_r]$. Let $H_1 = \{u_r : r \in \Lambda\}$, then H_1 is a totally ordered subset of D and $Au_r \geq u_r$ $(r \in \Lambda)$. By substitute H by H_1 in the proof of part (a), we know there exist $z^* \in D$ and $\{z_n\} \subset A(H_1)$ such that $z_n \to z^*$ as $n \to \infty$ and

$$u_r \leq Au_r \leq z^*, \quad r \in \Lambda, \tag{2.13}$$

$$z^* \leq Az^*. \tag{2.14}$$

Since $u_r \leq x$, $\forall r \in \Lambda$, $x \in \mathrm{Fix}(A)$, then

$$u_r \leq Au_r \leq Ax = x, \quad r \in \Lambda, \ x \in \mathrm{Fix}(A),$$

so

$$z_n \leq x, \quad x \in \mathrm{Fix}(A) \quad (n = 1, 2, 3, \ldots). \tag{2.15}$$

Because $z_n \to z^*$, the above inequality implies that

$$z^* \leq x, \quad x \in \mathrm{Fix}(A). \tag{2.16}$$

Similarly considering $G_1 = \{v_r : r \in \Lambda\}$, we can obtain there exists $w^* \in D$ that satisfies

$$v_r \geq w^*, \quad r \in \Lambda, \tag{2.17}$$

$$w^* \geq Aw^*, \tag{2.18}$$

$$w^* \geq x, \quad x \in \mathrm{Fix}(A). \tag{2.19}$$

Combining (2.16), (2.19) and $\mathrm{Fix}(A) \neq \emptyset$, it has $z^* \leq w^*$. Let $I^* = [z^*, w^*]$. Then $I^* \in S$. Since (2.13) and (2.17), I^* is a lower bound for S_1. So by Zorn's Lemma, we find S has a minimal element $I_* = [x_*, x^*]$. It must have

$$x_* \leq x \leq x^*, \quad x \in \mathrm{Fix}(A). \tag{2.20}$$

By the definition of S and $I^* \in S$, it is easy to know

$$x_* \leq Ax_* \leq Ax^* \leq x^*, \tag{2.21}$$

$$Ax_* \leq Ax = x \leq Ax^*, \quad \forall x \in \mathrm{Fix}(A). \tag{2.22}$$

Since A is increasing,

$$x_* \leq Ax_* \leq A\left(Ax^*\right) \leq A(Ax_*) \leq Ax^* \leq x^*. \tag{2.23}$$

Through (2.22) and (2.23), we get $\bar{I} = \{Ax_*, Ax^*\} \in S$ and $\bar{I} \leq I_*$. Since I_* is a minimal element, it requires $\bar{I} - I_*$, which implies $Ax_* = x_*$ and $Ax^* = x^*$. (2.20) shows x_* and x^* is the minimal fixed point and maximal fixed point of A.

Finally, since A is increasing and $u_0 \leq x_* \leq x^* \leq v_0$, it is easy to obtain (2.2). $\square$

Remark 2.1.1 From the above proof, the condition that $A([u_0, v_0])$ is relatively compact can bc rclaxcd to $A(u_0, v_0)$ is relatively compact for every totally ordered subset. Additionally, by using the same method, the above theorem can be generalized to the following one: Suppose $u_0, v_0 \in E$, $u_0 < v_0$, $A : [u_0, v_0] \to E$ is an increasing operator satisfying (2.1). If $A = \sum_{i=1}^{m} C_i B_i$, where $B_i : [u_0, v_0] \to E_i$ (E_i is an ordered Banach space) and $C_i : [B_i u_0, B_i v_0] \to E$ is increasing operator and $B_i([u_0, v_0])$ is relatively compact for every totally ordered subset ($i = 1, 2, \ldots, m$), then the conclusion of Theorem 2.1.1 holds. (See Sun and Zhao [173].)

Definition 2.1.1 An operator $T : D(T) \subset E \to E$ is said to be convex if for x, $y \in D(T)$ with $x \leq y$ and every $t \in [0, 1]$, we have

$$T\left(tx + (1-t)y\right) \leq tT(x) + (1-t)T(y). \tag{2.24}$$

T is concave if $-T$ is convex.

Lemma 2.1.1 (Yihong Du, see [88, 89]) *Suppose P is a normal cone, $v > \theta$, $A : [\theta, v] \to E$ is a concave and increasing operator. If there exists $0 < \varepsilon < 1$ such that $A\theta \geq \varepsilon v$, $Av \leq v$, then A has minimal fixed point u^* in $[\theta, v]$, $u^* > \theta$. Let $u_0 = \theta$, $u_n = Au_{n-1}$ $(n = 1, 2, 3, \ldots)$, then we have*

$$\|u_n - u^*\| \leq N\|A\theta\|\varepsilon^{-2}(1 - \varepsilon)^n \quad (n = 1, 2, 3, \ldots). \tag{2.25}$$

Here N is the normal constant of P.

Proof Let $\tau = 1 - \varepsilon (0 < \tau < 1)$, $B = \tau^{-1}A$. Then $B : [\theta, v] \to E$ is a concave and increasing operator. Also $B\theta \geq \tau^{-1}\varepsilon v$, $Bv \leq \tau^{-1}v$, $u_n = \tau Bu_{n-1}$ $(n = 1, 2, 3, \ldots)$. Obviously $u_1 = A\theta \geq \varepsilon v > \theta$ and $Av \leq v$, since A is increasing,

$$\theta = u_0 \leq u_1 \leq \cdots \leq u_n \leq \cdots \leq v, \tag{2.26}$$

$$B\theta \geq \tau^{-1}\varepsilon v \geq \varepsilon Bv \geq \varepsilon Bu_n = (1 - \tau)Bu_n. \tag{2.27}$$

For $\theta \leq y \leq x \leq v$, $0 \leq t \leq 1$, we have

$$\theta \leq x - ty \leq (1 - t)x + t(x - y) \leq x \leq v,$$

since B is concave,

$$B(x - ty) \geq (1 - t)Bx + tB(x - y),$$

which means

$$Bx - B(x - ty) \leq t[Bx - B(x - y)], \quad \theta \leq y \leq x \leq v, \ 0 \leq t \leq 1. \tag{2.28}$$

Next we prove

$$u_{n+1} - u_n \leq \tau^n(Bu_n - B\theta) \quad (n = 1, 2, 3, \ldots). \tag{2.29}$$

Because $u_2 - u_1 = \tau(Bu_1 - B\theta)$, (2.29) holds when $n = 1$. Suppose it also holds when $n = k$, that is

$$u_{k+1} - u_k \leq \tau^k(Bu_k - B\theta). \tag{2.30}$$

Let $x = u_{k+1}$, $y = Bu_k - B\theta$. We have by (2.27), (2.28), and (2.30)

$$\theta \leq y \leq Bu_k - (1 - \tau)Bu_k = \tau Bu_k = u_{k+1} = x \leq v, \tag{2.31}$$

$$u_k \geq x - \tau^k y, \tag{2.32}$$

$$u_{k+2} - u_{k+1} = \tau Bx - \tau Bu_k \leq \tau\left[Bx - B\left(x - \tau^k y\right)\right] \tag{2.33}$$

$$\leq \tau^{k+1}\left[Bx - B(x - y)\right] \leq \tau^{k+1}(Bx - B\theta), \tag{2.34}$$

so (2.29) holds when $n = k + 1$. By induction, for every natural number n, (2.29) holds.

Since (2.29) and (2.27), we have

$$u_{n+1} - u_n \le \tau^n \left(\frac{B\theta}{1-\tau} - B\theta \right) = \frac{\tau^{n+1}}{1-\tau} B\theta,$$

so if $m > n$,

$$\theta \le u_m - u_n \le (u_{n+1} - u_n) + \cdots + (u_m - u_{m-1})$$

$$\le \frac{B\theta}{1-\tau} \left(\tau^{n+1} + \cdots + \tau^m \right) \le \frac{\tau^{n+1} B\theta}{(1-\tau)^2}. \tag{2.35}$$

Since P is normal, it must have $\{u_n\}$ is Cauchy sequence. So there exists $u^* \in E$ such that $u_n \to u^*$ as $n \to \infty$. Obviously $u_n \le u^* \le v$, then $\tau Bu^* \ge \tau Bu_n = u_{n+1}$. Let $n \to \infty$, we get

$$\tau Bu^* \ge u^*. \tag{2.36}$$

In (2.35), let $m \to \infty$, we have

$$\theta \le u^* - u_n \le \frac{\tau^{n+1} B\theta}{(1-\tau)^2} \quad (n = 1, 2, 3, \ldots). \tag{2.37}$$

When n is large enough, we have $\tau^n (1-\tau)^{-2} < 1$. Combing (2.36), (2.37) with (2.28), it leads to

$$\theta \le \tau Bu^* - u^* \le \tau Bu^* - u_{n+1} = \tau \left(Bu^* - Bu_n \right)$$

$$\le \tau \left[Bu^* - B\left(u^* - \frac{\tau^n}{(1-\tau)^2} \cdot \tau B\theta \right) \right]$$

$$\le \frac{\tau^{n+1}}{(1-\tau)^2} \left[Bu^* - B\left(u^* - \tau B\theta \right) \right]$$

$$\le \frac{\tau^{n+1}}{(1-\tau)^2} \left(Bu^* - B\theta \right) \to \theta \quad (n \to \infty),$$

since P is normal, then $\tau Bu^* - u^* = \theta$, that is $u^* = Au^*$. Also, (2.37) implies (2.25). Clearly $u^* > \theta$ for $A\theta \ge \varepsilon v > \theta$. Finally, u^* is the minimal fixed point of A in $[\theta, v]$. Suppose $\theta \le \bar{x} \le v$ such that $A\bar{x} = \bar{x}$, since A is increasing, by induction we get $u_n \le \bar{x}$ $(n = 0, 1, 2, \ldots)$. As $n \to \infty$, it shows $u^* \le \bar{x}$. $\qquad\square$

Theorem 2.1.2 (Yihong Du [88]) *Suppose that the cone P is normal, $u_0, v_0 \in E$ and $u_0 < v_0$. Moreover, $A : [u_0, v_0] \to E$ is a increasing operator. Let $h_0 = v_0 - u_0$. If one of the following assumptions holds*:

(i) *A is a concave operator, $Au_0 \ge u_0 + \varepsilon h_0$, $Av_0 \le v_0$ where $\varepsilon \in (0, 1)$ is a constant*;

(ii) *A is a convex operator, $Au_0 \geq u_0$, $Av_0 \leq v_0 - \varepsilon h_0$ where $\varepsilon \in (0, 1)$ is a constant, then A has a unique fixed point x^* in $[u_0, v_0]$. Moreover, for any $x_0 \in [u_0, v_0]$, the iterative sequence $\{x_n\}$ given by $x_n = Ax_{n-1}$ $(n = 1, 2, \ldots)$ satisfying that*

$$\|x_n - x^*\| \to 0 \quad (n \to \infty),$$

$$\|x_n - x^*\| \leq M(1 - \varepsilon)^n \quad (n = 1, 2, \ldots),$$

with M a positive constant independent of x_0.

Proof Firstly, we assume that (i) holds. Let $Bx = A(x + u_0) - u_0$. Clearly, $B : [\theta, h_0] \to E$ is concave and increasing. Moreover, $B\theta \geq \varepsilon h_0$ and $Bh_0 \leq h_0$. Lemma 2.1.1 implies that B has a minimal fixed point u^* in $[\theta, h_0]$ and

$$\|u_n - u^*\| \leq M_0(1 - \varepsilon)^n \quad (n = 1, 2, \ldots),$$

where $u_0 = \theta$, $u_n = Bu_{n-1}$ $(n = 1, 2, \ldots)$, M_0 is a positive constant Let $h_n = Bh_{n-1}$ $(n = 1, 2, \ldots)$. Clearly, we have

$$h_0 \geq h_1 \geq \cdots \geq h_n \geq \cdots \geq u^*.$$

Let $t_n = \sup\{t > 0 : u^* \geq th_n\}$. Since $u^* = Bu^* \geq B\theta \geq \varepsilon h_0 \geq \varepsilon h_n$,

$$0 < \varepsilon \leq t_1 \leq \cdots \leq t_n \leq \cdots \leq 1, \qquad u^* \geq t_n h_n.$$

It follows that

$$u^* = Bu^* \geq B(t_n h_n) \geq (1 - t_n)B\theta + t_n b h_n$$

$$\geq (1 - t_n)\varepsilon h_0 + t_n h_{n+1} \geq \big[(1 - t_n)\varepsilon + t_n\big]h_{n+1}.$$

Thus, by the definition of t_{n+1}, we get $t_{n+1} \geq (1 - t_n)\varepsilon + t_n$, and hence,

$$1 - t_{n+1} \leq (1 - t_n)(1 - \varepsilon) \quad (n = 1, 2, \ldots).$$

It follows that

$$1 - t_n \leq (1 - t_1)(1 - \varepsilon)^{n-1} \leq (1 - \varepsilon)^n \quad (n = 1, 2, \ldots).$$

Notice that

$$\theta \leq h_n - u^* \leq h_n - t_n h_n \leq (1 - t_n)h_0 \leq (1 - \varepsilon)^n h_0.$$

This yields $\|h_n - u^*\| \leq N\|h_0\|(1 - \varepsilon)^n$ $(n = 1, 2, \ldots)$.

For any $y_0 \in [\theta, h_0]$, let $y_n = By_{n-1}$ $(n = 1, 2, \ldots)$. Then $u_n \leq y_n \leq h_n$ $(n = 1, 2, \ldots)$. Therefore,

$$\|y_n - u^*\| \leq \|y_n - u_n\| + \|u_n - u^*\|$$

$$\leq N\|h_n - u_n\| + \|u_n - u^*\|$$

$$\leq N\|h_n - u^*\| + (N+1)\|u_n - u^*\|$$

$$\leq M(1-\varepsilon)^n \quad (n=1,2,\ldots), \tag{2.38}$$

where $M = N^2\|h_0\| + (N+1)M_0$ is a constant. We can easily conclude from (2.38) that the fixed point of B in $[\theta, h_0]$ is unique. In fact, (2.38) yields

$$\|\bar{x} - u^*\| \leq M(1-\varepsilon)^n \to 0 \quad (n \to \infty),$$

and hence, $\bar{x} = u^*$. If we denote $x^* = u^* + u_0$, $x_n = y_n + u_0$ $(n=1,2,\ldots)$, then we see that the conclusions of this theorem hold.

On the other hand, when (ii) holds. Let $Bx = v_0 - A(v_0 - x)$. It can be easily checked that $B : [\theta, h_0] \to E$ is a concave and increasing operator satisfying $B\theta \geq \varepsilon h_0$ and $Bh_0 \leq h_0$. By similar arguments as in the case of (i), we get the conclusions of this theorem. $\qquad\square$

Corollary 2.1.1 (Yihong Du [88]) *Assume that P is a normal solid cone, $u_0, v_0 \in E$, $u_0 < v_0$, $A : [u_0, v_0] \to E$ is an increasing operator. Suppose one of the following assumptions holds*:

(i) *A is concave, $Au_0 \gg u_0$, $Av_0 \leq v_0$;*
(ii) *A is convex, $Au_0 \geq u_0$, $Av_0 \ll v_0$.*

Then A has a unique fixed point x_* in $[u_0, v_0]$. Moreover, for any $x_0 \in [u_0, v_0]$, the iterative sequence $\{x_n\}$ defined by $x_n = Ax_{n-1}$ $(n=1,2,\ldots)$ satisfying that $\|x_n - x^*\| \to 0$ $(n \to \infty)$ and

$$\|x_n - x^*\| \leq Mr^n \quad (n=1,2,\ldots), \tag{2.39}$$

where $r \in (0,1)$, $M > 0$ are constants.

Proof Let $h_0 = v_0 - u_0$. Since $Au_0 \gg u_0$ (or $Av_0 \ll v_0$), we can choose $\varepsilon \in (0,1)$ small enough such that $Au_0 \geq u_0 + \varepsilon h_0$ (or $Av_0 \leq v_0 - \varepsilon h_0$). Then, we complete the proof by applying Theorem 2.1.2. $\qquad\square$

Corollary 2.1.2 (Yihong Du [88]) *Suppose that P is a normal solid cone, $u_0, v_0 \in E$, $u_0 < v_0$ and $A : [u_0, v_0] \to E$ is a strongly positive operator (i.e., $x, y \in [u_0, v_0]$, $x < y \Rightarrow Ax \ll Ay$). Suppose one of the following two conditions holds*:

(i) *A is concave, $Au_0 > u_0$, $Av_0 \leq v_0$;*
(ii) *A is convex, $au_0 \geq u_0$, $Av_0 < v_0$.*

Then A has a unique fixed point in $[u_0, v_0]$. Moreover, for any $x_0 \in [u_0, v_0]$ the iterative sequence $\{x_n\}$ given by $x_n = Ax_{n-1}$ $(n=1,2,\ldots)$ satisfying that $\|x_n - x^*\| \to 0$ $(n \to \infty)$ and

$$\|x_n - x^*\| \leq Mr^n \quad (n=1,2,\ldots),$$

where $r \in (0,1)$, $M > 0$ are constants.

Proof Assume that (i) holds (the proof is similar if condition (ii) holds). Let $u_1 = Au_0$, then $u_1 > u_0$. Since A is a strongly increasing operator, we have $Au_1 \gg Au_0 = u_1$. Applying Corollary 2.1.1 to $[u_1, v_0]$, we see that A has a unique fixed point x^* in $[u_1, v_0]$ and if we define $x_n = Ax_{n-1}$ then (2.39) holds. Suppose $\bar{x}$ is a fixed point of A in $[u_1, v_0]$, then $\bar{x} > u_0$. Hence, $\bar{x} = A\bar{x} \gg Au_0 = u_1$. Therefore, there is no fixed point for A in $[u_0, u_1]$. Moreover, for any $x_0 \in [u_0, v_0]$, let $x_n = Ax_{n-1}$ $(n = 1, 2, \ldots)$. It follows that $x_1 \in [u_1, v_0]$, and hence, (2.39) also holds. $\square$

Example 2.1.1 Consider the following Hammerstein integral equation:

$$x(t) = \int_{\mathbb{R}^n} k(t, s) f\big(s, x(s)\big)\, ds,$$

where $k(t, s)$ is nonnegative and measurable in $\mathbb{R}^n \times \mathbb{R}^n$,

$$\lim_{t \to t_0} \int_{\mathbb{R}^n} \big|k(t, s) - k(t_0, s)\big|\, ds = 0, \quad \forall t_0 \in \mathbb{R}^n,$$

and there exist constants $M > m > 0$ such that

$$m \leq \int_{\mathbb{R}^n} k(t, s)s \leq M, \quad \forall t \in \mathbb{R}^n.$$

Moreover, for any $x \geq 0$, $f(\cdot, x)$ is measurable in $\mathbb{R}^n$; and for any $t \in \mathbb{R}^n$, $f(t, \cdot)$ is continuous in $[0, \infty)$. Furthermore, there exists $R > r > 0$ such that one of the following two conditions holds:

(i) for any $t \in \mathbb{R}^n$, $f(t, \cdot) : [r, R] \to \mathbb{R}$ is a concave increasing function satisfying that

$$f(t, r) \geq \left(\frac{1}{m} + \varepsilon_1\right) r, \quad f(r, R) \leq \frac{1}{M} R, \quad \forall t \in \mathbb{R}^n,$$

with ε_1 is a positive constant;

(ii) for any $t \in \mathbb{R}^n$, $f(t, \cdot) : [r, R] \to \mathbb{R}$ is a convex increasing function satisfying that

$$f(t, r) \geq \frac{1}{m} r, \quad f(t, R) \leq \left(\frac{1}{M} - \varepsilon_2\right) R, \quad \forall t \in \mathbb{R}^n,$$

with ε_2 is a positive constant.

We will show that the integral equation has a unique continuous solution $x^*(t)$ satisfying that $\forall t \in \mathbb{R}^n$, $r \leq x^*(t) \leq R$ and for any continuous function $x_0(t)$ with $r \leq x_0(t) \leq R$ the sequence $\{x_n(t)\}$ given by

$$x_n(t) = \int_{\mathbb{R}^n} k(t, s) f\big(s, x_{n-1}(s)\big)\, ds, \quad \forall t \in \mathbb{R}^n \ (n = 1, 2, \ldots),$$

has the following property:

$$\sup_{t \in \mathbb{R}^n} \big|x_n(t) - x^*(t)\big| \leq M_0 \tau^n \to 0 \quad (n \to \infty),$$

where $M_0 > 0$ and $\tau \in (0, 1)$ are constants independent of $x_0(t)$.

Proof Let

$$E = C_B(\mathbb{R}^n) = \left\{ x \in C(\mathbb{R}^n) : \sup_{t \in \mathbb{R}^n} |x(t)| < \infty \right\},$$

with its norm $\|x\|_{C_B} = \sup_{t \in \mathbb{R}^n} |x(t)|$, and $P = \{ s \in C_B(\mathbb{R}^n) : x(t) \geq 0, \ \forall t \in \mathbb{R}^n \}$. Then P is a normal solid cone and $\overset{\circ}{P} = \{ x \in C_B(\mathbb{R}^n) : \inf_{t \in \mathbb{R}^n} x(t) > 0 \}$. Assume that condition (i) holds (the proof is much similar in the case of (ii)). Consider the following operator:

$$(Ax)(t) = \int_{\mathbb{R}^n} k(t, s) f(s, x(s)) \, ds,$$

and let $u_0(t) \equiv r$ ($\forall t \in \mathbb{R}^n$), $v_0(t) \equiv R$ ($\forall t \in \mathbb{R}^n$). It can be easily checked that $A : [u_0, v_0] \to E$ is a concave increasing operator and $Au_0 \gg u_0$, $Av_0 \leq v_0$. Then the conclusions follows from Corollary 2.1.1. $\qquad\square$

Note the results on concave increasing operators and the example above are obtained by Du Yihong, see [88, 89].

Theorem 2.1.3 (Dajun Guo [106]) *Suppose P is a solid and normal cone, $A : P \to P$ is an increasing operator. Assume there exist $v \in \overset{\circ}{P}$ and $c > 0$ such that $\theta < Av \leq v$, $A\theta \geq cAv$. Suppose for arbitrary $0 < a < b < 1$ and every bounded set $B \subset P$, there exists $\eta = \eta(a, b, B) > 0$ such that*

$$A(tx) \geq t(1 + \eta)Ax, \quad x \in B, \ t \in [a, b]. \tag{2.40}$$

Then A has a unique fixed point x^* in P and $x^* \in (\theta, v]$. Furthermore, for every initial point $x_0 \in P$, constructing a sequence as

$$x_n = Ax_{n-1} \quad (n = 1, 2, 3, \ldots), \tag{2.41}$$

we have

$$\|x_n - x^*\| \to \infty. \tag{2.42}$$

Proof Firstly we prove A has no more than one fixed point in P. Suppose x^*, $\bar{x} \in P$ such that $x^* = Ax^*$, $\bar{x} = A\bar{x}$. We have

$$x^* = Ax^* \geq A\theta \geq cAv > \theta, \qquad \bar{x} = A\bar{x} \geq A\theta \geq cAv > \theta. \tag{2.43}$$

Since $v \in \overset{\circ}{P}$, we can find $0 < t_0 < 1$ such that $t_0 x^* \leq v$, $t_0 \bar{x} \leq v$, thus from (2.40), for certain $\eta_0 > 0$

$$Av \geq A(t_0 x^*) \geq t_0(1 + \eta_0)Ax^* = t_0(1 + \eta_0)x^*, \tag{2.44}$$

$$Av \geq A(t_0 \bar{x}) \geq t_0(1 + \eta_0)A\bar{x} = t_0(1 + \eta_0)\bar{x}. \tag{2.45}$$

Recall (2.43), there exists $0 < c_1 < 1$ such that $x^* > c_1 \bar{x}$ and $\bar{x} \geq c_1 x^*$. Define $t^* = \sup\{t > 0 : x^* \geq t\bar{x} \text{ and } \bar{x} \geq tx^*\}$. Then $c_1 \leq t \leq 1$, $x^* \geq t^* \bar{x}$, $\bar{x} \geq t^* x^*$. So if $0 < t^* < 1$, from (2.40), for certain $\eta^* > 0$

$$x^* = Ax^* \geq A(t^*\bar{x}) \geq t^*(1+\eta^*)A\bar{x} = t^*(1+\eta^*)\bar{x},$$
$$\bar{x} = A\bar{x} \geq A(t^*x^*) \geq t^*(1+\eta^*)Ax^* = t^*(1+\eta^*)x^*,$$

this contradicts the definition of t^*. So $t^* = 1$, which means $x^* = \bar{x}$.

If $x^* \in P$ such that $x^* = Ax^*$, we can prove $x \in (\theta, v]$. In fact, $x^* > \theta$ is from (2.43). Let $t' = \sup\{t > 0 : v \geq tx^*\}$. Then $0 < t' < \infty$, $v \geq t'x^*$. If $0 < t' < 1$, from (2.40), for certain $\eta' > 0$ such that

$$v \geq Av \geq A(t'x^*) \geq t'(1+\eta')Ax^* = t'(1+\eta')x^*,$$

this contradicts the definition of t'. So $t' > 1$, which means $v \geq x^*$.

Finally, we prove A has a fixed point in P and (2.42) holds. Since $v \in \mathring{P}$, we can find $0 < t_* < 1$ such that

$$t_* x_0 \leq v. \tag{2.46}$$

Let $v_0 = t_*^{-1} v$. From (2.40), there exists $\eta_* > 0$ such that

$$Av = A(t_* v_0) \geq t_*(1+\eta_*)Av_0, \tag{2.47}$$

consequently

$$\theta < Av \leq Av_0 \leq \left[t_*(1+\eta_*)\right]^{-1} Av \leq t_*^{-1} v = v_0. \tag{2.48}$$

Also

$$A\theta \geq cAv \geq c_* Av_0, \tag{2.49}$$

where $c_* = ct_*(1+\eta_*) > 0$. Now let

$$u_0 = \theta, \qquad u_n = Au_{n-1}, \qquad v_n = A(v_{n-1}) \quad (n = 1, 2, 3, \ldots). \tag{2.50}$$

Using (2.48) and the increase property of A, we get

$$u_0 \leq u_1 \leq \cdots \leq u_n \leq \cdots \leq v_n \leq \cdots \leq v_1 \leq v_0. \tag{2.51}$$

(2.49) implies $u_1 \geq c_* v_1 > \theta$. Define $t_n = \sup\{t > 0 : u_n \geq tv_n\}$ $(n = 1, 2, 3, \ldots)$, then $u_n \geq t_n v_n$. From (2.51),

$$0 < c_* \leq t_1 \leq t_2 \leq \cdots \leq t_n \leq \cdots \leq 1. \tag{2.52}$$

Suppose $t_n \to t^*$ $(n \to \infty)$, then $c_* < t^* \leq 1$. If $t^* < 1$, (2.40) implies that, for certain $\eta > 0$,

$$A(t_n v_n) \geq t_n(1+\eta)Av_n = t_n(1+\eta)v_{n+1} \quad (n = 1, 2, 3, \ldots),$$

so

$$u_{n+1} = Au_n \geq A(t_n v_n) \geq t_n(1+\eta)v_{n+1} \quad (n = 1, 2, 3, \ldots),$$

which means $t_{n+1} \geq t_n(1+\eta)$. So $t_{n+1} \geq t_1(1+\eta)^n \geq c_*(1+\eta)^n \to \infty \ (n \to \infty)$. This contradicts (2.52). Thus $\lim_{n\to\infty} t_n = t^* = 1$.

From (2.51) and $u_n \geq t_n v_n$, we get

$$\theta \leq u_{n+p} - u_n \leq v_n - u_n \leq (1 - t_n)v_n \leq (1 - t_n)v_0, \tag{2.53}$$

since P is normal,

$$\|u_{n+p} - u_n\| \leq N(1 - t_n)\|v_0\| \to 0, \quad n \to \infty,$$

where N is the normal constant in P. The above inequality means $\{u_n\}$ is a Cauchy sequence in E. Since E is complete, there exists u^* in P such that $\lim_{n\to\infty} u_n = u^*$.

Similarly by

$$\theta \leq v_n - v_{n+p} \leq v_n - u_n \leq (1 - t_n)v_n \leq (1 - t_n)v_0,$$

there exists v^* in P such that $\lim_{n\to\infty} v_n = v^*$. Obviously,

$$u_n \leq u^* \leq v^* \leq v_n, \tag{2.54}$$

and

$$\theta \leq v^* - u^* \leq v_n - u_n \leq (1 - t_n)v_n \leq (1 - t_n)v_0,$$

so $u^* = v^*$. Let $x^* = u^* = v^*$, (2.54) implies

$$u_{n+1} \leq Au_n \leq Ax^* \leq Av_n \leq v_{n+1}.$$

Let $n \to \infty$, we get $x^* \leq Ax^* \leq x^*$, so $x^* = Ax^*$. From (2.46), it can obtain $u_0 \leq x_0 \leq v_0$. Since A is increasing, by induction, $u_n \leq x_n \leq v_n$, where x_n is given by (2.41). Since $u_n \to x^*$ and $v_n \to x^*$ and P is normal, (2.42) must hold. $\qquad\Box$

2.2 Decreasing Operators

Decreasing operators are a class of important operators, but it is difficult to obtain fixed point theorems (see [100, 106]). In this section, without assuming operators to be continuous or compact, we first study decreasing operators with convexity or concavity and give existence and uniqueness theorems, then we study eigenvalue problems and structure of solution set, finally apply them to nonlinear integral equations on unbounded regions and differential equations in Banach spaces.

Suppose that E is a real Banach space, θ is the zero element of E, $P \subset E$ is a cone. $A : D \to E$ is called a decreasing operator, if $\forall x_1, x_2 \in D$, $x_1 \leq x_2 \Rightarrow Ax_1 \geq Ax_2$. $A : D \to E$ is called a strictly decreasing operator, if $\forall x_1, x_2 \in D$, $x_1 < x_2 \Rightarrow Ax_1 > Ax_2$.

Theorem 2.2.1 (Dajun Guo, see [106]) *Assume that P is a normal cone and $A : P \to P$ is a completely continuous operator. Moreover, A is a decreasing operator satisfying that $A\theta > \theta$, $A^2\theta \geq \varepsilon_0 A\theta$, with $\varepsilon_0 > 0$, and for any $(x, t) \in (\theta, A\theta] \times (0, 1)$, there exists $\eta = \eta(x, t) > 0$, such that*

$$A(tx) \leq \left[t(1 + \eta) \right]^{-1} Ax.$$

Then A has a unique fixed point $x^* > \theta$, and for any $x_0 \in P$ the iterative sequence $\{x_n\}$ given by $x_n = Ax_{n-1}$ $(n = 1, 2, \ldots)$ satisfying that

$$\left\| x_n - x^* \right\| \to 0.$$

Proof Let $u_0 = \theta$, $u_n = Au_{n-1}$ $(n = 1, 2, \ldots)$, then we have $u_0 \leq A^2 u_0$, $A^2 u_1 \leq u_1$. It follows from A is a continuous completely operator that $A^2([\theta, A\theta])$ is relatively compact. Theorem 2.1.1 implies that A^2 has a maximal fixed point u^* and a minimal fixed point u_*. Moreover,

$$\lim_{n \to \infty} u_{2n} = u_*, \qquad \lim_{n \to \infty} u_{2n+1} = u^*,$$

$$\theta = u_0 \leq u_2 \leq \cdots \leq u_{2n} \leq \cdots \leq u_*$$

$$\leq u^* \leq \cdots \leq u_{2n+1} \leq \cdots \leq u_3 \leq u_1 = A\theta. \tag{2.55}$$

Since $u_2 = A^2\theta \geq \varepsilon_0 A\theta > \theta$, $u^* \geq u_* > \theta$. Letting $n \to \infty$ in the following equalities:

$$u_{2n} = Au_{2n-1}, \qquad u_{2n+1} = u_{2n},$$

we have $u_* = Au^*$ and $u^* = Au_*$, and hence, $u_* \geq u_2 \geq \varepsilon_0 u_1 \geq \varepsilon_0 u^*$. Let $t_0 = \sup\{t > 0 | u_* \geq tu^*\}$, then $\varepsilon_0 \leq t_0 \leq 1$, $u_* \geq t_0 u^*$. If $t_0 < 1$, then there exists $\eta_0 > 0$ such that

$$A\left(t_0 u^*\right) \leq \left[t_0(1 + \eta_0) \right]^{-1} Au^* = \left[t_0(1 + \eta_0) \right]^{-1} u_*.$$

Therefore, we have

$$u^* = Au_* \leq A\left(t_0 u^*\right) \leq \left[t_0(1 + \eta_0) \right]^{-1} u_*,$$

that is $u_* \geq t_0(1 + \eta_0)u^*$ in contradiction with the definition of t_0. Hence, $t_0 = 1$ and $u_* = u^*$. If we denote $x^* = u_* = u^*$, then $x^* > \theta$ and $x^* = Ax^*$, that is, x^* is a positive fixed point of A.

If $\bar{x} > \theta$ is a fixed point of A, then $\theta < \bar{x} \leq A\theta$, $\bar{x} = A^2\bar{x}$. In other words, A^2 has a fixed point in $[\theta, A\theta]$. By the minimal of u_* and the maximal of u^*, we conclude that $u_* \leq \bar{x} \leq u^*$, and therefore, $\bar{x} = x^*$. Thus, the positive fixed point of A is unique.

Finally, we show that $\|x_n - x^*\| \to 0$. We deduced from $x_0 \geq \theta$ that $\theta \leq Ax_0 \leq A\theta$, i.e., $u_0 \leq x_1 \leq u_1$. It follows from the decreasing of A that $u_2 \leq x_2 \leq u_1$. Moreover, we can get

$$u_{2n} \leq x_{2n} \leq u_{2n-1}, u_{2n} \leq x_{2n+1} \leq u_{2n+1} \quad (n = 1, 2, \ldots).$$

Noticing that $u_* = x^* = u^*$, we conclude from Theorem 1.3.1 and (2.55) that $\|x_n - x^*\| \to 0$.
$\qquad\qquad\qquad\qquad\qquad\qquad\qquad\qquad\qquad\qquad\qquad\qquad\qquad\qquad\qquad\Box$

Theorem 2.2.2 (Zhitao Zhang, see [198, 202]) *Suppose that P is a normal cone of a real Banach space E, N is the normal constant of P, $A : P \to P$ is convex and decreasing, and $\exists \varepsilon > \frac{1}{2}$, such that*

$$A^2\theta \geq \varepsilon A\theta > \theta. \tag{2.56}$$

Then A has a unique fixed point x^* in P, and $\forall x_0 \in P$, constructing the sequence $x_n = Ax_{n-1}$, we have $\|x_n - x^*\| \to 0$ $(n \to +\infty)$, with the convergence rate $\|x_{2n} - x^*\| \leq 4N^2\|A\theta\| \cdot (\frac{1-\varepsilon}{\varepsilon})^{n-1}$, $\|x_{2n+1} - x^*\| \leq 2N^2\|A\theta\| \cdot (\frac{1-\varepsilon}{\varepsilon})^n$ $(n = 1, 2, \ldots)$.

Proof Let

$$u_0 = \theta, \qquad u_n = Au_{n-1} \quad (n = 1, 2, \ldots). \tag{2.57}$$

Since A is decreasing, we have

$$\theta = u_0 \leq u_2 \leq \cdots \leq u_{2n} \leq \cdots \leq u_{2n+1} \leq \cdots \leq u_3 \leq u_1 = A\theta, \tag{2.58}$$

and by (2.56) we get $u_2 \geq \varepsilon u_1 > \theta$, thus $u_{2n} \geq \varepsilon u_{2n+1}$ $(n = 1, 2, 3, \ldots)$. Let

$$t_n = \sup\{t > 0, u_{2n} \geq tu_{2n+1}\} \quad (n = 1, 2, 3, \ldots), \tag{2.59}$$

we have $u_{2n+2} \geq u_{2n} \geq t_n u_{2n+1} \geq t_n u_{2n+3}$, thus

$$0 < \varepsilon \leq t_1 \leq t_2 \leq \cdots \leq t_n \leq \cdots \leq 1. \tag{2.60}$$

Now we prove $t_n \to 1$ $(n \to +\infty)$.

Noticing that A is convex decreasing, by (2.59) we have $u_{2n} \geq t_n u_{2n+1}$, and

$$u_{2n+1} = Au_{2n} \leq A(t_n u_{2n+1}) = A\big(t_n u_{2n+1} + (1 - t_n)\theta\big)$$
$$\leq t_n Au_{2n+1} + (1 - t_n)A\theta = t_n u_{2n+2} + (1 - t_n)A\theta.$$

Noticing that $\varepsilon A\theta \leq A^2\theta = u_2 \leq u_{2n+2}$, $u_{2n+1} \geq u_{2n+3}$, we know that

$$u_{2n+3} \leq u_{2n+1} \leq t_n u_{2n+2} + \frac{1 - t_n}{\varepsilon} u_{2n+2} = \left(t_n + \frac{1 - t_n}{\varepsilon}\right) u_{2n+2},$$

i.e., $u_{2n+2} \geq (t_n + \frac{1-t_n}{\varepsilon})^{-1} u_{2n+3}$. Thus we get $t_{n+1} \geq (t_n + \frac{1-t_n}{\varepsilon})^{-1}$, and

$$1 - t_{n+1} \leq 1 - \frac{1}{t_n + \frac{1-t_n}{\varepsilon}} = \frac{(\frac{1}{\varepsilon} - 1)(1 - t_n)}{t_n + \frac{1-t_n}{\varepsilon}} \leq \left(\frac{1}{\varepsilon} - 1\right)(1 - t_n). \tag{2.61}$$

Noticing that $\varepsilon > \frac{1}{2}$, $\frac{1}{\varepsilon} - 1 < 1$,

$$1 - t_{n+1} \leq \left(\frac{1}{\varepsilon} - 1\right)(1 - t_n) \leq \left(\frac{1}{\varepsilon} - 1\right)^2 (1 - t_{n-1}) \leq \left(\frac{1}{\varepsilon} - 1\right)^n (1 - \varepsilon)$$

$$\leq \left(\frac{1-\varepsilon}{\varepsilon}\right)^{n+1}, \tag{2.62}$$

we have

$$t_n \to 1 \quad (n \to +\infty). \tag{2.63}$$

Since

$$\theta \leq u_{2n+2p} - u_{2n} \leq u_{2n+1} - u_{2n} \leq (1 - t_n)u_{2n+1} \leq (1 - t_n)A\theta, \tag{2.64}$$

noticing that P is normal, we know $\|u_{2n+2p} - u_{2n}\| \leq N \cdot (1 - t_n) \cdot \|A\theta\|$. By (2.62) and (2.63), we find that $\exists u^*$ such that $\lim_{n \to \infty} u_{2n} = u^*$, and

$$\|u^* - u_{2n}\| \leq (1 - t_n) \cdot N \cdot \|A\theta\| \leq \left(\frac{1-\varepsilon}{\varepsilon}\right)^n \cdot N \cdot \|A\theta\|.$$

Similarly, $\exists v^*$ such that $\lim_{n \to \infty} u_{2n+1} = v^*$, and

$$\|v^* - u_{2n+1}\| \leq (1 - t_n) \cdot N \cdot \|A\theta\| \leq \left(\frac{1-\varepsilon}{\varepsilon}\right)^n \cdot N \cdot \|A\theta\|.$$

And it is clear that $\theta \leq v^* - u^* \leq u_{2n+1} - u_{2n} \leq (1 - t_n)A\theta$, thus $\|v^* - u^*\| \leq N \cdot (1 - t_n) \cdot \|A\theta\| \to 0$ $(n \to \infty)$, so we have $u^* = v^* := x^*$, and $u_{2n} \leq x^* \leq u_{2n+1}$, $u_{2n+1} = Au_{2n} \geq Ax^* \geq Au_{2n+1} = u_{2n+2}$ $(n = 1, 2, 3, \ldots)$. Let $n \to +\infty$, we have $x^* \geq Ax^* \geq x^*$, thus $Ax^* = x^*$.

Now we prove the uniqueness. Suppose $\bar{x}$ is an arbitrary fixed point of A in P, then $u_0 = \theta \leq \bar{x} = A\bar{x} \leq A\theta = u_1$. As $u_{2n} \leq \bar{x} \leq u_{2n+1}$ $(n = 1, 2, \ldots)$, letting $n \to \infty$ we have $\bar{x} = x^*$.

At last, $\forall x_0 \in P$, we have $u_0 = \theta \leq x_1 = Ax_0 \leq A\theta = u_1$, $u_2 \leq x_2 = Ax_1 \leq u_1$. It is easy to get

$$u_{2n} \leq x_{2n} \leq u_{2n-1}, \quad u_{2n} \leq x_{2n+1} \leq u_{2n+1} \quad (n = 1, 2, \ldots), \tag{2.65}$$

thus $x_n \to x^*$. By (2.64) we get $\|u_{2n+1} - u_{2n}\| \leq N(1 - t_n) \cdot \|A\theta\|$, thus

$$\|x^* - x_{2n}\| \leq \|x^* - u_{2n}\| + \|u_{2n} - x_{2n}\| \leq 2N\|u_{2n-1} - u_{2n}\|$$

$$\leq 2N\big(\|u_{2n-1} - x^*\| + \|x^* - u_{2n}\|\big) \leq 4N^2 \cdot \left(\frac{1-\varepsilon}{\varepsilon}\right)^{n-1} \cdot \|A\theta\|,$$

$$\|x^* - x_{2n+1}\| \leq \|x^* - u_{2n}\| + \|u_{2n} - x_{2n+1}\| \leq 2N\|u_{2n+1} - u_{2n}\|$$

$$\leq 2N^2 \cdot \left(\frac{1-\varepsilon}{\varepsilon}\right)^n \cdot \|A\theta\|. \qquad \square$$

Corollary 2.2.1 *Suppose that P is a normal cone of E, $A : [\theta, v] \to [\theta, v]$ is concave decreasing such that $A^2v \leq (1 - \varepsilon)v + \varepsilon Av$, where $v > \theta$, $\frac{1}{2} < \varepsilon \leq 1$. Then A has a unique fixed point $(v - x^*)$ in $[0, v]$; moreover, $\forall x_0 \in [\theta, v]$, constructing the sequence $x_n = v - A(v - x_{n-1})$, we have $x_n \to x^*$ $(n \to \infty)$.*

Proof Let $Bx = v - A(v - x)$, $\forall x \in [0, v]$, it is easy to get $B : [\theta, v] \to [\theta, v]$ is convex and decreasing, such that $B\theta = v - Av$, $B^2\theta = v - A(v - v + Av) = v - A^2v \geq v - (1-\varepsilon)v - \varepsilon Av = \varepsilon(v - Av) = \varepsilon B\theta$. Thus by the proof of Theorem 2.2.2, we know B has a unique fixed point x^* such that $x^* = v - A(v - x^*)$, i.e., $v - x^*$ is a unique fixed point of A. Moreover, we can get x^* by iterative method. $\square$

Corollary 2.2.2 *Suppose that* $A : [u, v] \to [u, v]$ *is concave and decreasing, and* $A^2v \leq (1 - \varepsilon)v + \varepsilon Av$, *where* $\frac{1}{2} < \varepsilon \leq 1$. *Then A has a unique fixed point in* $[u, v]$.

Proof Let $Bx = A(x + u) - u$, $\forall x \in [\theta, v - u]$, it is easy to know that $B : [\theta, v - u] \to [\theta, v - u]$ is concave decreasing, and $B(v - u) = Av - u$,

$$B^2(v - u) = A^2v - u \leq (1 - \varepsilon)v + \varepsilon Av - u = (1 - \varepsilon)(v - u) + \varepsilon B(v - u).$$

Thus B satisfies the conditions of Corollary 2.2.1, and it is easy to get the conclusion. $\square$

Theorem 2.2.3 (Zhitao Zhang, see [198]) *Suppose the hypotheses of Theorem 2.2.2 are satisfied. Let $\lambda_0 = \frac{1}{2(1-\varepsilon)}$ (as $1/2 < \varepsilon < 1$), $\lambda_0 = +\infty$ (as $\varepsilon = 1$). Then $\forall \lambda \in [0, \lambda_0)$, the operator equation $\lambda Au = u$ has a unique solution $u(\lambda)$ in P, and let $u_0(\lambda) \equiv \theta$, $u_n(\lambda) = \lambda Au_{n-1}(\lambda)$, we have $u_n(\lambda) \to u(\lambda)$ $(n \to +\infty)$,*

$$\left\| u_{2n}(\lambda) - u(\lambda) \right\| \leq 4N^2 \cdot \|A\theta\| \cdot \lambda \left(\frac{1 - \varepsilon_\lambda}{\varepsilon_\lambda} \right)^{n-1}, \tag{2.66}$$

$$\left\| u_{2n+1}(\lambda) - u(\lambda) \right\| \leq 2N^2 \cdot \|A\theta\| \cdot \lambda \left(\frac{1 - \varepsilon_\lambda}{\varepsilon_\lambda} \right)^{n}, \tag{2.67}$$

where $\varepsilon_\lambda = \min\{\varepsilon, 1 - (1 - \varepsilon)\lambda\}$.

Proof (i) As $\frac{1}{2} < \varepsilon < 1$, let $\lambda_0 = \frac{1}{2(1-\varepsilon)}$, then we know that $\lambda_0 > 1$. For any $\lambda \in [0, 1]$, by $\lambda A\theta \leq A\theta$, we get $A(\lambda A\theta) \geq A^2\theta \geq \varepsilon A\theta$, thus

$$\lambda A(\lambda A\theta) \geq \varepsilon(\lambda A\theta). \tag{2.68}$$

$\forall \lambda \in [1, \lambda_0)$, since $A^2\theta = A(\frac{1}{\lambda}\lambda A\theta) \leq \frac{1}{\lambda}A(\lambda A\theta) + (1 - \frac{1}{\lambda})A\theta$, we have

$$\lambda A(\lambda A\theta) \geq \lambda^2\left[A^2\theta - \left(1 - \frac{1}{\lambda}\right)A\theta \right] \geq \lambda^2\left[\varepsilon - 1 + \frac{1}{\lambda} \right]A\theta = \left[1 - (1 - \varepsilon)\lambda\right](\lambda A\theta). \tag{2.69}$$

Since $\lambda_0 = \frac{1}{2(1-\varepsilon)}$, we know

$$1 - (1 - \varepsilon)\lambda > 1 - (1 - \varepsilon)\lambda_0 = 1 - (1 - \varepsilon) \cdot \frac{1}{2(1 - \varepsilon)} = \frac{1}{2}, \quad \forall \lambda < \lambda_0. \tag{2.70}$$

By (2.68)–(2.70), we see that $\forall \lambda \in [0, \lambda_0)$, λA satisfies all the conditions of Theorem 2.2.2, thus λA has a unique fixed point in P, i.e., $\lambda Au = u$ has a unique positive solution $u(\lambda)$, and let $u_0(\lambda) \equiv \theta$, $u_n(\lambda) = \lambda A u_{n-1}(\lambda)$, then $u_n(\lambda) \to u(\lambda)$ ($n \to +\infty$); moreover,

$$\left\| u_{2n}(\lambda) - u(\lambda) \right\| \le 4\lambda \cdot N^2 \cdot \|A\theta\| \cdot \left(\frac{1 - \varepsilon_\lambda}{\varepsilon_\lambda} \right)^{n-1} \quad (n = 1, 2, \ldots),$$

$$\left\| u_{2n+1}(\lambda) - u(\lambda) \right\| \le 2\lambda \cdot N^2 \cdot \|A\theta\| \cdot \left(\frac{1 - \varepsilon_\lambda}{\varepsilon_\lambda} \right)^{n} \quad (n = 1, 2, \ldots),$$

where $\varepsilon_\lambda = \min\{\varepsilon, 1 - (1 - \varepsilon)\lambda\}$.

(ii) As $\varepsilon = 1$, then $A\theta$ is the unique fixed point of A. Let $\lambda_0 = +\infty$, $\forall \lambda \in [0, 1]$, (2.68) is still valid, and we get $\lambda A(\lambda A\theta) \ge \lambda A\theta$; $\forall \lambda \in (1, +\infty)$, (2.69) is replaced by $\lambda A(\lambda A\theta) \ge \lambda A\theta$. Thus, $\forall \lambda \in [0, +\infty)$, by Theorem 2.2.2, we know that $\lambda A u = u$ has a unique positive solution $u(\lambda)$ such that $u(\lambda) = \lambda A\theta$, (2.66) and (2.67) are still valid. $\qquad \square$

Theorem 2.2.4 (Zhitao Zhang, see [198]) *Suppose all the hypotheses of Theorem 2.2.3 are satisfied, Then $u(\cdot) : [0, \lambda_0) \to P$ is strictly increasing, i.e., for $0 \le \lambda_1 < \lambda_2 < \lambda_0$, we have $u(\lambda_1) < u(\lambda_2)$.*

Proof If $\varepsilon = 1$, by the proof (ii) of Theorem 2.2.3, we know the conclusion is valid.

We may suppose that $\frac{1}{2} < \varepsilon < 1$. If $\lambda_2 > \lambda_1 = 0$, then by the proof of Theorem 2.2.3, we get $u(\lambda_2) > \theta = u(\lambda_1)$. We may suppose that $\lambda_1 > 0$.

Now using proof by contradiction, we first get

$$u(\lambda_2) \nleq u(\lambda_1). \tag{2.71}$$

Suppose $u(\lambda_2) \le u(\lambda_1)$, since A is decreasing, we get $\lambda_1 \cdot Au(\lambda_2) \ge \lambda_1 Au(\lambda_1) = u(\lambda_1)$, but $\lambda_1 \cdot Au(\lambda_2) = \frac{\lambda_1}{\lambda_2}\lambda_2 Au(\lambda_2) = \frac{\lambda_1}{\lambda_2}u(\lambda_2) < u(\lambda_2)$, thus $u(\lambda_1) < u(\lambda_2)$, which is a contradiction. Therefore, (2.71) is valid.

Now under the two conditions (i) $0 < \lambda_1 \le 1$, $\frac{\lambda_1}{\lambda_2} \le \frac{\varepsilon}{1-\varepsilon}$; (ii) $\lambda_1 > 1$ we prove that $\lambda_2 A(\lambda_2 Ax) \ge \lambda_1 A(\lambda_1 Ax)$, $\forall x \in P$, respectively.

(i) As $0 < \lambda_1 \le 1$, $\frac{\lambda_1}{\lambda_2} \le \frac{\varepsilon}{1-\varepsilon}$, by $\lambda_2 > \lambda_1$, $\frac{1}{2} < \varepsilon \le 1$, we get $1 < \frac{\lambda_2}{\lambda_1} \le \frac{\varepsilon}{1-\varepsilon}$ (as $\lambda_1 = 1$, it is clear that $\lambda_2 < \frac{1}{2(1-\varepsilon)} < \frac{\varepsilon}{1-\varepsilon}$). Then we have $\lambda_1 \varepsilon \ge \lambda_2(1 - \varepsilon)$, thus $(\lambda_2 - \lambda_1)((\lambda_1 + \lambda_2)\varepsilon - \lambda_2) \ge 0$, i.e.,

$$\left(\lambda_2^2 - \lambda_1^2\right)\varepsilon - \lambda_2(\lambda_2 - \lambda_1) \ge 0. \tag{2.72}$$

Since $\lambda_1 \le 1$, we get

$$A(\lambda_1 A\theta) \ge A^2\theta \ge \varepsilon A\theta. \tag{2.73}$$

Notice the following formula (2.74):

$$\forall x \in P, \quad \lambda_1 Ax \le \lambda_1 A\theta, \quad A(\lambda_1 Ax) \ge A(\lambda_1 A\theta). \tag{2.74}$$

By (2.72)–(2.74), we have

$$\frac{\lambda_2^2 - \lambda_1^2}{\lambda_1 \lambda_2} A(\lambda_1 A\theta) - \frac{\lambda_2(\lambda_2 - \lambda_1)}{\lambda_1 \lambda_2} A(\theta) \geq \theta, \tag{2.75}$$

$$\frac{\lambda_2^2 - \lambda_1^2}{\lambda_1 \lambda_2} A(\lambda_1 Ax) - \frac{\lambda_2(\lambda_2 - \lambda_1)}{\lambda_1 \lambda_2} A(\theta) \geq \theta. \tag{2.76}$$

Since A is convex, we know $A(\frac{1}{\delta} \cdot \delta u) \leq \frac{1}{\delta} A(\delta u) + (1 - \frac{1}{\delta}) A\theta$, $\forall \delta \geq 1$, $\forall u \in P$. Thus $A(\delta u) \geq \delta A(u) - \delta(1 - \frac{1}{\delta}) A\theta$, therefore,

$$A(\lambda_2 Ax) = A\left(\frac{\lambda_2}{\lambda_1} \cdot \lambda_1 Ax\right) \geq \frac{\lambda_2}{\lambda_1} A(\lambda_1 Ax) - \frac{\lambda_2}{\lambda_1}\left(1 - \frac{\lambda_1}{\lambda_2}\right) A\theta. \tag{2.77}$$

Noticing that

$$\frac{\lambda_2}{\lambda_1} A(\lambda_1 Ax) - \frac{\lambda_1}{\lambda_2} A(\lambda_1 Ax) = \frac{\lambda_2^2 - \lambda_1^2}{\lambda_1 \lambda_2} A(\lambda_1 Ax), \tag{2.78}$$

by (2.77), we have

$$A(\lambda_2 Ax) - \frac{\lambda_1}{\lambda_2} A(\lambda_1 Ax) \geq \frac{\lambda_2}{\lambda_1} A(\lambda_1 Ax) - \frac{\lambda_1}{\lambda_2} A(\lambda_1 Ax) - \left(\frac{\lambda_2}{\lambda_1} - 1\right) A\theta. \tag{2.79}$$

By (2.76), (2.78), (2.79), we get $A(\lambda_2 Ax) - \frac{\lambda_1}{\lambda_2} A(\lambda_1 Ax) \geq \theta$ i.e.,

$$\lambda_2 A(\lambda_2 Ax) \geq \lambda_1 A(\lambda_1 Ax).$$

(ii) As $\lambda_1 > 1$, $\lambda_2 > \lambda_1$, since $\lambda_0 = \frac{1}{2(1-\varepsilon)}$, we know $\lambda_1 + \lambda_2 \leq \frac{1}{1-\varepsilon}$, thus $\frac{\lambda_1}{\lambda_2} \geq \frac{\lambda_1}{\frac{1}{1-\varepsilon} - \lambda_1} = \frac{(1-\varepsilon)\lambda_1}{1 + (\varepsilon-1)\lambda_1}$, therefore, $(\frac{\lambda_1}{\lambda_2} + 1)(\varepsilon\lambda_1 - \lambda_1 + 1) \geq 1$. Moreover, $(\lambda_2^2 - \lambda_1^2)(\varepsilon\lambda_1 - \lambda_1 + 1) - \lambda_2(\lambda_2 - \lambda_1) \geq 0$. By $\lambda_1 > 1$ and the convexity of A, we get

$$A(\lambda_1 A\theta) \geq \lambda_1 A^2\theta - \lambda_1\left(1 - \frac{1}{\lambda_1}\right) A\theta \geq (\varepsilon\lambda_1 + 1 - \lambda_1) A\theta.$$

Thus (2.75) is still valid. Notice that (2.74), (2.76) are still valid. So we can prove

$$\lambda_2 A(\lambda_2 Ax) \geq \lambda_1 A(\lambda_1 Ax).$$

By (i), (ii), $\forall x \in P$, as $0 < \lambda_1 \leq 1$, $1 < \frac{\lambda_2}{\lambda_1} \leq \frac{\varepsilon}{1-\varepsilon}$, or $1 \leq \lambda_1 < \lambda_2$, we have

$$(\lambda_2 A)^2 x \geq (\lambda_1 A)^2 x. \tag{2.80}$$

Noticing that $u_n(\lambda) = \lambda A u_{n-1}(\lambda)$, $u_0(\lambda) = \theta$, $u_1(\lambda_1) = \lambda_1 A\theta < \lambda_2 A\theta = u_1(\lambda_2)$ and that $(\lambda A)^2$ is a increasing operator for fixed $\lambda \in [0, \lambda_0)$, by (2.80) and the proof of Theorem 2.2.2, we get $u(\lambda_1) \leq u(\lambda_2)$ as $0 < \lambda_1 \leq 1$, $1 < \frac{\lambda_2}{\lambda_1} \leq \frac{\varepsilon}{1-\varepsilon}$, or as $1 \leq \lambda_1 < \lambda_2$. Since λ_1 is arbitrary and partial order has transitive relation, we get $u(\lambda_1) \leq u(\lambda_2)$, $\forall 0 < \lambda_1 < \lambda_2 < \lambda_0$. By (2.71) we have $u(\lambda_1) < u(\lambda_2)$, $\forall \lambda_1, \lambda_2 \in [0, \lambda_0)$, $0 < \lambda_1 < \lambda_2$. $\square$

Theorem 2.2.5 (Zhitao Zhang, see [198]) *Suppose all the hypotheses of Theorem 2.2.3 are satisfied. Then*

(i) *$u(\lambda) : [0, \lambda_0) \to P$ is continuous.*
(ii) *$u(t\lambda) \geq tu(\lambda)$, $\forall t \in [0, 1]$, $\lambda \in [0, \lambda_0)$.*
(iii) *If A is continuous, then there exists $u(\lambda_0)$ such that $\lambda_0 Au(\lambda_0) = u(\lambda_0)$.*

Proof (i) By Theorem 2.2.4 , we get $u(\lambda_1) < u(\lambda_2)$, $\forall 0 \leq \lambda_1 < \lambda_2$. Noticing $\lambda_1 Au(\lambda_1) = u(\lambda_1)$, $\lambda_2 Au(\lambda_2) = u(\lambda_2)$, we have

$$\frac{u(\lambda_1)}{\lambda_1} = Au(\lambda_1) \geq Au(\lambda_2) = \frac{u(\lambda_2)}{\lambda_2}$$

i.e., $u(\lambda_2) \leq \frac{\lambda_2}{\lambda_1} u(\lambda_1)$, thus

$$\theta \leq u(\lambda_2) - u(\lambda_1) \leq \left(\frac{\lambda_2}{\lambda_1} - 1 \right) u(\lambda_1) \leq \left(\frac{\lambda_2}{\lambda_1} - 1 \right) \lambda_1 A\theta = (\lambda_2 - \lambda_1)A\theta.$$

Since P is normal, it is easy to know

$$\left\| u(\lambda_2) - u(\lambda_1) \right\| \leq N \cdot |\lambda_2 - \lambda_1| \cdot \|A\theta\|. \tag{2.81}$$

Thus $u(\lambda) \to u(\lambda_1)$, as $\lambda \to \lambda_1^+$; $u(\lambda) \to u(\lambda_2)$, as $\lambda \to \lambda_2^-$. So we know $u(\lambda)$ is continuous on $[0, \lambda_0)$.

(ii) $\forall t \in [0, 1]$, $\lambda \in [0, \lambda_0)$, we have $u(t\lambda) = t\lambda Au(t\lambda) \geq t\lambda Au(\lambda) = tu(\lambda)$.

(iii) By (2.81), we know $\lim_{\lambda \to \lambda_0^-} u(\lambda)$ exists; let $u(\lambda_0) = \lim_{\lambda \to \lambda_0^-} u(\lambda)$. Since $\lambda Au(\lambda) = u(\lambda)$, $\lambda \in [0, \lambda_0)$ and A is continuous, we get $\lambda_0 Au(\lambda_0) = u(\lambda_0)$. $\square$

Remark 2.2.1 It should be pointed out that A need not to be compact. In Theorems 2.2.2, 2.2.3, 2.2.4 and (i), (ii) of Theorem 2.2.5, Corollaries 2.2.1, 2.2.2, we do not assume A to be continuous. About convexity, we only use the following formula: $\forall x \in P$, $t \in [0, 1]$, $A(tx) \leq tAx + (1 - t)A\theta$.

Remark 2.2.2 By the proof of Theorem 2.2.2, for $u_n = Au_{n-1}$, we know if $\exists k_0 \geq 1$ such that $u_{2k_0} \geq \varepsilon A\theta$, the conclusion is still valid.

Remark 2.2.3 If P is normal, $A : P \to P$ is convex and decreasing, $A^2\theta \geq \varepsilon A\theta$ ($\varepsilon > 0$), $A^3\theta \geq cA\theta$ ($c > \frac{1}{2}$). Then $\forall \delta \in [0, \varepsilon]$, $\delta Au = u$ has a unique solution in P.

Proof $\forall \delta \in [0, \varepsilon]$, $A^2\theta \geq \varepsilon A\theta \geq \delta A\theta$, thus $A(\delta A\theta) \geq A^3\theta \geq cA\theta$, i.e., δA satisfies all the conditions of Theorem 2.2.2, δA has a unique fixed point in P, i.e., $\delta Au = u$ has a unique solution in P. $\square$

Example 2.2.1 Consider nonlinear Hammerstein integral equations on $\mathbb{R}^N$:

$$x(t) = Ax(t) := \int_{\mathbb{R}^N} k(t,s) \frac{1}{1+x(s)} \, ds \qquad (2.82)$$

Proposition 2.2.1 *Suppose that* $k : \mathbb{R}^N \times \mathbb{R}^N \to \mathbb{R}$ *is nonnegative and continuous,* $k \not\equiv 0$, *and* $\exists r > 0$, *such that* $\int_{\mathbb{R}^N} k(t,s)\,ds \le r < 1$. *Then* (2.82) *has a unique positive solution* $x^*(t)$, *and* $\forall x_0(t) \ge 0$, $x_n(t) = Ax_{n-1}(t)$, *we have* $x_n(t) \to x^*(t)(n \to \infty)$ *(uniformly for* $t \in \mathbb{R}^N$).

Proof Let $C_B(\mathbb{R}^N)$ denote the space of bounded continuous functions on $\mathbb{R}^N$, $\|x(\cdot)\| = \sup_{t \in \mathbb{R}^N} |x(t)|$, $P = C_B^+(\mathbb{R}^N)$ denote the nonnegative functions in $C_B(\mathbb{R}^N)$, then P is a normal and solid cone of $C_B(\mathbb{R}^N)$, and $A : P \to P$ is convex and decreasing. Noticing

$$A\theta = \int_{\mathbb{R}^N} k(t,s)\,ds > \theta, \qquad A\theta \le r < 1,$$

$$A^2\theta \ge \left(\int_{\mathbb{R}^N} k(t,s)\,ds \right) \cdot \frac{1}{1+r} = \frac{1}{1+r} A\theta$$

and $\frac{1}{1+r} > \frac{1}{2}$, by Theorem 2.2.2 we get the conclusion. $\qquad\square$

Example 2.2.2 Consider the following systems of nonlinear differential equations:

$$\begin{cases} -x_n'' = \dfrac{1}{n} \cdot (1 + x_{n+2})^{-\frac{1}{2}}, \\[2mm] x_n(0) = x_n'(1) = 0, \quad n = 1, 2, \dots. \end{cases} \qquad (2.83)$$

Let $E = \{x | x = (x_1, x_2, \dots), |x_i| < +\infty\}$, $\|x\| = \sup_i |x_i|$, $P = \{x \in E | x_i \ge 0\}$ is a normal and solid cone of E, (2.83) is equivalent to two-point boundary value problem in E.

Let $C[I, E] = \{x \mid x : I \to E$ is abstract continuous function$\}$, $I = [0, 1]$, $\bar{P} = \{x \in C[I, E] \mid x_i(t) > 0, i = 1, 2, \dots, \forall t \in I\}$ is normal and solid cone of $C[I, E]$.

Proposition 2.2.2 *System of equations* (2.83) *has a unique positive solution in* $\bar{P}$, *and* $\forall x_0(t) \in \bar{P}$, *let* $x_n = Ax_{n-1}$, *we have* $x_n \to x^*$ $(n \to +\infty)$, *where* $Ax(t) = \int_0^1 G(t,s) f(x(s))\,ds$, $G(t,s) = \min\{t, s | (t,s) \in I \times I\}$, $f(x) = (f_1(x), f_2(x), \dots)$, $f_i(x) = \frac{1}{i}(1 + x_{i+2})^{-\frac{1}{2}}$, $i = 1, 2, \dots$.

Proof It is easy to know $x(t) \in C^2[I, E] \cap \bar{P}$ is a solution of (2.83) if and only if $x \in \bar{P}$ is a solution of $Ax = x$. It is easy to know $A : \bar{P} \to \bar{P}$ is convex and decreasing, and

$$(A\theta)_i(t) = \frac{1}{i} \int_0^1 G(t,s)\,ds = \frac{1}{i}\left(t - \frac{t^2}{2}\right) \ge 0,$$

$$\left(A^2\theta\right)_i(t) = \int_0^1 G(t,s)f_i(A\theta)\,ds = \frac{1}{i}\int_0^1 G(t,s)\left(1 + \frac{2s-s^2}{2(i+2)}\right)^{-\frac{1}{2}} ds,$$

since $s - \frac{s^2}{2} \geq 0$, $\forall s \in [0,1]$, we have $(1 + \frac{s-\frac{s^2}{2}}{i+2})^{-\frac{1}{2}} \leq 1$. And by $1 + \frac{2s-s^2}{2(i+2)} \leq 1 + s - s^2/2 \leq \frac{3}{2}$, we get $(1 + \frac{2s-s^2}{2(i+2)})^{-\frac{1}{2}} \geq \sqrt{\frac{2}{3}}$, thus $A^2\theta \geq \sqrt{\frac{2}{3}}A\theta > \theta$, $\varepsilon = \sqrt{\frac{2}{3}} > \frac{1}{2}$. By Theorem 2.2.2, we get the conclusion. $\qquad\square$

Example 2.2.3 Consider eigenvector problem:

$$\begin{cases} -x_n'' = \dfrac{\lambda}{n} \cdot (1 + x_{n+2})^{-\frac{1}{2}}, \\ x_n(0) = x_n'(1) = 0, \quad n = 1, 2, \ldots. \end{cases} \tag{2.84}$$

Proposition 2.2.3 *By Proposition 2.2.2 and Theorems 2.2.3, 2.2.4, we get $\varepsilon = \sqrt{\frac{2}{3}}$, $\lambda_0 = \frac{1}{2(1-\varepsilon)} = \frac{1}{2(1-\sqrt{\frac{2}{3}})}$, $\forall \lambda \in [0, \lambda_0)$, (2.84) has a unique positive solution. Moreover, $u(\lambda)$ is continuous and increasing for λ.*

2.3 Mixed Monotone Operators

Let the real Banach space E be partially ordered by a cone P of E, i.e., $x \leq y$ if and only if $y - x \in P$. Let $D \subset E$, operator $A : D \times D \to E$ is said to be mixed monotone if $A(x,y)$ is non-decreasing in x and non-increasing in y. Point $(x^*, y^*) \in D \times D$ is called a coupled fixed point of A if $A(x^*, y^*) = x^*$ and $A(y^*, x^*) = y^*$. Element $x^* \in D$ is called a fixed point of A if $A(x^*, x^*) = x^*$.

Theorem 2.3.1 (Dajun Guo, see [101]) *Let the cone P be normal and $A : \overset{\circ}{P} \times \overset{\circ}{P} \to \overset{\circ}{P}$ be a mixed monotone operator. Suppose that there exists $0 \leq a < 1$ such that*

$$A\left(tx, t^{-1}y\right) \geq t^a A(x,y), \quad t \in (0,1), \ x, y \in \overset{\circ}{P}. \tag{2.85}$$

Then A has exactly one fixed point x^* in $\overset{\circ}{P}$ and constructing successively the sequences

$$x_n = A(x_{n-1}, y_{n-1}), \quad y_n = A(y_{n-1}, x_{n-1}) \quad (n = 1, 2, 3, \ldots) \tag{2.86}$$

for any initial $x_0, y_0 \in [\overset{\circ}{P}, \overset{\circ}{P}]$, we have

$$\left\| x_n - x^* \right\| \to 0, \quad \left\| y_n - x^* \right\| \to 0 \quad (n \to \infty), \tag{2.87}$$

with the convergence rate

$$\left\| x_n - x^* \right\| = O\left(1 - r^{a^n}\right), \quad \left\| y_n - x^* \right\| = O\left(1 - r^{a^n}\right), \tag{2.88}$$

where $0 < r < 1$ and r depends on (x_0, y_0). Moreover, for any coupled fixed points $(\bar{x}, \bar{y}) \in \overset{\circ}{P} \times \overset{\circ}{P}$ of A, it must be $\bar{x} = \bar{y} = x^*$.

Proof From hypothesis (2.85) we know first

$$A(x, y) = A\big(tt^{-1}x, t^{-1}ty\big) \geq t^a A\big(t^{-1}x, ty\big),$$

and so

$$A\big(t^{-1}x, ty\big) \leq t^{-a} A(x, y), \quad x, y \in \overset{\circ}{P},\ 0 < t < 1. \tag{2.89}$$

Let $z_0 \in \overset{\circ}{P}$ be arbitrarily given. Since $A(z_0, z_0) \in \overset{\circ}{P}$, we can choose $0 < t_0 < 1$ sufficiently small such that

$$t_0^{(1-a)/2} z_0 \leq A(z_0, z_0) \leq t_0^{-(1-a)/2} z_0. \tag{2.90}$$

Let $u_0 = t_0^{\frac{1}{2}} z_0,\ v_0 = t_0^{-\frac{1}{2}} z_0$ and

$$u_n = A(u_{n-1}, v_{n-1}), \quad v_n = A(v_{n-1}, u_{n-1}) \quad (n = 1, 2, \ldots). \tag{2.91}$$

Clearly,

$$u_0, v_0 \in \overset{\circ}{P}, \qquad u_0 < v_0, \qquad u_0 = t_0 v_0 \tag{2.92}$$

and by virtue of (2.85), (2.89) and the mixed monotone property of A, we have

$$u_1 = A(u_0, v_0) \leq A(v_0, u_0) = v_1$$

$$u_1 = A\big(t_0^{\frac{1}{2}} z_0, t_0^{-\frac{1}{2}} z_0\big) \geq t_0^{\frac{a}{2}} A(z_0, z_0) \geq t_0^{\frac{1}{2}} z_0 = u_0,$$

$$v_1 = A\big(t_0^{-\frac{1}{2}} z_0, t_0^{\frac{1}{2}} z_0\big) \leq t_0^{-\frac{a}{2}} A(z_0, z_0) \leq t_0^{-\frac{1}{2}} z_0 = v_0.$$

Now, it is easy to show by induction that

$$u_0 \leq u_1 \leq \cdots \leq u_{n-1} \leq u_n \leq \cdots \leq v_n \leq v_{n-1} \leq \cdots \leq v_1 \leq v_0. \tag{2.93}$$

If $u_n \geq t_0^{a^n} v_n$, then $v_n \leq t_0^{-a^n} u_n$ and

$$u_{n+1} = A(u_n, v_n) \geq A\big(t_0^{a^n} v_n, t_0^{-a^n} u_n\big) \geq t_0^{a^{n+1}} A(v_n, u_n)$$

$$= t_0^{a^{n+1}} v_n,$$

hence, by (2.92) and induction, we get

$$u_n \geq t_0^{a^n} v_n \quad (n = 0, 1, 2, \ldots). \tag{2.94}$$

From (2.93) and (2.94) we find

$$0 \leq u_{n+p} - u_n \leq v_n - u_n \leq \big(1 - t_0^{a^n}\big) v_n \leq \big(1 - t_0^{a^n}\big) v_0,$$

and consequently

$$\|u_{n+p} - u_n\| \le N\big(1 - t_0^{a^n}\big)\|v_0\|,$$

which implies that $\{u_n\}$ converges (in norm) to some $u^* \in E$. Similarly, we can prove that $\{v_n\}$ also converges to some $v^* \in E$ and, by (2.93),

$$u_n \le u^* \le v^* \le v_n \quad (n = 0, 1, \ldots). \tag{2.95}$$

Hence $u^*, v^* \in \overset{\circ}{P}$. Now, (2.95), (2.93) and (2.94) imply

$$0 \le v^* - u^* \le v_n - u_n \le \big(1 - t_0^{a^n}\big)v_0 \quad (n = 1, 2, \ldots), \tag{2.96}$$

and therefore $u^* = v^*$. Let $v^* = u^* = x^*$. On account of (2.95),

$$A\big(x^*, x^*\big) \ge A(u_n, v_n) = u_{n+1} \quad (n = 0, 1, \ldots),$$
$$A\big(x^*, x^*\big) \le A(v_n, u_n) = v_{n+1} \quad (n = 0, 1, \ldots).$$

Taking limit as $n \to \infty$, we get

$$x^* = u^* \le A\big(x^*, x^*\big) \le v^* = x^*,$$

hence, $A(x^*, x^*) = x^*$, i.e. x^* is a fixed point of A.

For any coupled fixed point $(\bar{x}, \bar{y}) \in \overset{\circ}{P} \times \overset{\circ}{P}$ of A, let $t_1 = \sup\{0 < t < 1 | tx^* \le \bar{x} \le t^{-1}x^*, tx^* \le \bar{y} \le t^{-1}x^*\}$. Clearly, $0 < t_1 \le 1$ and $t_1 x^* \le \bar{x} \le t_1^{-1}x^*$, $t_1 x^* \le \bar{y} \le t_1^{-1}x^*$. If $0 < t_1 < 1$, then by virtue of (2.85) and (2.89), we have

$$\bar{x} = A(\bar{x}, \bar{y}) \ge A\big(t_1 x^*, t_1^{-1}x^*\big) \ge t_1^a A\big(x^*, x^*\big) = t_1^a x^*,$$
$$\bar{x} = A(\bar{x}, \bar{y}) \le A\big(t_1^{-1}x^*, t_1 x^*\big) \le t_1^{-a} A\big(x^*, x^*\big) = t_1^{-a} x^*$$

i.e.

$$t_1^a x^* \le \bar{x} \le t_1^{-a} x^*. \tag{2.97}$$

Similarly, we get

$$t_1^a x^* \le \bar{y} \le t_1^{-a} x^*. \tag{2.98}$$

(2.97) and (2.98) contradict the definition of t_1, since $t_1^a > t_1$. Hence $t_1 = 1$ and $\bar{x} = \bar{y} = x^*$. This at the same time proves the uniqueness of fixed point of A in $\overset{\circ}{P}$.

It remains to show that (2.87) and (2.88) hold. Let $(x_0, y_0) \in \overset{\circ}{P} \times \overset{\circ}{P}$ be given. We can choose t_0 $(0 < t_0 < 1)$ so small that (2.90) holds and $t_0^{\frac{1}{2}} z_0 \le x_0 \le t_0^{-\frac{1}{2}} z_0$, $t_0^{\frac{1}{2}} z_0 \le y_0 \le t_0^{-\frac{1}{2}} z_0$, i.e. $u_0 \le x_0 \le v_0$, $u_0 \le y_0 \le v_0$. Suppose $u_{n-1} \le x_{n-1} \le v_{n-1}$, then

$$x_n = A(x_{n-1}, y_{n-1}) \ge A(u_{n-1}, v_{n-1}) = u_n,$$
$$x_n = A(x_{n-1}, y_{n-1}) \le A(v_{n-1}, u_{n-1}) = v_n,$$

and similarly, $y_n \geq u_n$, $y_n \leq v_n$. Hence, by induction,

$$u_n \leq x_n \leq v_n, \quad u_n \leq y_n \leq v_n \quad (n = 0, 1, \ldots). \tag{2.99}$$

Now, from

$$\left\| x_n - x^* \right\| \leq \left\| x_n - u_n \right\| + \left\| u_n - x^* \right\|,$$

$$0 \leq x_n - u_n \leq v_n - u_n \leq \left(1 - t_0^{a^n}\right) v_0$$

$$0 \leq u^* - u_n \leq v_n - u_n \leq \left(1 - t_0^{a^n}\right) v_0,$$

it follows that

$$\left\| x_n - x^* \right\| \leq 2N\left(1 - t_0^{a^n}\right) \|v_0\| \quad (n = 0, 1, \ldots). \tag{2.100}$$

In the same way, we get

$$\left\| y_n - x^* \right\| \leq 2N\left(1 - t_0^{a^n}\right) \|v_0\| \quad (n = 0, 1, \ldots). \tag{2.101}$$

Finally, (2.100) and (2.101) imply (2.88) with $r = t_0$, and therefore (2.87) holds. The proof is complete. $\qquad\square$

Theorem 2.3.2 (Dajun Guo, see [101]) *Let the cone P be normal and $A : \overset{\circ}{P} \times \overset{\circ}{P} \to \overset{\circ}{P}$ be a mixed monotone operator. Suppose that there exists $0 \leq a < 1$ such that (2.85) holds. Let x_t^* be the unique solution in $\overset{\circ}{P}$ of the equation*

$$A(x, x) = tx \quad (x > 0). \tag{2.102}$$

Then x_t^* is continuous with respect to t, i.e., $\|x_t^* - x_{t_0}^*\| \to 0$ as $t \to t_0$ ($t_0 > 0$). If, in addition, $0 \leq a < \frac{1}{2}$, then x_t^* is strongly decreasing with respect to t, i.e.,

$$0 < t_1 < t_2 \quad \Longrightarrow \quad x_{t_1}^* \gg x_{t_2}^*, \tag{2.103}$$

and

$$\lim_{t \to \infty} \|x_t^*\| = 0, \qquad \lim_{t \to +0} \|x_t^*\| = +\infty. \tag{2.104}$$

Proof Since the operator $t^{-1}A$ satisfies all conditions of Theorem 2.3.1, (2.102) has exactly one solution $x_t^* \in \overset{\circ}{P}$. Given $t_2 > t_1 > 0$ arbitrarily and let $s_0 = \sup\{s > 0 | x_{t_1}^* \geq s x_{t_2}^*, x_{t_2}^* \geq s x_{t_1}^*\}$. Clearly, $0 < s_0 < +\infty$ and

$$x_{t_1}^* \geq s_0 x_{t_2}^*, \qquad x_{t_2}^* \geq s_0 x_{t_1}^*. \tag{2.105}$$

It is easy to see from (2.105) that $s_0 \geq 1$ is impossible. Hence $0 < s_0 < 1$. By (2.85) and (2.105), we find

$$t_1 x_{t_1}^* = A\left(x_{t_1}^*, x_{t_1}^*\right) \geq A\left(s_0 x_{t_2}^*, s_0^{-1} x_{t_2}^*\right) \geq s_0^a A\left(x_{t_2}^*, x_{t_2}^*\right) = t_2 s_0^a x_{t_2}^*,$$

$$t_2 x_{t_2}^* = A\left(x_{t_2}^*, x_{t_2}^*\right) \ge A\left(s_0 x_{t_1}^*, s_0^a x_{t_1}^*\right) \ge s_0^a A\left(x_{t_1}^*, x_{t_1}^*\right) = t_1 s_0^a x_{t_1}^*.$$

Consequently,

$$x_{t_1}^* \ge t_2 t_1^{-1} s_0^a x_{t_2}^*, \qquad x_{t_2}^* \ge t_1 t_2^{-1} s_0^a x_{t_1}^*. \tag{2.106}$$

Observing the definition of s_0 and $t_2 t_1^{-1} s_0^a > s_0$, we conclude $t_1 t_2^{-1} s_0^a \le s_0$, and so

$$s_0 \ge (t_1/t_2)^{1/(1-a)}. \tag{2.107}$$

It follows from (2.105) and (2.107) that

$$(t_1/t_2)^{1/(1-a)} x_{t_2}^* \le x_{t_1}^* \le (t_2/t_1)^{1/(1-a)} x_{t_2}^*, \tag{2.108}$$

$$(t_1/t_2)^{1/(1-a)} x_{t_1}^* \le x_{t_2}^* \le (t_2/t_1)^{1/(1-a)} x_{t_1}^*. \tag{2.109}$$

Inequalities (2.108) and (2.109), together with the normality of cone P, imply that

$$\left\| x_{t_1}^* - x_{t_2}^* \right\| \to 0, \quad \text{as } t_1 \to t_2 - 0,$$
$$\left\| x_{t_2}^* - x_{t_1}^* \right\| \to 0, \quad \text{as } t_2 \to t_1 + 0.$$

Hence, the continuity of $x_{t_1}^*$ with respect to t $(t > 0)$ is proved.

Now, assume $0 \le a < \frac{1}{2}$, by virtue of (2.106) and (2.107), we have

$$x_{t_1}^* \ge (t_2/t_1)^{(1-2a)/(1-a)} x_{t_2}^*, \tag{2.110}$$

which implies (2.103) since

$$(t_2/t_1)^{(1-2a)/(1-a)} > 1.$$

Letting $t_1 = 1$ and $t_2 = t$ in (2.110), we find

$$x_1^* \ge t^{(1-2a)/(1-a)} x_t^*,$$

and so

$$\left\| x_t^* \right\| \ge N t^{-(1-2a)/(1-a)} \left\| x_1^* \right\|, \quad t > 1$$

which implies $\|x_t^*\| \to 0$ as $t \to +\infty$. On the other hand, letting $t_1 = t$ and $t_2 = 1$ in (2.110), we get

$$x_t^* \ge t^{-(1-2a)/(1-a)} x_1^*,$$

and therefore

$$\left\| x_t^* \right\| \ge N t^{-(1-2a)/(1-a)} \left\| x_1^* \right\|, \quad 0 < t < 1,$$

which implies $\|x_t^*\| \to +\infty$ as $t \to +0$. Hence, (2.104) holds and our theorem is proved. $\square$

Remark 2.3.1 It should be pointed out that in Theorems 2.3.1 and 2.3.2 we do not require operator A to be continuous.

As above, an existence and uniqueness theorem was established for operator $A : \overset{\circ}{P} \times \overset{\circ}{P} \to \overset{\circ}{P}$ under the condition: there exists $0 \le a < 1$ such that

$$A\left(tx, t^{-1}y\right) \ge t^{a} A(x, y), \quad x, y \in \overset{\circ}{P}, \ 0 < t < 1. \tag{2.111}$$

This result was applied to the IVP of ordinary differential equations

$$\begin{cases} x' = \displaystyle\sum_{i=1}^{n} a_i(t)x^{r_i} + \left(\sum_{j=1}^{m} b_j(t)x^{s_j}\right)^{-1}, & t \in [0, T], \\[4mm] x(0) = x_0, \end{cases} \tag{2.112}$$

where $0 < r_i < 1$, $0 < s_j < 1$, $i = 1, 2, \ldots, n$; $j = 1, 2, \ldots, m$.

Next we only assume that $A : P \times P \to P$ and replace (2.111) by, in some sense, weaker condition: for any $0 < a < b < 1$ and bounded $B \subset P$, there exists an $r = r(a, b, B) > 0$ such that

$$A\left(tx, t^{-1}y\right) \ge t(1 + r)A(x, y), \quad x, y \in B, \ t \in [a, b].$$

The result obtained can be applied to the IVP (2.112) and also the two-point BVP

$$\begin{cases} -x'' = \displaystyle\sum_{i=1}^{n} a_i(t)x^{r_i} + \left(\sum_{j=1}^{m} b_j(t)x^{s_j}\right)^{-1}, & t \in [0, 1], \\[4mm] ax(0) - bx'(0) = x_0, & cx(1) + dx'(1) = x_1, \end{cases}$$

in the more general case that some one of the r_i and s_j may be equal to 1 and another one of the r_i and s_j may be equal to 0.

Theorem 2.3.3 (Dajun Guo, see [102]) *Let the cone P be normal and $A : P \times P \to P$ be a mixed monotone operator. Suppose that*

(a) *there exist $v > 0$ and $c > 0$ such that $0 < A(v, 0) \le v$ and $A(0, v) \ge cA(v, 0)$,*
(b) *for any a, b satisfying $0 < a < b < 1$, there exists $r = r(a, b) > 0$ such that*

$$A\left(tx, t^{-1}y\right) \ge t(1 + r)A(x, y), \quad t \in [a, b], \ 0 \le y \le x \le v.$$

Then A has exactly one fixed point x^* in $[0, v]$ and $x^* > 0$. Moreover, constructing successively the sequences

$$x_n = A(x_{n-1}, y_{n-1}), \quad y_n = A(y_{n-1}, x_{n-1}) \quad (n = 1, 2, 3, \ldots) \tag{2.113}$$

for any initial $x_0, y_0 \in [0, v]$, we have

$$\left\| x_n - x^* \right\| \to 0, \quad \left\| y_n - x^* \right\| \to 0 \quad (n \to \infty). \tag{2.114}$$

Proof Set $u_0 = 0$ and choose v and c to satisfy condition (a). Then $0 < A(v, 0) \leq v$ and $A(0, v) \geq cA(v, 0)$. If we set $v_0 = v$ and define $u_1 = A(u_0, v_0)$, $v_1 = A(v_0, u_0)$, then

$$u_0 = 0 \leq A(0, v) = u_1, \qquad v_1 = A(v, 0) \leq v = v_0,$$
$$u_1 = A(0, v) \leq A(v, 0) = v_1 \quad \text{and} \quad u_1 \geq cv_1. \tag{2.115}$$

Define $u_n = A(u_{n-1}, v_{n-1})$, $v_n = A(v_{n-1}, u_{n-1})$ for $n = 1, 2, 3, \ldots$ and assume that $u_{n-1} \leq u_n \leq v_n \leq v_{n-1}$. Then

$$u_n = A(u_{n-1}, v_{n-1}) \leq A(u_n, v_n) = u_{n+1} \leq A(v_n, u_n) = v_{n+1}$$
$$\leq A(v_{n-1}, u_{n-1}) = v_n.$$

Hence, by induction,

$$0 = u_0 < u_1 \leq \cdots \leq u_{n-1} \leq u_n \leq \cdots \leq v_n \leq v_{n-1} \leq \cdots \leq v_1 \leq v_0 = v. \tag{2.116}$$

It follows form (2.115) and (2.116) that

$$u_n \geq u_1 \geq cv_1 \geq cv_n \quad (n = 1, 2, 3, \ldots). \tag{2.117}$$

Let $t_n = \sup\{t > 0 : u_n \geq tv_n\}$. Then

$$u_n \geq t_n v_n \tag{2.118}$$

and, on account of (2.116) and (2.117) and the fact $u_{n+1} \geq u_n \geq t_n v_n \geq t_n v_{n+1}$, we have

$$0 < c \leq t_1 \leq t_2 \leq \cdots \leq t_n \leq \cdots \leq 1 \tag{2.119}$$

which implies that $\exists t^*$ such that

$$\lim_{n \to \infty} t_n = t^* \tag{2.120}$$

and $0 < t^* \leq 1$. We check that

$$t^* = 1. \tag{2.121}$$

In fact, if $t^* < 1$, then $t_n \in [c, t^*]$ $(n = 1, 2, \ldots)$, and so, by virtue of (2.116) and condition (b), there exists $r > 0$ such that

$$A\left(t_n v_n, t_n^{-1} u_n\right) \geq t_n(1 + r)A(v_n, u_n) \quad (n = 1, 2, 3, \ldots). \tag{2.122}$$

Since $u_n \geq t_n u_n$, $v_n \leq t_n^{-1} u_n$, we have by (2.122)

$$u_{n+1} = A(u_n, v_n) = A\left(t_n v_n, t_n^{-1} u_n\right) \geq t_n(1 + r)A(v_n, u_n) = t_n(1 + r)v_{n+1},$$

which implies that

$$t_{n+1} \geq t_n(1 + r) \quad (n = 1, 2, 3, \ldots)$$

and therefore

$$t_n \geq t_1(+r)^n \geq c(1+r)^n \quad (n=1,2,3,\ldots).$$

Hence $t_n \to \infty$, which contradicts (2.119), and so (2.121) is true. Now, (2.116) and (2.118) imply

$$0 \leq u_{n+p} - u_n \leq v_n - u_n \leq (1-t_n)v_n \leq (1-t_n)v,$$

and so

$$\|u_{n+p} - u_n\| \leq N(1-t_n)\|v\|, \tag{2.123}$$

where N is the normal constant of P. It follows from (2.123), (2.120) and (2.121) that $\lim_{n\to\infty} u_n = u^*$ exists. In the same way we can prove that $\lim_{n\to\infty} v_n = v^*$ also exists. Since

$$u_n \leq u^* \leq v^* \leq v_n,$$

we have

$$0 \leq v^* - u^* \leq v_n - u_n \leq (1-t_n)v,$$

so

$$\left\|v^* - u^*\right\| \leq N(1-t_n)\|v\| \to 0 \quad (n \to \infty).$$

Hence $u^* = v^*$. Let $x^* = u^* = v^*$, then $x^* \in [0, v]$, $x^* > 0$. Since

$$u_n \leq x^* \leq v_n \quad (n = 1,2,3,\ldots),$$

we have

$$u_{n+1} = A(u_n, v_n) \leq A\left(x^*, x^*\right) \leq A(v_n, u_n) = v_{n+1},$$

and, after taking the limit,

$$x^* \leq A\left(x^*, x^*\right) \leq x^*.$$

Hence $A(x^*, x^*) = x^*$, i.e., x^* is a fixed point of A.

Let $\bar{x}$ be any fixed point of A in $[0, v]$. Then $u_0 = 0 \leq \bar{x} \leq v = v_0$, $A(\bar{x}, \bar{x}) = \bar{x}$, so

$$u_1 = A(u_0, v_0) \leq A(\bar{x}, \bar{x}) = \bar{x} \leq A(v_0, u_0) = v_1.$$

It is easy to see by induction that

$$u_n \leq \bar{x} \leq v_n \quad (n = 1,2,3,\ldots), \tag{2.124}$$

which implies by taking limit that $\bar{x} = x^*$.

Finally, we verify that (2.114) holds. Let $x_0, y_0 \in [0, v]$. Then, similar to (2.124), we get easily from (2.113) that

$$u_n \leq x_n \leq v_n, \quad u_n \leq y_n \leq v_n \quad (n = 1,2,3,\ldots). \tag{2.125}$$

Consequently, (2.114) follows from (2.125) and the fact that P is normal and $u_n \to x^*$ and $v_n \to x^*$. The proof is complete. $\qquad\square$

Theorem 2.3.4 (Dajun Guo [102]) *Let the cone P be normal and solid and $A : P \times P \to P$ be a mixed monotone operator. Suppose that*

(a') *there exist $v \in \overset{\circ}{P}$ and $c > 0$ such that $0 < A(v, 0) \le v$ and $A(0, v) \ge cA(v, 0)$,*
(b') *for any a, b satisfying $0 < a < b < 1$, and any bounded set $B \subset P$, there exists an $r = r(a, b, B) > 0$ such that*

$$A(tx, t^{-1}y) \ge t(1 + r)A(x, y), \quad t \in [a, b], \ x, y \in B. \tag{2.126}$$

Then A has exactly one fixed point x^* in P and $0 < x^* \le v$. Moreover, constructing successively the sequences (2.113) for any initial $x_0, y_0 \in P$, we have (2.114) holds.

Proof By Theorem 2.3.3, A has a fixed point x^* in P and $0 < x^* \le v$. Let $\bar{x}$ be any fixed point of A in P and $x_0, y_0 \in P$ be given. Since $v \in \overset{\circ}{P}$, we can choose $0 < t_0 < 1$ sufficiently small such that

$$v \ge t_0\bar{x}, \qquad v \ge t_0 x_0, \qquad v \ge t_0 y_0. \tag{2.127}$$

Observing that $A(x, y) = A(tt^{-1}x, t^{-1}ty)$, it is easy to see from condition (b') that for any $0 < a < b < 1$ and any bounded $B \subset P$, there exists $h = h(a, b, B) > 0$ such that

$$A(t^{-1}x, ty) \le \left(t(1 + h)\right)^{-1}A(x, y), \quad t \in [a, b], \ x, y \in B. \tag{2.128}$$

Now, (2.128) and conditions (a') and (b') imply that there exist $h_1 > 0$, $r_1 > 0$, $r_2 > 0$ such that

$$0 < A(v, 0) \le A\left(t_0^{-1}v, 0\right) = A\left(t_0^{-1}v, t_0 0\right) \le \left(t_0(1 + h_1)\right)^{-1}A(v, 0) \le t_0^{-1}v$$

and

$$A\left(0, t_0^{-1}v\right) = A\left(t_0, t_0^{-1}v\right) \ge t_0(1 + r_1)A(0, v) \ge t_0(1 + r_1)cA(v, 0)$$

$$= t_0(1 + r_1)cA\left(t_0 t_0^{-1}v, t_0^{-1}0\right) \ge t_0(1 + r_1)ct_0(1 + r_2)A\left(t_0^{-1}v, 0\right)$$

$$= c_1 A\left(t_0^{-1}v, 0\right),$$

where $c_1 = ct_0^2(1 + r_1)(1 + r_2) > 0$. Hence, conditions (a) and (b) of Theorem 2.3.3 are satisfied for $t_0^{-1}v$ instead of v. So, Theorem 2.3.3 implies that A has exactly one fixed point in $[0, t_0^{-1}v]$ and (2.114) holds for any initial $x_0, y_0 \in [0, t_0^{-1}v]$. Since, by (2.127), $x^*, \bar{x}, x_0$ and y_0 all belong to $[0, t_0^{-1}v]$, it must be $x^* = \bar{x}$ and (2.114) is true. The proof is complete. $\qquad\square$

Remark 2.3.2 It should be pointed out that in Theorems 2.3.3 and 2.3.4 we do not assume operator A to be continuous or compact.

Convex (concave) operators are a class of important ones , which are extensively used in nonlinear differential and integral equations (see [8, 110]). We first study mixed monotone operators with convexity and concavity. Next, not assuming operators to be continuous or compact, we give existence and uniqueness theorems, then we study existence and uniqueness and continuity of eigenvectors, finally offering some applications to nonlinear integral equations on unbounded regions and differential equations in Banach spaces.

Theorem 2.3.5 (Zhitao Zhang [196]) *Let P be a normal cone of E, $A : P \times P \to P$ be a mixed monotone operator, suppose that*

(i) *for fixed y, $A(\cdot, y) : P \to P$ is concave; for fixed x, $A(x, \cdot) : P \to P$ is convex;*
(ii) *$\exists v > \theta$, $c > \frac{1}{2}$ such that $\theta < A(v, \theta) \leq v$ and*

$$A(\theta, v) \geq cA(v, \theta). \tag{2.129}$$

Then A has exactly one fixed point $x^* \in [\theta, v]$, and constructing successively sequences

$$x_n = A(x_{n-1}, y_{n-1}), \quad y_n = A(y_{n-1}, x_{n-1}) \quad (n = 1, 2, \ldots), \tag{2.130}$$

for any initial $(x_0, y_0) \in [\theta, v] \times [\theta, v]$, we have

$$\left\| x_n - x^* \right\| \to 0, \quad \left\| y_n - x^* \right\| \to 0 \quad (n \to \infty),$$

with convergence rate

$$\left\| x_n - x^* \right\| \leq N^2 \left(\frac{1-c}{c} \right)^n \cdot \|v\|, \qquad \left\| y_n - x^* \right\| \leq N^2 \left(\frac{1-c}{c} \right)^n \cdot \|v\|. \tag{2.131}$$

Proof Let $u_0 = \theta$, $v_0 = v$, then

$$u_0 < v_0. \tag{2.132}$$

Let

$$u_n = A(u_{n-1}, v_{n-1}), \quad v_n = A(v_{n-1}, u_{n-1}) \quad (n = 1, 2, \ldots). \tag{2.133}$$

It is easy to show

$$\theta = u_0 < u_1 \leq u_2 \leq \cdots \leq u_n \leq \cdots \leq v_n \leq \cdots \leq v_2 \leq v_1 \leq v_0 = v, \tag{2.134}$$

hence, by (2.129), (2.134), we get

$$u_n \geq u_1 \geq cv_1 \geq cv_n. \tag{2.135}$$

Let

$$t_n = \sup\{t > 0, u_n \geq tv_n\} \quad (n = 1, 2, \ldots), \tag{2.136}$$

then

$$u_n \geq t_n v_n, \tag{2.137}$$

and on account of (2.135) and the fact $u_{n+1} \geq u_n \geq t_n v_n \geq t_n v_{n+1}$, we have

$$0 < c \leq t_1 \leq t_2 \leq \cdots \leq t_n \leq \cdots \leq 1, \tag{2.138}$$

which implies that $\lim_{n \to \infty} t_n = t^*$ exists and $0 < t^* \leq 1$, we check that

$$t^* = 1. \tag{2.139}$$

From the hypothesis (i), we get the following (2.140)–(2.142), $\forall x_1 \leq x_2, \ y_1 \leq y_2, \ t \in [0, 1]$:

$$A\big(tx_1 + (1-t)x_2, y\big) \geq tA(x_1, y) + (1-t)A(x_2, y), \tag{2.140}$$

$$A\big(x, ty_1 + (1-t)y_2\big) \leq tA(x, y_1) + (1-t)A(x, y_2), \tag{2.141}$$

$$A(x, y) = A\big(x, t \cdot t^{-1}y\big) \leq tA\big(x, t^{-1}y\big) + (1-t)A(x, \theta), \quad \forall t \in (0, 1]. \tag{2.142}$$

Then by (2.142), we get

$$A\big(x, t^{-1}y\big) \geq t^{-1}\big[A(x, y) - (1-t)A(x, \theta)\big], \quad \forall t \in (0, 1]. \tag{2.143}$$

By (2.133), (2.134), (2.137), (2.140)–(2.143), and the fact that A is a mixed monotone operator, we have

$$
\begin{aligned}
u_{n+1} &= A(u_n, v_n) \geq A(t_n v_n, v_n) \geq t_n A(v_n, v_n) + (1 - t_n)A(\theta, v_n) \\
&\geq t_n A\big(v_n, t_n^{-1}u_n\big) + (1 - t_n)A(\theta, v) \\
&\geq t_n \cdot \big[t_n^{-1} \cdot A(v_n, u_n) - t_n^{-1} \cdot (1 - t_n)A(v_n, \theta)\big] + (1 - t_n)A(\theta, v) \\
&= A(v_n, u_n) + (1 - t_n)\big[A(\theta, v) - A(v_n, \theta)\big] \\
&\geq v_{n+1} + (1 - t_n)[u_1 - v_1] \\
&\geq v_{n+1} + (1 - t_n)\Big(1 - \frac{1}{c}\Big)u_1 \\
&\geq v_{n+1} + (1 - t_n)\Big(1 - \frac{1}{c}\Big)v_{n+1} \\
&= \Big[1 + (1 - t_n)\Big(1 - \frac{1}{c}\Big)\Big]v_{n+1}
\end{aligned}
\tag{2.144}
$$

which implies that

$$t_{n+1} \geq 1 + (1 - t_n)\Big(1 - \frac{1}{c}\Big), \tag{2.145}$$

therefore

$$1 - t_{n+1} \leq (1 - t_n)\left(\frac{1}{c} - 1\right). \tag{2.146}$$

By the hypothesis (ii), we get $\frac{1}{2} < c \leq 1$ and $\frac{1}{c} - 1 < 1$. Thus (2.146) implies that

$$1 - t_{n+1} \leq (1 - t_n)\left(\frac{1}{c} - 1\right) \leq \left(\frac{1}{c} - 1\right)^n (1 - t_1) \leq \left(\frac{1 - c}{c}\right)^{n+1}. \tag{2.147}$$

Hence

$$t_n \to 1 \quad (n \to \infty). \tag{2.148}$$

Now from (2.134) and (2.148), we have

$$\theta \leq u_{n+p} - u_n \leq v_n - u_n \leq (1 - t_n)v_n \leq (1 - t_n)v. \tag{2.149}$$

Since P is normal, we get

$$\|u_{n+p} - u_n\| \leq N(1 - t_n) \cdot \|v\| \leq N \cdot \left(\frac{1 - c}{c}\right)^n \cdot \|v\|, \tag{2.150}$$

$$\|v_n - u_n\| \leq N(1 - t_n) \cdot \|v\| \leq N \cdot \left(\frac{1 - c}{c}\right)^n \cdot \|v\|, \tag{2.151}$$

where N is the normal constant of P. So by (2.150), we know that $\lim_{n \to \infty} u_n = u^*$ exists. In the same way we can prove that $\lim_{n \to} v_n = v^*$ exists. By (2.150), we get

$$\|u_n - u^*\| < N \cdot \left(\frac{1 - c}{c}\right)^n \cdot \|v\|. \tag{2.152}$$

Similarly,

$$\|v_n - v^*\| \leq N \cdot \left(\frac{1 - c}{c}\right)^n \cdot \|v\|. \tag{2.153}$$

Since $u_n \leq u^* \leq v^* \leq v_n$, we have

$$\theta \leq v^* - u^* \leq v_n - u_n \leq (1 - t_n)v. \tag{2.154}$$

It follows that $\|v^* - u^*\| \leq N \cdot (1 - t_n)\|v\| \to 0 \ (n \to \infty)$, hence $u^* = v^*$. Let $x^* = u^* = v^*$, then $x^* \in [\theta, v]$, $x^* > \theta$. Since

$$u_n \leq x^* \leq v_n, \qquad u_{n+1} = A(u_n, v_n) \leq A(x^*, x^*) \leq A(v_n, v_n) = v_{n+1}, \tag{2.155}$$

after taking the limit, we have

$$x^* \leq A(x^*, x^*) \leq x^*. \tag{2.156}$$

Hence $A(x^*, x^*) = x^*$, i.e., x^* is a fixed point of A. Let $\bar{x}$ be any fixed point of A in $[\theta, v]$, then

$$u_0 = \theta \le \bar{x} \le v = v_0, \qquad A(\bar{x}, \bar{x}) = \bar{x},$$

so

$$u_1 = A(u_0, v_0) \le A(\bar{x}, \bar{x}) = \bar{x} \le A(v_0, u_0) = v_1.$$

It is easy to see by induction that

$$u_n \le \bar{x} \le v_n \quad (n = 1, 2, \ldots), \tag{2.157}$$

which implies by taking limit that $\bar{x} = x^*$. Finally, $\forall x_0, y_0 \in [\theta, v]$, then, similar to (2.157), we get

$$u_n \le x_n \le v_n, \quad u_n \le y_n \le v_n \quad (n = 1, 2, \ldots). \tag{2.158}$$

Consequently, by (2.151) we get

$$\|x_n - x^*\| \le N\|v_n - u_n\| \le N^2 \cdot \left(\frac{1-c}{c}\right)^n \|v\|, \tag{2.159}$$

$$\|y_n - x^*\| \le N\|v_n - u_n\| \le N^2 \cdot \left(\frac{1-c}{c}\right)^n \|v\|, \tag{2.160}$$

therefore $x_n \to x^*$, $y_n \to x^*$ $(n \to \infty)$. $\qquad\square$

Theorem 2.3.6 (Zhitao Zhang [196]) *Let the cone P be normal and $A : [\theta, v] \times [\theta, v] \to [\theta, v]$ be a mixed monotone operator. Suppose that*

(i) *for fixed $y \in [\theta, v]$, $A(\cdot, y) : [\theta, v] \to [\theta, v]$ is convex; for fixed $x \in [\theta, v]$, $A(x, \cdot) : [\theta, v] \to [\theta, v]$ is concave;*
(ii) *there is a constant c, $\frac{1}{2} < c \le 1$ such that*

$$A(v, \theta) \le cA(\theta, v) + (1 - c)v. \tag{2.161}$$

Then A has exactly one fixed point $x^* \in [\theta, v]$. Moreover, constructing successively the sequences

$$x_n = v - A(v - x_{n-1}, v - y_{n-1}), \quad y_n = v - A(v - y_{n-1}, v - x_{n-1})$$

$$(n = 1, 2, \ldots) \tag{2.162}$$

for any initial $x_0, y_0 \in [\theta, v]$, we have

$$v - x_n \to x^*, \quad v - y_n \to x^* \quad (n \to \infty). \tag{2.163}$$

Proof Let

$$B(x, y) = v - A(v - x, v - y), \quad \forall x, y \in [\theta, v - u] \tag{2.164}$$

then B is a mixed monotone operator. Moreover, for fixed $y \in [\theta, v]$, $B(\cdot, y):$ $[\theta, v] \to [\theta, v]$ is concave, and for fixed $x \in [\theta, v]$, $B(x, \cdot): [\theta, v] \to [\theta, v]$ is convex. If $A(\theta, v) = v$, then $A(x, y) = v$, $\forall x, y \in [\theta, v]$, and $A(v, v) = v$, all the results are obvious. If $A(\theta, v) < v$, then $\theta < B(v, \theta) = v - A(\theta, v) \le v$. By (2.161), we know that

$$A(v, \theta) < cv + (1 - c)v = v, \tag{2.165}$$

and

$$v - A(v, \theta) \ge v - \left[cA(\theta, v) + (1 - c)v \right] = c\left(v - A(\theta, v) \right), \tag{2.166}$$

i.e.,

$$B(\theta, v) \ge cB(v, \theta). \tag{2.167}$$

So B satisfies all the conditions of Theorem 2.3.5 and B has exactly one fixed point $y^* > \theta$, i.e.

$$y^* = B\left(y^*, y^* \right) = v - A\left(v - y^*, v - y^* \right), \tag{2.168}$$

thus $A(v - y^*, v - y^*) = v - y^*$. Let $x^* = v - y^*$, then $\forall x_0, y_0 \in [\theta, v]$ by (2.162), (2.164) and Theorem 2.3.5, we get

$$x_n \to y^*, \quad y_n \to y^* \quad (n \to \infty). \tag{2.169}$$

Hence

$$v - x^n \to x^*, \quad v - y_n \to x^* \quad (n \to \infty). \tag{2.170}$$

$\square$

Theorem 2.3.7 (Zhitao Zhang [196]) *Suppose all the conditions of Theorem 2.3.5 are satisfied, then $\exists \lambda_0 \ge 1$, such that $\lambda_0 A(v, \theta) \le v$, and $\forall \lambda \in [0, \lambda_0]$, the equation*

$$u = \lambda A(u, u) \tag{2.171}$$

has exactly one solution $u(\lambda)$. Let $u_0(\lambda) = \theta$, $v_0(\lambda) = v$, $u_n(\lambda) = \lambda A(u_{n-1}(\lambda), v_{n-1}(\lambda))$, $v_n(\lambda) = \lambda A(v_{n-1}(\lambda), u_{n-1}(\lambda))$, we have

$$\left\| u_n(\lambda) - u(\lambda) \right\| \le N \cdot \left(\frac{1 - c}{c} \right)^n \| v \| \to 0 \quad (n \to \infty), \tag{2.172}$$

$$\left\| v_n(\lambda) - u(\lambda) \right\| \le N \cdot \left(\frac{1 - c}{c} \right)^n \| v \| \to 0 \quad (n \to \infty). \tag{2.173}$$

Proof If $\lambda = 0$, then the conclusion is obvious and $u(0) = \theta$.

Suppose $\lambda \in (0, \lambda_0]$, where $\lambda_0 = \sup\{t > 0, tA(v, \theta) \le v)\}$. From $A(v, \theta) \le v$, we get $\lambda_0 \ge 1$, and from $\theta < \lambda A(v, \theta) \le \lambda_0 A(v, \theta) \le v$, $\lambda A(\theta, v) \ge c\lambda A(v, \theta)$, we see that λA satisfies all conditions of Theorem 2.3.5, so λA has exactly one fixed point $u(\lambda) \in [\theta, v]$, and $u(\lambda) > \theta$; obviously, other results are valid. $\square$

Lemma 2.3.1 (See [88]) *Let E be an ordered Banach space with positive cone P such that $\overset{\circ}{P} \neq \emptyset$. Let X be an ordered Banach space with positive cone K, K is normal. Suppose $A : D(A) \subset E \to X$ is a concave or convex operator, $x_0 \in \overset{\circ}{D}(A)$, the interior of $D(A)$ in E. Then A is continuous at x_0 if and only if A is locally bounded at x_0, i.e., there is a $\delta > 0$ such that A is bounded on the δ-neighborhood $N_\delta(x_0)$ of x_0.*

Theorem 2.3.8 (Zhitao Zhang [196]) *Let P be a normal and solid cone of E, $A : P \times P \to P$ is a mixed monotone operator, $A(v, \theta) \gg \theta$, and the hypotheses* (i), (ii) *of Theorem 2.3.5 are satisfied. Then the equation*

$$\lambda A(u, u) = u, \quad \lambda \in [0, \lambda_0], \tag{2.174}$$

where $\lambda_0 = \sup\{t > 0, t A(v, \theta) \leq v\}$, has exactly one solution $u(\lambda)$ satisfying

(i) $u(\cdot) : [0, \lambda_0] \to [\theta, v]$ *is continuous;*
(ii) $\forall 0 < \lambda_1 < \lambda_2 \leq \lambda_0$, *we have*

$$u(\lambda_2) \geq \frac{\lambda_2}{\lambda_1} c \cdot u(\lambda_1), \tag{2.175}$$

$$u(\lambda_1) \geq \frac{\lambda_1}{\lambda_2} c \cdot u(\lambda_2). \tag{2.176}$$

Proof (i) Set $u_0(\lambda_1) = \theta$, $v_0(\lambda) = v$,

$$u_n(\lambda) = \lambda A\big(u_{n-1}(\lambda), v_{n-1}(\lambda)\big), \quad v_n(\lambda) = \lambda A\big(v_{n-1}(\lambda), u_{n-1}(\lambda)\big)$$

$$(n = 1, 2, \ldots) \tag{2.177}$$

by Theorem 2.3.7, we know that the convergences of $u_n(\lambda) \to u(\lambda)$, $v_n(\lambda) \to v(\lambda)$ $(n \to \infty)$ are both uniform for $\lambda \in [0, \lambda_0]$. Hence $u(\lambda)$ is continuous on $[0, \lambda_0]$ if each $u_n(\lambda)$ $v_n(\lambda)$ is.

In fact, $\forall x_0, y_0 \in \overset{\circ}{P} \cap [\theta, v]$, $x, y \in [\theta, v]$, then

$$\big\|A(x, y) - A(x_0, y_0)\big\| \leq \big\|A(x, y) - A(x_0, y)\big\| + \big\|A(x_0, y) - A(x_0, y_0)\big\|, \tag{2.178}$$

and for fixed y, $A(\cdot, y)$ is bounded on $[\theta, v]$, so $A(\cdot, y)$ is continuous at x_0, similarly $A(x_0, \cdot)$ is continuous at y_0. By Lemma 2.3.1 and (2.178), we get A is continuous at (x_0, y_0). since x_0, y_0 are arbitrary, A is continuous in $\overset{\circ}{P} \cap [\theta, v]$. Obviously,

$$\lim_{\lambda \to 0} u(\lambda) = \lim_{\lambda \to 0} \lambda A\big(u(\lambda), u(\lambda)\big) = \theta = u(0) \tag{2.179}$$

by $A(v, \theta) \gg \theta$ and (2.129), (2.177), we get $\forall \lambda \in (0, \lambda_0]$,

$$u_1(\lambda) = \lambda A(\theta, v) \gg \theta, \qquad v_1(\lambda) = \lambda A(v, \theta) \gg \theta \tag{2.180}$$

and $u_1(\lambda)$, $v_1(\lambda)$ are continuous, hence we can easily prove by induction that $u_n(\lambda)$, $v_n(\lambda)$ are continuous on $[0, \lambda_0]$, therefore, $u(\lambda)$ is continuous on $[0, \lambda_0]$.

(ii) Since $u(\lambda) \in [\theta, v]$, by (2.129) we know that

$$u(\lambda_1) = \lambda_1 A(u(\lambda_1), u(\lambda_1)) \geq \lambda_1 A(\theta, v) \geq \lambda_1 c \cdot A(v, \theta)$$

$$\geq \frac{\lambda_1}{\lambda_2} c \cdot \lambda_2 \cdot A(u(\lambda_2), u(\lambda_2)) = \frac{\lambda_1}{\lambda_2} c \cdot u(\lambda_2). \tag{2.181}$$

Similarly

$$u(\lambda_2) \geq \frac{\lambda_2}{\lambda_1} c \cdot u(\lambda_1). \tag{2.182}$$

$\square$

Corollary 2.3.1 *Let P be a normal cone of E, $A : [u, v] \times [u, v] \to [u, v]$ is a mixed monotone operator, suppose that the hypothesis* (i) *of Theorem 2.3.5 is satisfied, and $\exists c$ such that $\frac{1}{2} < c \leq 1$,*

$$A(u, v) \geq cA(v, u) + (1 - c)u. \tag{2.183}$$

Then A has exactly one fixed point $\bar{x} \in [u, v]$.

Proof Let

$$B(x, y) = A(x + u, y + u) - u, \quad \forall x, y \in [\theta, v - u] \tag{2.184}$$

then $B : [\theta, v - u] \times [\theta, v - u]$ is a mixed monotone operator satisfying the hypothesis (i) of Theorem 2.3.5. Moreover,

$$B(v - u, \theta] = A(v, u) - u, \tag{2.185}$$

$$B(\theta, v - u) = A(u, v) - u. \tag{2.186}$$

By (5.56), (5.58) and (5.59), we get

$$B(v - u, \theta) \leq v, A(u, v) - u \geq cA(v, u) - cu, \tag{2.187}$$

i.e.,

$$B(\theta, v - u) \geq cB(v - u, \theta). \tag{2.188}$$

We may suppose that $B(v - u) \geq \theta$, (since if $B(v - u, \theta) = \theta$ then $A(v, u) = u$, therefore, $\forall x, y \in [u, v]$, $A(x, y) = u$, and u is the unique fixed point of A). Thus the hypothesis (ii) of Theorem 2.3.5 is satisfied, so B has exactly one fixed point $x^* \in [\theta, v - u]$, i.e.,

$$A(x^* + u, x^* + u) - u = x^*. \tag{2.189}$$

Obviously, A has the unique fixed point $\bar{x} = u + x^* \in [u, v]$. $\square$

Corollary 2.3.2 *Let P be a normal cone of E, $A : [u, v] \times [u, v] \to [u, v]$ is a mixed monotone operator. Suppose that the hypothesis* (i) *of Theorem 2 is satisfied, and $\exists c$ such that*

$$\frac{1}{2} < c \le 1, \qquad A(v, u) \le cA(u, v) + (1 - c)v. \tag{2.190}$$

Then A has exactly one fixed point $x^* \in [u, v]$.

Proof Let $B(x, y) = A(x + u, y + u) - u$, $\forall x, y \in [\theta, v - u]$. Similarly to the proof of Corollary 2.3.1, we can verify that $B : [\theta, v - u] \times [\theta, v - u] \to [\theta, v - u]$ satisfies all the hypotheses of Theorem 2.3.6 and B has exactly one fixed point $x^* \in [\theta, v - u]$, i.e.,

$$x^* = B\left(x^*, x^*\right) = A\left(x^* + u, x^* + u\right) - u, \tag{2.191}$$

thus A has the unique fixed point $\bar{x} = x^* + u \in [u, v]$. $\square$

Remark 2.3.3 It should be pointed out that we do not assume operator A to be continuous or compact in Theorems 2.3.5–2.3.8 too.

Definition 2.3.1 Operator $A : \overset{\circ}{P} \to \overset{\circ}{P}$. If there exists $0 \le \alpha < 1$, such that

$$A(tx) \ge t^\alpha A(x), \quad \text{or} \quad A(tx) \le t^{-\alpha} Ax, \quad \forall x \in \overset{\circ}{P}, \ 0 < t < 1, \tag{2.192}$$

then A is called α concave or $-\alpha$ convex, respectively.

We usually assume $A(x, y)$ has the same type of convex-concave property, namely $A(x, y)$ is convex in x, concave in y (refer to [196]), or α concave in x, $-\alpha$ convex in y (refer to [101]). Here $A(x, y)$ has different convex-concave type, such as A is concave in x and $-\alpha$ convex in y, or α concave in x and convex in y.

Theorem 2.3.9 (Zhitao Zhang [197]) *Let the cone P be normal and solid and $A : \overset{\circ}{P} \times \overset{\circ}{P} \to \overset{\circ}{P}$ be a mixed monotone operator.*

(i) *for fixed y, $A(\cdot, y) : P \to \overset{\circ}{P}$ is concave; for fixed x, $A(x, \cdot) : \overset{\circ}{P} \to \overset{\circ}{P}$ is $-\alpha$ convex.*

(ii) *$\exists u_0, v_0 \in P$, $\varepsilon > 0$, $\varepsilon \ge \alpha$ such that such that*

$$0 \ll u_0 \le v_0, \qquad u_0 \le A(u_0, v_0), \qquad A(v_0, u_0) \le v_0, \tag{2.193}$$

and

$$A(\theta, v_0) \ge \varepsilon A(v_0, u_0). \tag{2.194}$$

Then A has a unique fixed point x^* in $[u_0, v_0]$. For any $(x_0, y_0) \in [u_0, v_0] \times [u_0, v_0]$, constructing successively the sequences

$$x_n = A(x_{n-1}, y_{n-1}), \quad y_n = A(y_{n-1}, x_{n-1}) \quad (n = 1, 2, 3, \ldots) \tag{2.195}$$

we have

$$\|x_n - x^*\| \to 0, \quad \|y_n - x^*\| \to 0 \quad (n \to \infty). \tag{2.196}$$

Proof Define

$$u_n = A(u_{n-1}, v_{n-1}), \quad v_n = A(v_{n-1}, u_{n-1}), \quad n = 1, 2, \ldots. \tag{2.197}$$

It is easy to show

$$0 \ll u_0 \leq u_1 \leq \cdots \leq u_{n-1} \leq u_n \leq \cdots \leq v_n \leq v_{n-1} \leq \cdots \leq v_1 \leq v_0 \leq v. \tag{2.198}$$

By (2.194),

$$u_n \geq u_1 \geq \varepsilon v_1 \geq \varepsilon v_n. \tag{2.199}$$

Let

$$t_n = \sup\{t > 0 | u_n \geq t v_n\}, \quad n = 1, 2, \ldots \tag{2.200}$$

then $u_n \geq t_n v_n$.

From $u_{n+1} \geq u_n \geq t_n v_n \geq t_n v_{n+1}$, we have

$$0 < \varepsilon \leq t_1 \leq t_2 \leq \cdots \leq t_n \leq \cdots \leq 1 \tag{2.201}$$

which implies $\lim_{n \to \infty} t_n = t^*$ exists and $\varepsilon \leq t^* \leq 1$. We check that $t^* = 1$.

From (i) we know

$$A\left(x, t^{-1}y\right) \geq t^\alpha A(x, y), \quad \forall x \in P, \ y \in \overset{\circ}{P}, \ 0 < t < 1. \tag{2.202}$$

Therefore, through (2.194), we have

$$u_{n+1} = A(u_n, v_n) \geq A(t_n v_n, v_n) \geq t_n A(v_n, v_n) + (1 - t_n)A(\theta, v_n)$$

$$\geq t_n A\left(v_n, t^{-1} u_n\right) + (1 - t_n)A(\theta, v_n)$$

$$\geq t_n t_n^\alpha A(v_n, u_n) + (1 - t_n)A(\theta, v_0)$$

$$\geq \left(t_n^{1+\alpha} + (1 - t_n)\varepsilon\right)v_{n+1}. \tag{2.203}$$

So

$$t_{n+1} \geq t_n^{1+\alpha} + (1 - t_n)\varepsilon, \tag{2.204}$$

let $n \to \infty$, then t^* satisfies

$$t^* \geq \left(t^*\right)^{1+\alpha} + \left(1 - t^*\right)\varepsilon. \tag{2.205}$$

Observing

$$f(t) = t^{1+\alpha} - (1 + \varepsilon)t + \varepsilon, \quad t \in [0, \infty) \tag{2.206}$$

it is easy to know

$$f(0) = \varepsilon, \qquad f(1) = 0, \tag{2.207}$$

$$f'(t) = (1+\alpha)t^\alpha - (1+\varepsilon). \tag{2.208}$$

We divide the problem into two cases:

(i) $\alpha = 0$. $\forall t \in (0, 1]$, $f'(t) = -\varepsilon < 0$. So $f(t)$ is strictly decreasing in $[\varepsilon, 1]$, from (2.207),

$$f(t) > 0, \quad \forall t \in [\varepsilon, 1).$$

(ii) $0 < \alpha < 1$. $f''(t) = (1+\alpha)\alpha t^{\alpha-1} > 0$, $\forall t \in (0, \infty)$, also

$$f'(0) = -(1+\varepsilon) < 0, \qquad f'\left(\left(\frac{1+\varepsilon}{1+\alpha}\right)^{\frac{1}{\alpha}}\right) = 0.$$

Thus $f'(t) < 0$, $\forall t \in [0, (\frac{1+\varepsilon}{1+\alpha})^{\frac{1}{\alpha}})$.

Since $\varepsilon \geq \alpha$, $(\frac{1+\varepsilon}{1+\alpha})^{\frac{1}{\alpha}} \geq 1$, we get $f'(t) < 0$, $\forall t \in [0, 1)$, which means $f(t)$ is strictly decreasing in $[0, 1)$, by (2.207), $f(t) > 0$, $\forall t \in [\varepsilon, 1)$.

From the above discussion (i) and (ii), if $t^*[\varepsilon, 1)$, then $f(t^*) > 0$, which is

$$\left(t^*\right)^{1+\alpha} + \left(1 - t^*\right)\varepsilon > t^*. \tag{2.209}$$

This contradicts (2.205). So $t^* = 1$. The rest of proof is routine. Readers can refer to Theorem 2.3.5 or [196]. $\qquad\square$

Theorem 2.3.10 (Zhitao Zhang [197]) *Let the cone P be normal and solid and $A : \mathring{P} \times \mathring{P} \to \mathring{P}$ be a mixed monotone operator. Assume that*

(i) *for fixed y, $A(\cdot, y) : \mathring{P} \to \mathring{P}$ is α concave; for fixed x, $A(x, \cdot) : P \to \mathring{P}$ is convex;*
(ii) *$\exists u_0, v_0 \in P$, $\varepsilon \geq \frac{1}{2-\alpha}$ such that*

$$0 \ll u_0 \leq v_0, \qquad u_0 \leq A(u_0, v_0), \qquad A(v_0, u_0) \leq v_0;$$

and

$$A(u_0, v_0) \geq \varepsilon A(v_0, \theta). \tag{2.210}$$

Then A has a unique fixed point x^* in $[u_0, v_0]$. For any $(x_0, y_0) \in [u_0, v_0] \times [u_0, v_0]$, (2.195) and (2.196) still hold.

Proof From the proof of Theorem 2.3.9, we only need to prove $t^* = 1$ when $\frac{1}{2-a} \leq \varepsilon < 1$ (the notations follow that of Theorem 2.3.9).

By condition (i), for fixed x, we have

$$A\left(x, t \cdot t^{-1}y\right) \leq t A\left(x, t^{-1}y\right) + (1-t)A(x, \theta), \quad \forall t \in [0, 1].$$

Therefore,

$$A\left(x, t^{-1}y\right) \geq t^{-1}\left[A(x, y) - (1-t)A(x, \theta)\right], \tag{2.211}$$

consequently,

$$\begin{aligned}
u_{n+1} = A(u_n, v_n) &\geq A(t_n v_n, v_n) \geq t_n^\alpha A\left(v_n, t_n^{-1}u_n\right) \\
&\geq t_n^\alpha t_n^{-1}\left[A(v_n, u_n) - (1-t_n)A(v_n, \theta)\right] \\
&\geq t_n^{\alpha-1}\left[v_{n+1} - (1-t_n)A(v_0, \theta)\right] \\
&\geq t_n^{\alpha-1}\left[v_{n+1} - (1-t_n)\frac{1}{\varepsilon}u_1\right] \\
&\geq t_n^{\alpha-1}\left[v_{n+1} - (1-t_n)\frac{1}{\varepsilon}u_{n+1}\right],
\end{aligned}$$

equivalently,

$$u_{n+1} \geq t_n^{\alpha-1}\left[1 + (1-t_n)t_n^{\alpha-1}\cdot\frac{1}{\varepsilon}\right]^{-1}v_{n+1}. \tag{2.212}$$

So

$$t_{n+1} \geq t_n^{\alpha-1}\left[1 + (1-t_n)t_n^{\alpha-1}\cdot\frac{1}{\varepsilon}\right]^{-1}. \tag{2.213}$$

Let $n \to \infty$, then

$$t^* \geq \left(t_n^*\right)^{\alpha-1}\left[1 + \left(1-t^*\right)\left(t_n^*\right)^{\alpha-1}\cdot\frac{1}{\varepsilon}\right]^{-1}. \tag{2.214}$$

Consider the function

$$g(t) = t^{2-\alpha} + t\cdot\frac{1-t}{\varepsilon} - 1, \quad t \in [\varepsilon, 1]. \tag{2.215}$$

Then it is easy to know

$$g(1) = 0, \qquad g'(t) = (2-\alpha)t^{1-\alpha} + \frac{1-t}{\varepsilon} - \frac{t}{\varepsilon}, \tag{2.216}$$

$$g''(t) = (2-\alpha)(1-\alpha)t^{-\alpha} - \frac{2}{\varepsilon}. \tag{2.217}$$

Since $\varepsilon \geq \frac{1}{2-\alpha}$, $g'(1) = 2 - \alpha - \frac{1}{\varepsilon} \geq 0$. At the same time, from $\varepsilon < 1, 0 \leq \alpha < 1$,

$$\varepsilon^{1-\alpha} < \frac{2}{(2-\alpha)(1-\alpha)}, \tag{2.218}$$

which means $\varepsilon < \frac{2\varepsilon^\alpha}{(2-\alpha)(1-\alpha)}$. If $t \in [\varepsilon, 1]$, $\varepsilon < \frac{2t^\alpha}{(2-\alpha)(1-\alpha)}$, that is

$$(2-\alpha)(1-\alpha) \cdot t^{-\alpha} < \frac{2}{\varepsilon}. \tag{2.219}$$

From (2.217), we get $g''(t) < 0$, $\forall t \in [\varepsilon, 1]$. Also by $g'(1) \geq 0$, we get $g'(t) > 0$, $\forall t \in [\varepsilon, 1]$. Therefore $g(1) = 0$ leads to

$$g(t) < 0, \quad t \in [\varepsilon, 1). \tag{2.220}$$

This means $t^{2-\alpha} + t \cdot \frac{1-t}{\varepsilon} < 1$, $t \in [\varepsilon, 1)$ in (2.215). This means

$$t^2 + t^{1+\alpha}(1-t)\frac{1}{\varepsilon} < t^\alpha, \qquad t + t^\alpha(1-t)\frac{1}{\varepsilon} < t^{\alpha-1},$$

$$t < t^{\alpha-1}\left[1 + t^{\alpha-1}(1-t)\frac{1}{\varepsilon}\right]^{-1}, \qquad \forall t \in [\varepsilon, 1). \tag{2.221}$$

If $t^* \in [\varepsilon, 1)$, then t^* satisfies (2.221), which contradicts (2.214). Thus $t^* = 1$. The rest of the proof is routine. $\qquad\square$

2.4 Applications of Mixed Monotone Operators

We give applications of Theorem 2.3.1 and 2.3.2 to the following initial value problem:

$$\begin{cases} x' = \sum_{i=1}^{n} a_i(t)x^{r_i} + \left(\sum_{j=1}^{m} b_j(t)x^{s_j}\right)^{-1}, & \text{a.e. on } J \\ x(0) = x_0, \end{cases} \tag{2.222}$$

where $J = [0, T]$ $(T > 0)$, $0 < r_i < 1$, $0 < s_j < 1$ $(i = 0, 1, \ldots, n;\ j = 0, 1, \ldots, m)$, $x_0 > 0$, $a_i(t)$ are nonnegative bounded measurable functions (on J) and $b_j(t)$ are nonnegative measurable functions such that

$$\inf_{t \in J} \sum_{j=1}^{m} b_j(t) > 0.$$

The set of all absolutely continuous functions from J into $\mathbb{R}$ is denoted by $AC[J, \mathbb{R}]$. A function $x(t)$ on J is said to be a solution of the initial value problem (2.222) if $x(t) \in AC[J, \mathbb{R}]$ and satisfies (2.222).

Theorem 2.4.1 (Dajun Guo [101]) *Under conditions mentioned above, initial value problem (2.222) has exactly one positive solution $x^*(t)$. Moreover, constructing a* *successive sequence of functions*

$$x_n(t) = x_0 + \int_0^T \left\{ \sum_{i=1}^n a_i(s)\big(x_{n-1}(s)\big)^{r_i} \right\} ds + \int_0^T \left\{ \sum_{j=1}^m b_j(s)\big(x_{n-1}(s)\big)^{s_j} \right\}^{-1} ds,$$

$$(2.223)$$

$(n = 1, 2, \ldots)$ *for any initial positive function* $x_0(t) \in AC[J, \mathbb{R}]$, *the sequence of functions* $\{x_n(t)\}$ *converges to* $x^*(t)$ *uniformly on* J.

Proof It is clear, $x(t) \in AC[J, \mathbb{R}]$ is a positive solution of (2.222) if and only if $x(t) \in C[J, \mathbb{R}]$ is a positive solution of the following integral equation:

$$x(t) = x_0 + \int_0^T \left\{ \sum_{i=1}^n a_i(s)\big(x(s)\big)^{r_i} \right\} ds + \int_0^T \left\{ \sum_{j=1}^m b_j(s)\big(x(s)\big)^{s_j} \right\}^{-1} ds,$$

$$t \in J. \tag{2.224}$$

Let $E = C[J, \mathbb{R}]$ and $P = \{x \in C[J, \mathbb{R}] : x(t) \geq 0 \text{ for } t \in J\}$, then P is a normal and solid cone in E and (2.224) can be written in the form

$$x = A(x, x), \tag{2.225}$$

where

$$A(x, y) = A_1(x) + A_2(y),$$

$$A_1(x) = x_0 + \int_0^T \left\{ \sum_{i=1}^n a_i(s)\big(x(s)\big)^{r_i} \right\} ds,$$

$$A_2(y) = \int_0^T \left\{ \sum_{j=1}^m b_j(s)\big(x(s)\big)^{s_j} \right\}^{-1} ds.$$

It is clear that $A_1 : \overset{\circ}{P} \to \overset{\circ}{P}$ is non-decreasing and $A_2 : \overset{\circ}{P} \to \overset{\circ}{P}$ is non-increasing, so $A : \overset{\circ}{P} \times \overset{\circ}{P} \to \overset{\circ}{P}$ is a mixed monotone operator. We now check that A satisfies all of the conditions of Theorem 2.3.1. For $x, y \in \overset{\circ}{P}$ and $0 < t < 1$, it is easy to see

$$A_1(tx) \geq t^{r_0} A_1(x),$$

$$A_2\big(t^{-1} y\big) \geq t^{s_0} A_2(y),$$

where $r_0 = \max\{r_1, \ldots, r_n\}$, $s_0 = \max\{s_1, \ldots, s_m\}$, $0 < r_0 < 1$, $0 < s_0 < 1$. Therefore

$$A\big(tx, t^{-1} y\big) \geq t^r A(x, y), \quad x, y \in \overset{\circ}{P}, \ 0 < t < 1,$$

where $r = \max\{r_0, s_0\}$, $0 < r < 1$. Hence, by Theorem 2.3.1, we conclude that A has exactly one fixed point x^* in $\mathring{P}$ and, for any initial $x_0 \in \mathring{P}$,

$$\|x_n - x^*\| = \max_{t \in J} |x_n(t) - x^*(t)| \to 0 \quad (n \to \infty),$$

where

$$x_n = A(x_{n-1}, x_{n-1}) \quad (n = 1, 2, \ldots).$$

The proof is complete. $\qquad\square$

Using Theorem 2.3.2, we get similarly the following.

Theorem 2.4.2 (Dajun Guo [101]) *Let the hypotheses of Theorem 2.4.1 be satisfied. Denote by $x^*(t)$ the unique positive solution of the initial value problem*

$$\begin{cases} rx' = \displaystyle\sum_{i=1}^{n} a_i(t)x^{r_i} + \left(\sum_{j=1}^{m} b_j(t)x^{s_j}\right)^{-1}, & a.e. \ on J \\ rx(0) = x_0. \end{cases} \tag{2.226}$$

Then $x_r^*(t)$ converges to $x_{r_0(t)}^*$ uniformly on $t \in J$ as $r \to r_0$ ($r_0 > 0$). If, in addition, $0 < r_i < \frac{1}{2}$, $0 < s_j < \frac{1}{2}$ ($i = 0, 1, \ldots, n$; $j = 0, 1, \ldots, m$), then

$$0 < r < r' \quad \Longrightarrow \quad x_r^*(t) > x_{r'}^*(t), \quad t \in J,$$

and

$$\max_{t \in J} x_r^*(t) \to 0 \quad as \ r \to +\infty, \qquad \max_{t \in J} x_r^*(t) \to +\infty \quad as \ r \to +0.$$

We apply Theorem 2.3.4 to the IVP

$$\begin{cases} x' = \displaystyle\sum_{i=0}^{n} a_i(t)x^{r_i} + \left(\sum_{j=0}^{m} b_j(t)x^{s_j}\right)^{-1}, & t \in [0, T], \\ x(0) = x_0, \end{cases} \tag{2.227}$$

and the two-point BVP

$$\begin{cases} -x'' = \displaystyle\sum_{i=0}^{n} a_i(t)x^{r_i} + \left(\sum_{j=0}^{m} b_j(t)x^{s_j}\right)^{-1}, & t \in [0, 1], \\ ax(0) - bx'(0) = x_0, \qquad cx(1) + dx'(1) = x_1, \end{cases} \tag{2.228}$$

where

$$0 = r_0 < r_1 < \cdots < r_{n-1} < r_n = 1, \qquad 0 = s_0 < s_1 < \cdots < s_{m-1} < s_n = 1,$$

$$T > 0, \qquad x_0 \geq 0, \qquad x_1 \geq 0, \qquad a \geq 0, \qquad b \geq 0, \qquad c \geq 0, \qquad d \geq 0,$$

$$Q = ac + ad + bc > 0$$

and $a_i(t), b_j(t)$ $(i = 0, 1, \ldots, n; \ j = 0, 1, \ldots, m)$ are nonnegative continuous functions on $J = [0, T]$ (for problem (2.227)) or $I = [0, 1]$ (for problem (2.228)).

Theorem 2.4.3 (Dajun Guo [102]) *Suppose that $a_0(t) > 0$, $b_0(t) > 0$ for $t \in J$ and*

$$\int_0^T a_n(t)\, dt < 1. \tag{2.229}$$

Then IVP (2.227) has exactly one nonnegative nontrivial C^1 solution $x^*(t)$ on J. Moreover, constructing a successive sequence of functions

$$x_n(t) = x_0 + \int_0^T \left\{ \sum_{i=0}^n a_i(s)\big(x_{n-1}(s)\big)^{r_i} \right\} ds + \int_0^T \left\{ \sum_{j=0}^m b_j(s)\big(x_{n-1}(s)\big)^{s_j} \right\}^{-1} ds$$

$$(n = 1, 2, \ldots)$$

for any initial nonnegative function $x_0(t) \in C(J, \mathbb{R})$, the sequence $\{x_n(t)\}$ converges to $x^*(t)$ uniformly on J.

Proof It is clear that $x(t) \in C^1[J, \mathbb{R}]$ is a nonnegative solution of (2.227) if and only if $x(t) \in C[J, \mathbb{R}]$ is a nonnegative solution of the following integral equation:

$$x(t) = x_0 + \int_0^T \left\{ \sum_{i=0}^n a_i(s)\big(x(s)\big)^{r_i} \right\} ds + \int_0^T \left\{ \sum_{j=0}^m b_j(s)\big(x(s)\big)^{s_j} \right\}^{-1} ds,$$

$$t \in J. \tag{2.230}$$

Let $E = C[J, \mathbb{R}]$ and $P = \{x \in C[J, \mathbb{R}] : x(t) \geq 0 \text{ for } t \in J\}$, then P is a normal and solid cone in E and (2.230) can be written in the form

$$x = A(x, x), \tag{2.231}$$

where

$$A(x, y) = A_1(x) + A_2(y),$$

$$A_1(x) = x_0 + \int_0^T \left\{ \sum_{i=0}^n a_i(s)\big(x(s)\big)^{r_i} \right\} ds,$$

$$A_2(y) = \int_0^T \left\{ \sum_{j=0}^m b_j(s)\big(x(s)\big)^{s_j} \right\}^{-1} ds.$$

Evidently, $A_1 : P \to P$ is non-decreasing and $A_2 : P \to P$ is non-increasing, so $A : P \times P \to P$ is a mixed monotone operator. We now check that A satisfies all of the conditions of Theorem 2.3.4.

By virtue of (2.229), we can choose a constant $R > 0$ sufficiently large such that

$$x_0 + \sum_{i=0}^{n} R^{r_i} \int_0^T a_i(s)\,ds + \int_0^T b_0(s)^{-1}\,ds < R. \tag{2.232}$$

Hence, putting $v(t) \equiv R \ (t \in J)$, we have $v \in \overset{\circ}{P}$ and $0 < A(v, 0) \le v$. On the other hand, since, by hypothesis,

$$\min_{t \in J} a_0(t) = \bar{a}_0 > 0, \qquad \min_{t \in J} b_0(t) = \bar{b}_0 > 0,$$

$$\max_{t \in J} a_i(t) = a_i^* \ge 0, \qquad \max_{t \in J} b_j(t) = b_j^* \ge 0,$$

$$(i = 0, 1, 2, \ldots, n; \ j = 0, 1, 2, \ldots, m),$$

we can choose $0 < c_0 < 1$ sufficiently small such that

$$c_0 \left(\frac{1}{\bar{b}_0} + \sum_{i=0}^{n} a_i^* R^{r_i} \right) < \bar{a}_0. \tag{2.233}$$

Let

$$F(t) = x_0 + \int_0^t a_0(s)\,ds + \int_0^t \left(\sum_{j=0}^{m} b_j(s) R^{s_j} \right)^{-1} ds$$

$$- c_0 \left(x_0 + \int_0^t a_i(s) R^{r_i}\,ds + \int_0^t \left(b_0(s) \right)^{-1} ds \right), \quad t \in J.$$

Then, (2.233) implies

$$F(t) \ge (1 - c_0) x_0 + \int_0^t \left\{ \bar{a}_0 - c_0 \left(\frac{1}{\bar{b}_0} + \sum_{i=0}^{n} a_i^* R^{r_i} \right) \right\} ds \ge 0, \quad t \in J$$

i.e., $A(0, v) \ge c_0 A(v, 0)$. Thus, condition (a$'$) of Theorem 2.3.4 is satisfied.

Let $0 < a < b < 1$ and a bounded $B \subset P$ be given. So, $\|x\| \le R_1$ for all $x \in B$, where $R_1 > 0$ is a constant. We now define

$$r = \min \left\{ \bar{a}_0 (1 - b) \left(\bar{a}_0 b + b \sum_{i=1}^{n} n a_i^* R_1^{r_i} \right)^{-1}, \bar{b}_0 (1 - b) \left(\bar{b}_0 b + \sum_{j=1}^{m} b_j^* R_1^{s_j} \right)^{-1} \right\} > 0, \tag{2.234}$$

and check that it satisfies (2.126). For any $t_1 \in [a, b]$, $x, y \in B$, we have

$$A_1(t_1 x) = x_0 + \int_0^t \left\{ \sum_{i=0}^{n} a_i(s) t_1^{r_i} \left(x(s) \right)^{r_i} \right\} ds$$

$$\geq t_1 \left(x_0 b^{-1} + \int_0^t \left\{ a_0(s) b^{-1} + \sum_{i=1}^n a_i(s) \big(x(s) \big)^{r_i} \right\} ds \right).$$

To show that $A_1(t_1 x) \geq t_1 (1 + r A_1(x))$, it is sufficient to show that

$$a_0(s) b^{-1} + \sum_{i=1}^n a_i(s) \big(x(s) \big)^{r_i} \geq (1 + r) \left\{ a_0(s) + \sum_{i=1}^n a_i(s) \big(x(s) \big)^{r_i} \right\}, \quad s \in J$$

since $b^{-1} \geq 1 + r$ by (2.234). This inequality can be written in the form

$$a_0(s) \big(b^{-1} - 1 - r \big) \geq r \sum_{i=1}^n a_i(s) \big(x(s) \big)^{r_i},$$

which will certainly be true if

$$\bar{a}_0 \big(1 - b(1 + r) \big) \geq br \sum_{i=1}^n a_i^* R_1^{r_i}.$$

This last inequality follows from (2.234). Similarly, we have

$$A_2\big(t_1^{-1} y \big) \geq t_1 \int_0^t \left\{ b b_0(s) + \sum_{j=1}^m b_j(s) \big(y(s) \big)^{s_j} \right\}^{-1} ds$$

$$\geq t_1 (1 + r) \int_0^t \left\{ b_0(s) + \sum_{j=1}^m b_j(s) \big(y(s) \big)^{s_j} \right\}^{-1} ds$$

$$= t_1 (1 + r) A_2(y).$$

Hence, $A(t_1 x, t_1^{-1} y) \geq t_1 (1 + r) A(x, y)$, and (2.126) is true.

Finally, our conclusions follow from Theorem 2.3.4 and the proof is complete. $\square$

Theorem 2.4.4 (Dajun Guo [102]) *Suppose that $a_0(t) > 0$, $b_0(t) > 0$ for $t \in I$ and*

$$\int_0^1 a_n(s) \, ds < \frac{Q}{(a + b)(c + d)}.$$

Then BVP (2.228) has exactly one nonnegative nontrivial C^2 solution $x^*(t)$ on I. Moreover, constructing successively the sequence of functions

$$x_k(t) = z_0(t) + \int_0^1 G(t, s) \left(\sum_{i=0}^n a_i(s) \big(x_{k-1}(s) \big)^{r_i} \left\{ \sum_{j=0}^m b_j(s) \big(x_{k-1}(s) \big)^{s_j} \right\}^{-1} \right) ds$$

$$(k = 1, 2, \ldots)$$

for any initial nonnegative function $x_0(t) \in C[I, \mathbb{R}]$, the sequence $\{x_k(t)\}$ converges to $x^*(t)$ uniformly on I, where

$$G(t, s) = \begin{cases} Q^{-1}(at + b)(c(1 - s) + d), & t \le s, \\ Q^{-1}(as + b)(c(1 - t) + d), & t > s \end{cases}$$

and

$$z_0(t) = Q^{-1}\{(c(1 - t) + d)x_0 + (at + b)x_1\}.$$

Proof It is well known that $x(t) \in C^2(I, \mathbb{R})$ is a nonnegative solution of (2.228) if and only if $x(t) \in C(I, \mathbb{R})$ is a nonnegative solution of the following integral equation:

$$x(t) = z_0(t) + \int_0^1 G(t, s)\left(\sum_{i=0}^n a_i(s)(x(s))^{r_i} + \left\{\sum_{j=1}^m b_j(s)(x(s))^{s_j}\right\}^{-1}\right) ds.$$

This integral equation can be regarded as an equation of the form (2.231), where $A(x, y) = A_1(x) + A_2(y)$ and

$$A_1(x) = z_0(t) + \int_0^1 G(t, s)\left\{\sum_{i=0}^n a_i(s)(x(s))^{r_i}\right\} ds,$$

$$A_2(y) = \int_0^1 G(t, s)\left\{\sum_{j=0}^m b_j(s)(x(s))^{s_j}\right\}^{-1} ds.$$

Let $E = C(I, \mathbb{R})$ and $P = \{x \in C(I, \mathbb{R}) : x(t) \ge 0 \text{ for } t \in I\}$. Then $A : P \times P \to P$ is a mixed monotone operator. In the same way as in the proof of Theorem 2.4.3, we can show the operator A satisfies condition (a′) and (b′) of Theorem 2.3.4. Hence our conclusion follow from Theorem 2.3.4. □

Next we use some of our results to several existence theorems for nonlinear integral equations on unbounded region and nonlinear differential equations in Banach spaces.

We first consider the following nonlinear integral equation on $\mathbb{R}^N$:

$$x(t) = Ax(t) = \int_{\mathbb{R}^N} K(t, s)\left[\frac{x^2(s) + x(s)}{2} + \sqrt{1 - x^2(s)}\right] ds. \tag{2.235}$$

Proposition 2.4.1 (Zhitao Zhang [196]) *Suppose that $K : \mathbb{R}^N \times \mathbb{R}^N \to \mathbb{R}$ is continuous and $K(t, s) \ge 0$, $K \not\equiv 0$, moreover, $\exists a > 0$ such that*

$$\int_{\mathbb{R}^N} K(t, s)\, ds \le a < \frac{1}{4}. \tag{2.236}$$

Then (2.235) has unique one solution $x^*(t)$ satisfying $0 \le x^*(t) < 1$, and $x^*(t) \not\equiv 0$.

Proof We use Theorem 2.3.6 to prove it. Let $C_B(\mathbb{R}^N)$ denote the set of all bounded continuous functions, we define $\|x\| = \sup_{t \in \mathbb{R}^N} |x(t)|$, then $C_B(\mathbb{R}^N)$ is a real Banach space. Let $P = C_B^+(\mathbb{R}^N)$ denote the set of nonnegative functions of $C_B(\mathbb{R}^N)$, then P is a normal and solid cone of $C_B(\mathbb{R}^N)$. (2.235) can be written in the form

$$x = A(x, x); \tag{2.237}$$

where

$$A(x, y) = A_1(x) + A_2(y); \tag{2.238}$$

$$A_1(x) = \int_{\mathbb{R}^N} K(t, s) \cdot \frac{x^2(s) + x(s)}{2} \, ds; \tag{2.239}$$

$$A_2(y) = \int_{\mathbb{R}^N} K(t, s) \cdot \sqrt{1 - y^2(s)} \, ds. \tag{2.240}$$

Let $v \equiv 1$, then

$$A(v, 0) = A_1(v) + A_2(0) = 2 \int_{\mathbb{R}^N} K(t, s) \, ds \le 2a < \frac{1}{2}, \tag{2.241}$$

and

$$A(x, y) \le A(v, 0), \quad \forall x, y \in [0, v]. \tag{2.242}$$

So $A : [0, v] \times [0, v] \to [0, v]$ is a mixed monotone operator, and $A(\cdot, y)$ is a convex operator for fixed y, $A(x, \cdot)$ is a concave operator for fixed x, thus the hypothesis (i) of Theorem 2.3.6 is satisfied. Moreover,

$$A(0, v) = A_1(0) + A_2(1) = 0. \tag{2.243}$$

Let $c = 1 - 2a > \frac{1}{2}$, then

$$A(v, 0) \le 2a \le cA(0, v) + (1 - c)v, \tag{2.244}$$

thus the hypothesis (ii) of Theorem 2.3.6 is satisfied. So A has exactly one fixed point $x^* \in [0, 1]$, moreover, $\forall x_0, y_0 \in [0, v]$, $x_n - 1 - A(1 - x_{n-1}, 1 - y_{n-1})$, $y_n = 1 - A(1 - y_{n-1}, 1 - x_{n-1})$, we have

$$1 - x_n \to x^*, \quad 1 - y_n \to x^* \quad (n \to \infty). \qquad \square$$

Now we consider the system of equations:

$$-x_n'' = \frac{1}{n} \cdot \left[\frac{1}{3} x_{2n}^{\frac{1}{2}} + (1 + x_{n+2})^{-\frac{1}{2}} \right]; \tag{2.245}$$

$$x_n(0) = x_n'(1) = 0, \quad n = 1, 2, \ldots. \tag{2.246}$$

Let $E = \{x \mid x = (x_1, x_2, \ldots), \sup_i |x_i| < \infty\}$, $\|x\| = \sup_i |x_i|$. $P = \{x \mid x \in E, x_i \ge 0\} \subset E$ is a normal and solid cone. Let $I = [0, 1]$, $C[I, E] = \{x \mid x(\cdot) : I \to$

E is continuous}, $\|x\| = \max_{t\in I} \|x(t)\|$. $\bar{P} = \{x \in C[I, E] | x(t) \in P, \ \forall t \in I\} \subset C[I, E]$ is a normal and solid cone. Then (2.245)–(2.246) is equivalent to the two-point Boundary Value Problem in E. $\square$

Proposition 2.4.2 (Zhitao Zhang [196]) *The system* (2.245)–(2.246) *has a unique positive solution* $x^*(t) \in [0, v]$, *where* $v = (1, 1, \ldots, 1, \ldots)$. *Moreover,* $\forall x_0, y_0 \in [0, v]$, $x_n = A(x_{n-1}, \ y_{n-1}), y_n = A(y_{n-1}, x_{n-1})$, *we have* $x_n \to x^*$, $y_n \to x^* \ (n \to \infty)$,

$$\left\| x_n - x^* \right\| \le \left(\frac{4\sqrt{2}}{3} - 1 \right)^n, \qquad \left\| y_n - x^* \right\| \le \left(\frac{4\sqrt{2}}{3} - 1 \right)^n, \tag{2.247}$$

where

$$A(x, y) = A_1(x) + A_2(y) \tag{2.248}$$

(in the following $i = 1, 2, \ldots$)

$$\left(A_1(x) \right)_i = \frac{1}{3i} \int_0^1 G(t, s) \cdot x_{2i}^{\frac{1}{2}}(s)\, ds;$$

$$\left(A_2(y) \right)_i = \frac{1}{i} \int_0^1 G(t, s) \cdot \left(1 + x_{i+2}(s) \right)^{-\frac{1}{2}} ds; \tag{2.249}$$

$$G(t, s) = \begin{cases} t, & 0 \le t \le s \le 1, \\ s, & 1 \ge t > s \ge 0. \end{cases}$$

Proof It is easy to know $x(t) \in C^2[I, E] \cap \bar{P}$ is a solution of (2.245)–(2.246) iff $x \in \bar{P}$ is a solution of $A(x, x) = x$. We shall prove A has exactly one fixed point in $[0, v]$. Obviously, $A : \bar{P} \times \bar{P} \to \bar{P}$ is a mixed monotone operator, and $A(\cdot, y)$ is a concave operator for fixed y, $A(x, \cdot)$ is a convex one for fixed x.

$$\left(A(v, 0) \right)_i = \left(A_1(v) \right)_i + \left(A_2(0) \right)_i = \frac{4}{3i} \int_0^1 G(t, s)\, ds = \frac{4}{3i}\left(t - t^2 \right)$$

$$\le \frac{4}{3i} \cdot \frac{1}{2} \le 1, \tag{2.250}$$

and since $\forall x, y \in [0, v]$, $A(x, y) \le A(v, 0)$, we get $A : [0, v] \times [0, v] \to [0, v]$. Moreover,

$$\left(A(0, v) \right)_i = \left(A_1(0) \right)_i + \left(A_2(v) \right)_i = 0 + \frac{1}{i} \int_0^1 G(t, s)(1 + 1)^{-\frac{1}{2}}\, ds$$

$$= \frac{1}{i} \cdot 2^{-\frac{1}{2}} \int_0^1 G(t, s)\, ds \tag{2.251}$$

by (2.250) and (2.251), we get $A(v, 0) > 0$,

$$\left(A(0, v)\right)_i = \frac{1}{i} \cdot 2^{-\frac{1}{2}} \cdot \frac{3i}{4}\left(A(v, 0)\right)_i$$

$$= \frac{3}{4\sqrt{2}}\left(A(v, 0)\right)_i. \tag{2.252}$$

Since $c = \frac{3}{4\sqrt{2}} > \frac{1}{2}$ such that $A(0, v) \geq cA(v, 0)$, we know that all the hypotheses of Theorem 2.3.5 are satisfied and A has a unique fixed point $x^* \in [0, v]$, i.e., (2.245)–(2.246) has a unique solution in $[0, v]$. Moreover, $\forall x_0, y_0 \in [0, v]$, let

$$(x_n)_i = \frac{1}{i}\int_0^1 G(t, s)\left[\frac{1}{3}\left(x_{n-1}(s)\right)_{2i}^{\frac{1}{2}} + \left(1 + y_{n-1}(s)\right)_{i+2}^{-\frac{1}{2}}\right]ds, \tag{2.253}$$

$$(y_n)_i = \frac{1}{i}\int_0^1 G(t, s)\left[\frac{1}{3}\left(y_{n-1}(s)\right)_{2i}^{\frac{1}{2}} + \left(1 + x_{n-1}(s)\right)_{i+2}^{-\frac{1}{2}}\right]ds, \tag{2.254}$$

we have

$$(x_n)_i \to x_i^*, \quad (y_n)_i \to x_i^* \quad (n \to \infty)$$

uniformly convergent for $i = 1, 2, \ldots$. Since the normal constant of $\bar{P}$ is 1, we get

$$\|x_n - x^*\| \leq \left(\frac{1 - \frac{3}{4\sqrt{2}}}{\frac{3}{4\sqrt{2}}}\right)^n \cdot \|v\| = \left(\frac{4\sqrt{2}}{3} - 1\right)^n;$$

$$\|y_n - x^*\| \leq \left(\frac{1 - \frac{3}{4\sqrt{2}}}{\frac{3}{4\sqrt{2}}}\right)^n \cdot \|v\| = \left(\frac{4\sqrt{2}}{3} - 1\right)^n. \qquad \square$$

Consider the Hammerstein type integral equation on $\mathbb{R}^N$

$$x(t) = (Ax)(t) = \int_{\mathbb{R}^N} K(t, s)\left[\left(1 + x(s)\right) + x^{-\frac{1}{3}}(s)\right]ds. \tag{2.255}$$

Proposition 2.4.3 (Zhitao Zhang [197]) *Suppose* $K : \mathbb{R}^N \times \mathbb{R}^N \to \mathbb{R}$ *is continuous,* $K(t, s) \geq 0$, *and*

$$\frac{1}{21} \leq \int_{\mathbb{R}^N} K(t, s)\,ds \leq \frac{1}{5} \tag{2.256}$$

then (2.255) *has a unique positive solution* $x^*(t)$, *satisfying* $10^{-1} \leq x^*(t) \leq 1$, *and* $\forall(x_0(t), y_0(t)) \in [10^{-1}, 1] \times [10^{-1}, 1]$, (2.195) *and* (2.196) *hold.*

Proof Let $E = C_B(\mathbb{R}^N)$ is the space of bounded continuous functions on $\mathbb{R}^N$, define norm $\|x\| = \sup_{t \in R^N} |x(t)|$, then E is a Banach space.

Let $P = C_B^+(\mathbb{R}^N)$ denotes the set of all nonnegative functions in E, then P is a normal and solid cone. Equation (2.255) can be read as $x = A(x, x)$, where

$$A(x, y) = A_1(x) + A_2(y),$$

$$A_1(x) = \int_{R^N} K(t, s)\big(1 + x(s)\big)\,ds,$$

$$A_2(y) = \int_{R^N} K(t, s)y^{-\frac{1}{3}}(s)\,ds.$$

We aim to apply Theorem 2.3.9. Let $u_0 = 10^{-1}$, $v_0 = 1$, and it is easy to know $A : P \times \overset{\circ}{P} \to \overset{\circ}{P}$ is mixed monotone operator. For fixed y, $A(\cdot, y) : P \to \overset{\circ}{P}$ is concave; for fixed x, $A(x, \cdot) : \overset{\circ}{P} \to \overset{\circ}{P}$ is $-\frac{1}{3}$ convex.

Obviously $\theta \ll u_0 \le v_0$, and by (2.256),

$$A(u_0, v_0) = \int_{\mathbb{R}^N} K(t, s)\big[\big(10^{-1} + 1\big)\big]\,ds \ge 10^{-1}, \tag{2.257}$$

$$A(v_0, u_0) = \int_{\mathbb{R}^N} K(t, s)\big[(1 + 1) + 10^{\frac{1}{3}}\big]\,ds \le 1, \tag{2.258}$$

$$A(\theta, v_0) = \int_{\mathbb{R}^N} K(t, s)\big[(1 + 0) + 1\big]\,ds \ge \frac{2}{5}A(v_0, u_0). \tag{2.259}$$

Let $\varepsilon = \frac{2}{5}$, then $\varepsilon > 0$, $\varepsilon > \frac{1}{3}$. Then by Theorem 2.3.9, this proposition is proved. $\square$

2.5 Further Results on Cones and Partial Order Methods

Let E be a real Banach space, P be a cone in E. Let $P^* = \{f \in E^* | f(x) \ge 0, \forall x \in P\}$, if $\overline{P - P} = E$, then $P^* \subset E^*$ is a cone; by Theorem 1 of [13], we know P is normal $\Leftrightarrow P^*$ is generating.

Theorem 2.5.1 (Zhitao Zhang [199]) *Let P be a cone in E, then the following assertions are equivalent*:

(i) *P is normal*;
(ii) *$x_n \le z_n \le y_n$ $(n = 1, 2, \ldots)$, $x_n \to x$ weakly and $y_n \to x$ weakly imply $z_n \to x$ weakly*.

Proof (ii) $\Rightarrow$ (i) Suppose (i) is not true, then $\exists \theta \le x_n \le y_n$ such that $\|x_n\| > n^2\|y_n\|$, $n = 1, 2, \ldots$. Let $z_n = \frac{x_n}{n\|y_n\|}$, $y_n' = \frac{y_n}{n\|y_n\|}$, then $\theta \le z_n \le y_n'$, and $\|y_n'\| \to 0$, thus $y_n' \to \theta$ weakly, but we know that $\|z_n\| > n$, so $\{z_n\}$ is unbounded, $z_n \to \theta$ weakly is impossible, which contradicts (ii).

(i) $\Rightarrow$ (ii) $\forall x_n \le z_n \le y_n$, $x_n \to x$ weakly, $y_n \to x$ weakly. By (i), we know P^* is generating, i.e., $\forall f \in E^*$, $\exists f_1, f_2 \in P^*$ such that $f = f_1 - f_2$. We have

$$f_i(x_n) \leq f_i(z_n) \leq f_i(y_n), \quad f_i(x_n) \to f_i(x), \quad f_i(y_n) \to f_i(x)$$

$$(i = 1, 2, \ n \to +\infty)$$

thus we get $f_i(z_n) \to f_i(x)$ and $f(z_n) = f_1(z_n) - f_2(z_n) \to f_1(x) - f_2(x) = f(x)$, i.e., $z_n \to x$ weakly. $\qquad\square$

Theorem 2.5.2 (Zhitao Zhang [199]) *Let P be a strongly minihedral and solid cone, then every bounded (in norm) set D has a least upper bound $\sup D$.*

Proof We know $\exists M > 0$ such that $\forall x \in D$, $\|x\| < M$. Since P is a solid cone, we find that $\exists u_0 \in \overset{\circ}{P}$, $\delta > 0$ such that $u_0 - \delta \frac{x}{\|x\|} \in P$. So $x \leq \frac{\|x\|}{\delta} u_0 < \frac{M}{\delta} u_0$, thus D is a bounded set (in order), by the definition of strongly minihedral cone, we get D has a least upper bound $\sup M$. $\qquad\square$

Next we consider two operators form equations:

$$A(x, x) + Bx = x. \tag{2.260}$$

Definition 2.5.1 $B : D \to E$ satisfies

$$B\big(tx + (1 - t)y\big) = tBx + (1 - t)By$$

for $x, y \in D$, and $t \in [0, 1]$, then B is said to be affine.

Theorem 2.5.3 (Li, Liang and Xiao [129]) *Let P be normal, N be the normal constant of P. Let $u, v \in P \cap \mathcal{D}(B)$, $u < v$, operator $A : [u, v] \times [u, v] \to E$ be mixed monotone and $B : \mathcal{D}(B) \to E$ be affine on $[u, v]$, where $[u, v] = \{x \in E \,|\, u \leq x \leq v\}$. Assume that:*

(i) *for fixed y, $A(\cdot, y) : [u, v] \to E$ is concave; for fixed x, $A(x, \cdot) : [u, v] \to E$ is convex;*
(ii) *$(I - B)^{-1} : E \to \mathcal{D}(B)$ exists and is an increasing operator on $[u - Bu, v - Bv]$, i.e., for $x, y \in [u - Bu, v - Bv]$, $x \geq y$ implies that $(I - B)^{-1}x \geq (I - B)^{-1}y$, where I is the identity operator on E;*
(iii) *$A(u, v) \geq u$, $A(v, u) \leq v$, $Bu \geq \theta$ and $Bv \leq \theta$;*
(iv) *there exists some $m_0 \in \mathbb{N} \cup \{0\}$ such that*

$$u_{m_0+1} \geq \frac{1}{2}(v_{m_0+1} + u_{m_0}), \tag{2.261}$$

where

$$u_0 = u, \qquad v_0 = v,$$

$$u_n = (I - B)^{-1} A(u_{n+1}, v_{n+1}) \quad (n = 1, 2, \ldots), \tag{2.262}$$

$$v_n = (I - B)^{-1} A(v_{n+1}, u_{n+1}) \quad (n = 1, 2, \ldots). \tag{2.263}$$

Then (2.260) has a unique positive solution x^* in $[u, v]$. Moreover, constructing successively the sequences

$$x_n = (I - B)^{-1} A(x_{n-1}, y_{n-1}) \quad (n = 1, 2, 3, \ldots), \qquad (2.264)$$

$$y_n = (I - B)^{-1} A(y_{n-1}, x_{n-1}) \quad (n = 1, 2, 3, \ldots) \qquad (2.265)$$

for any initial $(x_0, y_0) \in [0, v] \times [u, v]$, we have

$$x_n \to x^*, \quad y_n \to x^* \quad (n \to \infty),$$

and the convergence rates are

$$\left\| x_{m_0+n} - x^* \right\| \le \frac{2N^2}{n+1} \| v - u \| \quad (n \to \infty), \qquad (2.266)$$

$$\left\| y_{m_0+n} - x^* \right\| \le \frac{2N^2}{n+1} \| v - u \| \quad (n \to \infty). \qquad (2.267)$$

Proof For convenience of presentation, we denote $C = (I - B)^{-1} A$. From (ii) and (iii) we have

$$C(u, v) = (I - B)^{-1} A(u, v) \ge (I - B)^{-1} u \ge u, \qquad (2.268)$$

and

$$C(v, u) = (I - B)^{-1} A(v, u) \ge (I - B)^{-1} v \ge v. \qquad (2.269)$$

Hence $\mathcal{R}(C|_{[u,v] \times [u,v]}) \subset [u, v]$. Since B is affine on $[u, v]$, if follows that $(I - B)^{-1}$ is affine on $[u - Bu, v - Bv]$. So by (ii) we know that C is mixed monotone, and for fixed y, $C(\cdot, y) : [u, v] \to E$ is concave, for fixed x, $C(x, \cdot) : [u, v] \to E$ is convex. By (2.262), (2.263), (2.268) and (2.269), we show easily that for $n \in \mathbb{N}$,

$$u = u_0 \le u_1 \le \cdots \le u_{m_0} \le u_{m_0+1} \le \cdots \le u_{m_0+n} \le u_{u+m_0+1} \le \cdots$$

$$\le v_{m_0+n+1} \le v_{m_0+n} \le \cdots \le v_{m_0+1} \le v_{m_0} \le \cdots \le v_1$$

$$\le v_0 = v. \qquad (2.270)$$

Thus, it follows from (2.261) and (2.270) that

$$\theta \le \frac{1}{2}(v_{m_0+n} - u_{m_0}) \le \cdots \le \frac{1}{2}(v_{m_0+1} - u_{m_0}) \le u_{m_0+1} - u_{m_0} \le \cdots \le u_{m_0+n} - u_{m_0}.$$

We set

$$t_n = \sup\left\{ t > 0 : u_{m_0+n} \ge t v_{m_0+n} + (1 - t) u_{m_0} \right\} \quad (n = 1, 2, \ldots).$$

Obviously,

$$u_{m_0+n} \ge t_n v_{m_0+n} + (1 - t_n) u_{m_0}, \qquad (2.271)$$

and

$$\frac{1}{2} \le t_1 \le t_2 \le \cdots \le t_n \le \cdots \le 1.$$

Next, we prove $t_n \to 1$ as $n \to \infty$. For $n \in \mathbb{N}$, making use of (2.271) we obtain

$$u_{m_0+n} \ge t_n u_{m_0+n} + (1 - t_n)u_{m_0}$$

$$\ge t_n v_{m_0+n} + (1 - t_n)\frac{2u_{m_0} - t_n v_{m_0+n}}{2 - t_n}$$

$$\ge \frac{t_n}{2 - t_n} v_{m_0+n} + \frac{2(1 - t_n)}{2 - t_n} u_{m_0}.$$

So,

$$u_{m_0+n+1} = C(u_{m_0+n}, v_{m_0+n})$$

$$\ge \frac{t_n}{2 - t_n} C(v_{m_0+n}, v_{m_0+n}) + \frac{2(1 - t_n)}{2 - t_n} C(u_{m_0}, v_{m_0+n}),$$

$$\ge \frac{t_n}{2 - t_n} C(v_{m_0+n}, v_{m_0+n}) + \frac{2(1 - t_n)}{2 - t_n} C(u_{m_0}, v_{m_0+1}),$$

i.e.,

$$t_n C(v_{m_0+n}, v_{m_0+n}) + 2(1 - t_n)C(u_{m_0}, v_{m_0+1}) \le (2 - t_n)u_{m_0+n+1}. \qquad (2.272)$$

From (2.261), (2.271) and (2.272), we have

$$v_{m_0+n+1} = C(v_{m_0+n}, u_{m_0+n})$$

$$\le t_n C(v_{m_0+n}, v_{m_0+n}) + (1 - t_n)C(v_{m_0+n}, u_{m_0})$$

$$\le t_n C(v_{m_0+n}, v_{m_0+n}) + (1 - t_n)v_{m_0+1}$$

$$\le t_n C(v_{m_0+n}, v_{m_0+n}) + (1 - t_n)(2u_{m_0+1-u_{m_0}})$$

$$= t_n C(v_{m_0+n}, v_{m_0+n}) + 2(1 - t_n)u_{m_0+1} - (1 - t_n)u_{m_0}$$

$$\le t_n C(v_{m_0+n}, v_{m_0+n}) + 2(1 - t_n)C(u_{m_0}, v_{m_0+1}) - (1 - t_n)u_{m_0}$$

$$\le (2 - t_n)u_{m_0+n+1} - (1 - t_n)u_{m_0}.$$

Consequently,

$$u_{m_0+n+1} \ge \frac{1}{2 - t_n} v_{m_0+n+1} - (1 - t_n)u_{m_0}.$$

This means that

$$t_{n+1} \ge \frac{1}{2 - t_n},$$

i.e.,

$$1 - t_{n+1} \le \frac{1 - t_n}{2 - t_n}. \tag{2.273}$$

For convenience, we set $s_n = 1 - t_n \neq 0$. Then (2.273) can be rewritten as follows:

$$s_{n+1} \le \frac{s_n}{1 + s_n} = \frac{1}{1 + \frac{1}{s_n}}. \tag{2.274}$$

Therefore,

$$\frac{1}{s_{n+1}} \ge 1 + \frac{1}{s_n}$$

and

$$\frac{1}{s_n} \ge 1 + \frac{1}{s_n}. \tag{2.275}$$

Combining (2.274) and (2.275) gives

$$1 - t_{n+1} = s_{n+1} \le \frac{1}{2 + \frac{1}{s_{n-1}}} \le \cdots \le \frac{1}{n + \frac{1}{s_1}} \le n + 2. \tag{2.276}$$

Hence, $s_n \to 0$ ($n \to \infty$), i.e., $t_n \to 1$ as $n \to \infty$. In addition, from (2.270) and (2.271) we obtain, for $n, p \in \mathbb{N}$,

$$\theta \le u_{m_0+n+p} - u_{m_0+n} \le v_{m_0+n} - u_{m_0+n}$$

$$\le (1 - t_n)(v_{m_0+n} - u_{m_0}) \le (1 - t_n)(v - u). \tag{2.277}$$

Since P is normal we get

$$\|u_{m_0+n+p} - u_{m_0+n}\| \le N(1 - t_n)\|v - u\| \tag{2.278}$$

and

$$\|v_{m_0+n} - u_{m_0+n}\| \le N(1 - t_n)\|v - u\|. \tag{2.279}$$

From (2.278) we know that $\{u_{m_0+n}\}_{n=1}^{\infty}$ is a Cauchy sequence. Hence there exists $x^* \in [u, v]$ such that $u_n \to x^*$ ($n \to \infty$). In combination with (2.270), (2.277), and (2.279), this implies

$$\lim_{n \to \infty} u_n = \lim_{n \to \infty} v_n = x^*, \tag{2.280}$$

and

$$u_{m_0+n} \le x^* \le v_{m_0+n}. \tag{2.281}$$

Therefore,

$$u_{m_0+n+1} = C(u_{m_0+n}, v_{m_0+n}) \le C(x^*, x^*) \le C(v_{m_0+n}, u_{m_0+n}) = v_{m_0+n+1}.$$

Taking limit and using (2.280), we conclude that

$$C(x^*, x^*) = x^*,$$

i.e.,

$$(I - B)^{-1} A(x^*, x^*) = x^*.$$

So,

$$A(x^*, x^*) + Bx^* = x^*,$$

that is, (2.260) has a positive solution x^*.

Now for each $(x_0, y_0) \in [0, v] \times [u, v]$, considering the sequences (2.264) and (2.265) we have

$$u_n \le x_n \le v_n \quad (n = 0, 1, \ldots), \tag{2.282}$$

and

$$u_n \le y_n \le v_n \quad (n = 0, 1, \ldots). \tag{2.283}$$

Hence by (2.280) we get

$$\|x_n - x^*\| \to 0 \quad (n \to \infty), \tag{2.284}$$

and

$$\|y_n - x^*\| \to 0 \quad (n \to \infty). \tag{2.285}$$

Moreover, using (2.276), (2.279), (2.281), and (2.282) we obtain the following inequality:

$$\begin{aligned}
\|x_{m_0+n} - x^*\| &\le \|x_{m_0+n} - u_{m_0+n}\| + \|x^* - u_{m_0+n}\| \\
&\le 2N \|v_{m_0+n} - u_{m_0+n}\| \\
&\le 2N^2 (1 - t_n) \|v - u\| \\
&\le \frac{2N^2}{n+1} \|v - u\| \quad (n = 1, 2, \ldots).
\end{aligned}$$

In the same way we can get (2.267).

Finally, let $y^* \in [u, v]$, $y^* \neq x^*$ and $C(y^*, y^*) = y^*$. We take initial value $(x_0, y_0) = (y^*, y^*)$. Then,

$$x_1 = C(x_0, y_0) = C(y^*, y^*) = y^*$$

and

$$y_1 = C(y_0, x_0) = C(y^*, y^*) = y^*.$$

By induction we have $x_n = y_n = y^*$ $(n = 1, 2, \ldots)$. Noting that (2.284) and (2.285), we get $y^* = x^*$. This implies the uniqueness of positive solution of (2.260). $\square$

Remark 2.5.1 In Theorem 2.5.3 they do not require operators A and B to be compact or continuous.

In the special case $B = 0$, we have the following corollary, which improved Corollary 2.3.1.

Corollary 2.5.1 (Li, Liang and Xiao [129]) *Let P be normal, $u, v \in P$, $u < v$, operator $A : [u, v] \times [u, v] \to E$ is mixed monotone and satisfy the following condition* (i) *of Theorem 2.5.3. Suppose that*

$$A(u, v) \geq u, \quad A(v, u) \leq v \quad and \quad A(u, v) \geq \frac{1}{2}\big[A(v, u) + u\big].$$

Then A has a unique fixed point x^* in $[u, v]$.

Remark 2.5.2 In Corollary 2.3.1, the constant c must be strictly larger than $\frac{1}{2}$. But Corollary 2.5.1 works even for $c = \frac{1}{2}$.

Example 2.5.1 (Zhitao Zhang and Liming Sun) *An example to show that constant $c = \frac{1}{2}$ is the best one*:
According to the Theorem 2.3.5 and Corollary 2.5.1, it requires that the constant $c \in [\frac{1}{2}, 1]$. Then a natural question comes out that whether this result holds when $c < \frac{1}{2}$. The answer is negative. We will give an example to show mixed monotone operator A may has no fixed point when $c < \frac{1}{2}$.

We consider Banach space

$$l^\infty = \big\{x = (x_1, x_2, \ldots) \mid x_i \in \mathbb{R}, i = 1, 2, \ldots \text{ and } |x_i| \text{ is uniformly bounded}\big\}$$

with norm $\|x\| = \sup_{i \in \mathbb{N}} |x_i|$. Define subspace

$$V = \Big\{x \in l^\infty \mid \text{there exists } \bar{x} \in \mathbb{R}, \text{s.t. } \lim_{i \to \infty} x_i = \bar{x} < \infty\Big\}.$$

Then V is a Banach space. Let $P = \{x \in V \mid x_i \geq 0, \forall i \geq 1\}$, then it is easy to verify that P is a normal cone of V.

First, we need a function $S : [0, 1] \to [0, 1]$ which is convex, decreasing and continuous in $[0, 1]$, but $S^n(0)$ and $S^n(1)$ do not converge to the fixed point of S. For example, suppose $c < \frac{1}{2}$, choose $\varepsilon > 0$ small enough such that $c + \varepsilon < \frac{1}{2}$, let

$$S(x) = \begin{cases} -\frac{1-c}{c+\varepsilon}x + 1, & x \in [0, c + \varepsilon], \\ c, & x \in (c + \varepsilon, 1]. \end{cases} \tag{2.286}$$

The graph of S is presented in Fig. 2.1. It is easy to know that S has a unique fixed point $x = \frac{c+\varepsilon}{1+\varepsilon}$ and a couple fixed points $x' = \frac{c^2+\varepsilon}{c+\varepsilon}$, $x'' = c$, which means $S(x') = x''$ and $S(x'') = x'$. We can verify that

$$S(0) = 1, \qquad S(1) = c,$$

Fig. 2.1 The graph of S

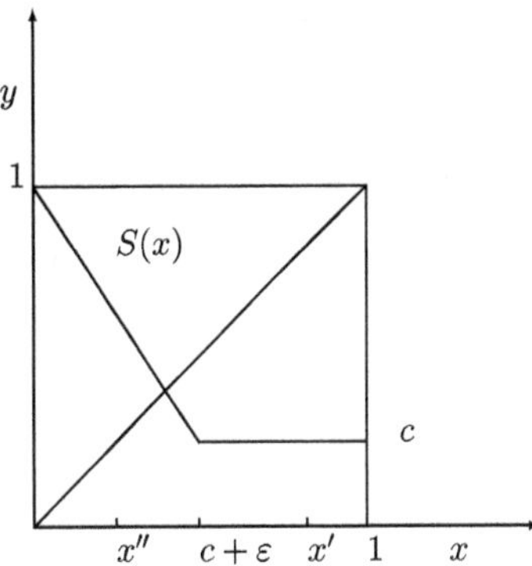

$$S^{2n}(0) = S^{2n+1}(1) = x'', \quad n \geq 1,$$

$$S^{2n+1}(0) = S^{2n}(1) = x', \quad n \geq 1.$$

Let $\bar{v} = (1, 1, 1, \ldots)$, $\theta = (0, 0, \ldots)$,

$$A : P \to P$$

$$x \mapsto Ax = \big(1, S(x_1), S(x_2), \ldots\big)$$

A maps P to P, in fact suppose $x \in P$ with $\lim_{i \to \infty} x_i = \bar{x}$, then $\lim_{i \to \infty} S(x_i) = S(\bar{x})$, since S is continuous.

Then it is easy to verify A is decreasing and convex,

$$A(\bar{v}) = \big(1, S(1), S(1), \ldots\big),$$

$$A(\theta) = \big(1, S(0), S(0), \ldots\big) = \bar{v}.$$

So if there exists k such that $A(\bar{v}) \geq k A(\theta)$, then $k \leq \frac{S(1)}{S(0)} = c < \frac{1}{2}$, we can just let $k = c < \frac{1}{2}$.

We next show A has no fixed point in P. If there exists $x \in V$ such that $Ax = x$ then

$$x_1 = 1,$$

$$x_2 = S(x_1),$$

$$x_3 = S(x_2),$$

$$\vdots$$

$$x_n = S(x_{n-1}),$$

$$\vdots$$

So $x_i = S^{i-1}(1)$, but we know that this consequence diverges, thus A has no fixed point in P.

Operator A is a special mixed monotone operator, A only has decreasing part. $\square$

Definition 2.5.2 Let P be a cone of Banach space E, $A : P \to P$, $u_0 > \theta$ (i.e., $u_0 \in P$, $u \neq \theta$), A is called a u_0-concave operator, if

(i) $\forall x > \theta$ there exist $\alpha = \alpha(x) > 0$, $\beta = \beta(x) > 0$ such that

$$\alpha x \leq Ax \leq \beta x; \tag{2.287}$$

(ii) $\forall x \in P$ such that $\alpha_1 u_0 \leq x \leq \beta_1 u_0$ for some $\alpha_1 = \alpha_1(x) > 0$, $\beta_1 = \beta_1(x) > 0$, and $0 < t < 1$, there exists $\eta = \eta(x, t) > 0$ such that

$$A(tx) \geq (1 + \eta) t Ax.$$

Definition 2.5.3 Let P be a cone of Banach space E, $A : P \to P$, $u_0 > \theta$ (i.e. $u_0 \in P, u \neq \theta$), A is called a u_0-convex operator, if

(i) $\forall x > \theta$ there exist $\alpha = \alpha(x) > 0$, $\beta = \beta(x) > 0$ such that (2.287) is satisfied;
(ii) $\forall x \in P$ such that $\alpha_1 u_0 \leq x \leq \beta_1 u_0$ for some $\alpha_1 = \alpha_1(x) > 0$, $\beta_1 = \beta_1(x) > 0$, and $0 < t < 1$, there exists $\eta = \eta(x, t) > 0$ such that

$$A(tx) \leq (1 - \eta) t Ax.$$

Theorem 2.5.4 (Zhitao Zhang [199]) *Let $P \subset E$ be a normal and solid cone, $A :$ $P \to P$ is a condensing map, $A^2\theta > \theta$, A is strongly decreasing, i.e., $\theta \leq x < y$ implies $Ax \gg Ay$, and $A(tx) < t^{-1}Ax$ for $x > \theta$, $t \in (0, 1)$. Then A has a unique fixed point $x^* > \theta$, and $\forall x_0 \in P$, let $x_n = Ax_{n-1}$ $(n = 1, 2, \ldots)$, we have $\|x_n - x^*\| \to 0(n \to \infty)$.*

Proof Since P is normal we know $[\theta, A\theta]$ is bounded. Moreover, $[\theta, A\theta]$ is convex closed subset. Since A is decreasing, we get $A([\theta, A\theta]) \subset [\theta, A\theta]$. By the Sadovskii Theorem, we see that A has at least one positive fixed point. Obviously, $A^2 : P \to P$ is a condensing and increasing map. Let $u_0 = \theta$, $u_n = Au_{n-1}$, it is easy to know A^2 has a minimal fixed point u_* and a maximal fixed point u^* in $[\theta, A\theta]$. Moreover,

$$\theta = u_0 \leq u_2 \leq \cdots \leq u_{2n} \leq \cdots \leq u_* \leq \cdots \leq u^* \leq \cdots$$

$$\leq u_{2n+1} \leq \cdots \leq u_3 \leq u_1 = A\theta, \tag{2.288}$$

$$\lim_{n \to \infty} u_{2n} = u_*, \qquad \lim_{n \to \infty} u_{2n+1} = u^*.$$

Since $A(tx) < t^{-1}Ax$, $\forall t \in (0, 1)$, we get

$$A\left(t \cdot t^{-1}x\right) < t^{-1} \cdot A\left(t^{-1}x\right),$$

i.e., $A(t^{-1}x) > t \cdot Ax$. Since A is strongly decreasing, we get

$$A^2(tx) \gg A\left(t^{-1}Ax\right) > tA^2x.$$

Thus A^2 is strongly sublinear, and $\exists e \in \overset{\circ}{P}$ such that A^2 is e-concave (see [110]). Therefore, by Theorem 2.2.2 of [110], we see that A^2 has at most one positive (i.e., $> \theta$) fixed point. Let $u_* = u^* = x^*$, then $A^2 x^* = x^*$, and $A(A^2 x^*) = A^2(Ax^*) = Ax^*$, thus $Ax^* = x^*$. Moreover, $\forall x_0 \in P$, $\theta \leq Ax_0 \leq A\theta$, $A^2\theta \leq A^2 x_0 \leq A\theta$, i.e., $u_0 \leq x_1 \leq u_1$, $u_2 \leq x_2 \leq u_1$, by induction, we get

$$u_{2n} \leq x_{2n+1} \leq u_{2n+1}, \qquad u_{2n} \leq x_{2n} \leq u_{2n-1}.$$

By taking the limit, we get $x_n \to x^*$ $(n \to \infty)$. $\qquad\qquad\qquad\qquad\qquad\qquad$ $\square$

Theorem 2.5.5 (Zhitao Zhang [199]) *Suppose $P \subset E$ is a normal and solid cone, $A : P \times P \to P$ is completely continuous, and $A(x, y)$ is strongly increasing in x, $A(x, y)$ is decreasing in y. Moreover, $\exists \theta < u_0 < v_0$ such that $A([u_0, v_0] \times [u_0, v_0]) \subset [u_0, v_0]$; $A(tx, y) \ll tA(x, y)$, for $x, y \in P$, $t \in (0, 1)$. Then A has a unique fixed point $x^* \in [u_0, v_0]$ and*

$$\forall x_0, y_0 \in [u_0, v_0], \quad x_n = A(x_{n-1}, y_{n-1}), \quad y_n = A(y_{n-1}, x_{n-1}),$$

we have $x_n \to x^*$, $y_n \to x^*$ $(n \to \infty)$.

Proof By Theorem 2.1.7 of [110], we know A has a couple quasi-fixed point $(x^*, y^*) \in [u_0, v_0] \times [u_0, v_0]$, i.e., $x^* = A(x^*, y^*)$, $y^* = A(y^*, x^*)$, which is minimal and maximal in the sense that $x^* \leq \bar{x} \leq y^*$ and $x^* \leq \bar{y} \leq y^*$ for any coupled quasi-fixed point $(\bar{x}, \bar{y}) \in [u_0, v_0] \times [u_0, v_0]$. Moreover, we have $x^* = \lim_{n \to \infty} u_n$, $y^* = \lim_{n \to \infty} v_n$, where $u_n = A(u_{n-1}, v_{n-1})$, $v_n = A(v_{n-1}, u_{n-1})$ $(n = 1, 2, \ldots)$ which satisfy

$$u_0 \leq u_1 \leq \cdots \leq u_n \leq x^* \leq \cdots \leq y^* \leq v_n \leq \cdots \leq v_1 \leq v_0. \qquad (2.289)$$

Now we prove $x^* = y^*$.

If $x^* \neq y^*$ then $x^* < y^*$. Let $t_0 = \inf\{t \mid x^* \leq ty^*\}$, then since A is strongly increasing in x, we get

$$x^* = A(x^*, y^*) \ll A(y^*, y^*) \leq A(y^*, x^*) = y^*, \qquad (2.290)$$

thus by $x^* \geq u_0 > \theta$ and (2.290), we have $0 < t_0 < 1$ and

$$x^* = A(x^*, y^*) \leq A(t_0 y^*, x^*) \ll t_0 A(y^*, x^*) = t_0 y^*, \qquad (2.291)$$

which contradicts the definition of t_0.

Thus we have $x^* = y^*$ and

$$x^* = A(x^*, x^*), \qquad u_n \to x^*, \qquad v_n \to x^* \quad (n \to \infty).$$

Moreover, $\forall x_0, y_0 \in [u_0, v_0]$, let

$$x_n = A(x_{n-1}, y_{n-1}), \qquad y_n = A(y_{n-1}, x_{n-1}).$$

It is obvious that $u_n \leq x_n \leq v_n$, $u_n \leq y_n \leq v_n$. Following the normality of P, we get $x_n \to x^*$, $y_n \to x^*$ $(n \to \infty)$. $\qquad\qquad\qquad\qquad\qquad\qquad\qquad\qquad\qquad$ $\square$

Proposition 2.5.1 *There is no operator $A : P \to P$ which is decreasing and e-convex, where $e > \theta$.*

Proof If $A : P \to P$ is decreasing and e-convex, then for $x \in P$ such that $\alpha_1(x)e \leq x \leq \beta_1(x)e$ $(\alpha_1(x) > 0, \; \beta_1(x) > 0)$, and $\forall t \in (0, 1)$, $\exists \eta = \eta(x, t) > 0$ (see [110]) such that

$$Ax \leq A(tx) \leq (1 - \eta)tAx,$$

so we have $(1 - (1 - \eta)t)Ax \leq \theta$, thus $Ax \leq \theta$, but by the definition of e-convex, we have $\exists \alpha = \alpha(x) > 0$ such that $Ax \geq \alpha e$. So we know $e \leq \theta$, which contradicts $e > \theta$. $\qquad\square$

Proposition 2.5.2 *Let E be weakly complete and P be a normal and solid cone in E, $A : P \to P$ is continuous and strongly decreasing, $A^2\theta > \theta$; $A(tx) < t^{-1}Ax$, for $t \in (0, 1)$, $x \in P$. Then A has a unique fixed point $x^* \in P$, and $\forall x_0 \in P$, let $x_n = Ax_{n-1}$, we have $x_n \to x^*$ $(n \to \infty)$.*

Proof Let

$$u_0 = \theta, \qquad u_n = Au_{n-1}.$$

We know $A^2 : [\theta, A\theta] \to [\theta, A\theta]$ is continuous and strongly increasing. Moreover,

$$\theta \leq u_0 \leq u_2 \leq \cdots \leq u_{2n} \leq \cdots \leq u_{2n+1} \leq \cdots \leq u_1 = A\theta. \tag{2.292}$$

Since E is weakly complete and P is normal, by Theorem 2.2 of [89], we get P is regular. By Theorem 2.1.1 of [110], we know A^2 has a minimal fixed point u_* and a maximal fixed point u^*, and $u_{2n} \to u_*$, $u_{2n+1} \to u^*$. Similarly to the proof of Theorem 2.5.4, we can prove $u^* = u_*$, and let $x^* = u_* = u^*$, then $Ax^* = x^*$, and $\forall x_0 \in P$, $x_n = Ax_{n-1}$, we have $x_n \to x^*$ $(n \to \infty)$. $\qquad\square$

On Differentiable Maps   Let (E, P) be an OBS with open unit ball B, a subset $D \subset E$ is called a right (or left) neighborhood of point $x \in E$ if there exists a positive number ε such that $x + \varepsilon B^+ \subset D$ (or $x - \varepsilon B^+ \subset D$). The set D is called P-open (or $-P$-open), if D is a right (or left) neighborhood of each of its points (see [8]).

Let D be a right neighborhood of some $x \in E$ and let F be an arbitrary Banach Space. A map $A : D \to F$ is said to be right differentiable in x, if there exists a bounded linear operator $T \in L(\overline{P - P}, F)$ such that

$$\lim_{h \to \theta, h \in P} \frac{\|A(x + h) - A(x) - Th\|}{\|h\|} = 0. \tag{2.293}$$

T is denoted by $A'_+(x)$ and called the right derivative of A at x. Let $D \subset E$ be P-open, then $A : D \to F$ is called right differentiable if A has the right derivative at every $x \in D$. In this case, the map $A'_+ : D \to L(\overline{P - P}, F)$ is called the right derivative of A. If A'_+ maps D continuously into the Banach space $L(\overline{P - P}, F)$ then A is said to be continuously right differentiable. Left derivatives and left differentiable maps are defined in the obvious way (see [8]).

Theorem 2.5.6 (Zhitao Zhang, see [199]) *Suppose that $P \subset E$ is a normal cone, $A : P \times P \to P$ is a mixed monotone operator, and $A(x, y) = A_1(x) + A_2(y)$, where A_1 is increasing and A_2 is decreasing. Moreover, $\exists v \in P$ such that $A : [\theta, v] \times [\theta, v] \to [\theta, v]$; $A_1 : P \to P$ and $A_2 : P \to P$ are both continuously right differentiable, and $\sup_{x \in [\theta, v]} \|A'_{1+}(x)\| + \sup_{y \in [\theta, v]} \|A'_{2+}(y)\| = \delta < 1$. Then A has a unique fixed point $x^* \in [\theta, v]$.*

Proof Let

$$u_0 = \theta, \qquad v_0 = v, \qquad u_n = A(u_{n-1}, v_{n-1}), \qquad v_n = A(v_{n-1}, u_{n-1}).$$

Since A is mixed monotone, we know easily

$$\theta = u_0 \le u_1 \le u_2 \le \cdots \le u_n \le \cdots \le v_n \le \cdots \le v_3 \le v_1 \le v_0 = v. \qquad (2.294)$$

Moreover, we know that $x + t(y - x) \in P$, $\forall t \in [0, 1]$, $\forall \theta \le x \le y \le v$. So $A_i(x + t(y - x)) : [0, 1] \to P$, $i = 1, 2$ are right differentiable on $[0, 1]$, and by mean value theorem (see [11]), we get $A_i(y) - A_i(x) \in (1 - 0)\overline{co}\{A'_{i+}(x + t(y - x))(y - x), t \in [0, 1]\} = \overline{co}\{A'_{i+}(x + t(y - x))(y - x), t \in [0, 1]\}$ $(i = 1, 2)$. So we have

$$\begin{aligned}
\|A(y, x) - A(x, y)\| &= \|A_1(y) + A_2(x) - A_1(x) - A_2(y)\| \\
&\le \|A_1(y) - A_1(x)\| + \|A_2(x) - A_2(y)\| \\
&\le \sup_{x \in [\theta, v]} \|A'_{1+}(x)\| \cdot \|y - x\| + \sup_{y \in [\theta, v]} \|A'_{2+}(y)\| \cdot \|y - x\| \\
&= \delta \|y - x\|.
\end{aligned} \qquad (2.295)$$

So by standard argument, we know that $\exists u^*$ such that $u_n \to u^*$, $v_n \to u^*$, and $A(u^*, u^*) = u^*$. $\qquad \square$

Theorem 2.5.7 (Zhitao Zhang [199]) *Let $P \subset E$ be a normal cone, $A : P \to P$ be a continuous, continuously right differentiable, twice right differentiable decreasing map. Suppose that one of the following hypotheses is satisfied:*

(i) *A is order convex, i.e., $\forall \theta \le x \le y$, $t \in [0, 1]$, $A(tx + (1 - t)y) \le tAx + (1 - t)Ay$, and $A'_+(0)u \ge -Nu$, $\forall u \in P$;*

(ii) *A is order concave on $[\theta, A\theta]$, i.e., $\forall x, y \in [\theta, A\theta]$, $x \le y$, $t \in [0, 1]$, $A(tx + (1 - t)y) \ge tAx + (1 - t)Ay$, and $A'_+(A\theta)u \ge -Nu$, $\forall u \in P$, where N is a positive constant.*

Then A has a unique fixed point $x^* \in P$.

Proof (i) Since A is order convex, in virtue of Theorem 23.1, Theorem 23.3 in [8] and Proposition 3.2 in [7], we know

$$\forall \theta \le x \le y, \quad A'_+(y)h \ge A'_+(x)h, \quad \forall h \in P \qquad (2.296)$$

and

$$Ay - Ax \geq A'_+(x)(y - x) \geq A'_+(0)(y - x) \geq -N(y - x). \qquad (2.297)$$

So we get $(A + N)y - (A + N)x \geq \theta$. Let

$$Bx = \frac{Ax + Nx}{1 + N},$$

then $B : P \to P$ is increasing and $\forall x \in [\theta, A\theta]$,

$$\theta \leq B\theta = \frac{A\theta}{1 + N} \leq A\theta, \qquad B(A\theta) = \frac{A^2\theta + NA\theta}{1 + N} \leq \frac{A\theta + NA\theta}{1 + N} \leq A\theta. \qquad (2.298)$$

Let

$$u_0 = \theta, \qquad v_0 = A\theta, \qquad u_n = Bu_{n-1}, \qquad v_n = Bv_{n-1},$$

we have

$$\theta = u_0 \leq u_1 \leq u_2 \leq \cdots \leq u_n \leq \cdots \leq v_n \leq \cdots \leq v_2 \leq v_1 \leq v_0 = A\theta, \qquad (2.299)$$

and

$$\theta \leq v_n - u_n = \frac{Av_{n-1} + Nv_{n-1}}{1 + N} - \frac{Au_{n-1} + Nu_{n-1}}{1 + N}$$

$$= \frac{Av_{n-1} - Au_{n-1} + N(v_{n-1} - u_{n-1})}{1 + N} \leq \frac{N}{1 + N}(v_{n-1} - u_{n-1})$$

$$\leq \left(\frac{N}{1 + N}\right)^n (v_0 - u_0) = \left(\frac{N}{1 + N}\right)^n \cdot A\theta. \qquad (2.300)$$

Following the normality of P, there exists a unique $u^* \in P$ such that $u_n \to u^*$, $v_n \to u^*$ $(n \to \infty)$ and $Bu^* = u^*$, i.e., $Au^* = u^*$. Moreover, $\forall x_0 \in [\theta, A\theta]$, $x_n = Bx_{n-1}$, we get $u_n \leq x_n \leq v_n$, $x_n \to u^*$ $(n \to \infty)$. The assertion (i) is valid.

(ii) Since A is order concave, in virtue of Theorem 23.1, Theorem 23.3 in [8] and Proposition 3.2 in [7], we have

$$A'_+(y) \leq A'_+(x), \qquad \forall \theta \leq x \leq y \leq A\theta,$$

i.e.,

$$A'_+(y)h \leq A'_+(x)h, \qquad \forall h \in P,$$

and

$$Ay - Ax \geq A'_+(y)(y - x) \geq A'_+(A\theta)(y - x) \geq -N(y - x), \qquad (2.301)$$

i.e.,

$$(Ay + Ny) - (Ax + Nx) \geq \theta.$$

Similar to the proof of (i), we can prove the assertion (ii). $\square$

Example 2.5.2 Consider the following integral equation:

$$x(t) = Ax(t) = \int_0^1 k(t,s) \frac{1}{(1+x(s))^p} \, ds \quad (0 < p < 1). \tag{2.302}$$

Proposition 2.5.3 *If $k : [0,1] \times [0,1] \to \mathbb{R}$ is continuous and $k(t,s) > 0$, then (2.302) has a unique positive solution $x^*(t)$, $\forall x_0(t) \geq 0$, $x_n = Ax_{n-1}$, we have $x_n \to x^*$ $(n \to \infty)$.*

Proof Let $E = C[0,1]$, $P = \{x \mid x(t) \geq 0, x \in E\}$ is a normal and solid cone in E. We know that $A : P \to P$ is completely continuous, $A^2\theta > \theta$. If $0 \leq x_1(t) < x_2(t)$, then

$$Ax_1(t) - Ax_2(t) = \int_0^1 k(t,s) \left(\frac{1}{(1+x_1(s))^p} - \frac{1}{(1+x_2(s))^p} \right) ds. \tag{2.303}$$

Since $k(t,s) > 0$, and $\frac{1}{(1+x_1(s))^p} - \frac{1}{(1+x_2(s))^p} > \theta$, there exists a constant $\delta > 0$ such that $Ax_1(t) - Ax_2(t) \geq \delta > 0$, so A is strongly decreasing. Moreover, $\forall x > \theta$, $\forall \tau \in (0,1)$, we have

$$A(\tau x) = \int_0^1 k(t,s) \frac{1}{(1+\tau x(s))^p} \, ds = \int_0^1 k(t,s) \frac{1}{\tau^p(\tau^{-1}+x(s))^p} \, ds$$

$$\leq \frac{1}{\tau^p} \int_0^1 k(t,s) \frac{1}{(1+x(s))^p} \, ds < \tau^{-1} Ax. \tag{2.304}$$

By Theorem 2.5.4, we get A has a unique fixed point $x^* \in P$, and (2.302) has a unique positive solution $x^*(t)$. Moreover, $\forall x_0(t) \in P$, let $x_n(t) = Ax_{n-1}(t)$, then $x_n(t) \to x^*(t)$ uniformly on $[0,1]$. $\qquad\square$

Chapter 3
Minimax Methods

3.1 Mountain Pass Theorem and Minimax Principle

Theorem 3.1.1 (Mountain Pass Theorem, Ambrosetti–Rabinowitz [11], see also [159]) *Let E be a real Banach space and $I \in C^1(E, \mathbb{R})$ satisfying the* (PS) *condition. Suppose $I(0) = 0$ and*

(I_1) *there are constants $\rho, \alpha > 0$ such that $I|_{\partial B_\rho} \geq \alpha$, and*
(I_2) *there is an $e \in E \setminus B_\rho$ such that $I(e) \leq 0$.*

Then I possesses a critical value $c \geq \alpha$. Moreover, c can be characterized as

$$c = \inf_{g \in \Gamma} \max_{u \in g([0,1])} I(u), \tag{3.1}$$

where $\Gamma = \{g \in C([0, 1], E) \mid g(0) = 0, g(1) = e\}$.

Proof The definition of c shows that $c < \infty$. If $g \in \Gamma$, $g([0, 1]) \cap \partial B_\rho \neq \emptyset$. Therefore,

$$\max_{u \in g([0,1])} I(u) \geq \inf_{w \in \partial B_\rho} I(w) \geq \alpha$$

via (I_1). Consequently $c \geq \alpha$. Suppose that c is not a critical value of I. Then Theorem 1.2.7 with $\bar{\varepsilon} = \alpha/2$ yields $\varepsilon \in (0, \bar{\varepsilon})$ and η as in that result. Choose $g \in \Gamma$ such that

$$\max_{u \in g([0,1])} I(u) \leq c + \varepsilon \tag{3.2}$$

and consider $h(t) \equiv \eta(1, g(t))$. Clearly $h \in C([0, 1], E)$. Also $g(0) = 0$ and $I(0) = 0 < \alpha/2 \leq c - \bar{\varepsilon}$ imply $h(0) = 0$ by (2) of Theorem 1.2.7. Similarly $g(1) = e$ and $I(e) \leq 0$ imply that $h(1) = e$. Consequently $h \in \Gamma$ and by (3.1)

$$c \leq \max_{u \in h([0,1])} I(u). \tag{3.3}$$

Z. Zhang, *Variational, Topological, and Partial Order Methods with Their Applications*,
Developments in Mathematics 29,
DOI 10.1007/978-3-642-30709-6_3, © Springer-Verlag Berlin Heidelberg 2013

But by (3.2), $g([0, 1]) \subset I_{c+\varepsilon}$, so (7) of Theorem 1.2.7 implies $h([0, 1]) \subset I_{c-\varepsilon}$, i.e.,

$$\max_{u \in h([0,1])} I(u) \leq c - \varepsilon, \tag{3.4}$$

contrary to (3.3). Thus c is a critical value of I. $\square$

Theorem 3.1.2 (See [159]) *Let E be a real Banach space and $I \in C^1(E, \mathbb{R})$ satisfying* (PS) *condition. If I is bounded from below, then*

$$c \equiv \inf_E I$$

is a critical value of I.

Proof Clearly c is finite. Set $S = \{\{x\}|x \in E\}$, i.e., S is a collection of sets, each consisting of one point. Trivially we have

$$c = \inf_{M \in S} \max_{u \in M} I(u)$$

Now for any choice of $\bar{\varepsilon}$, e.g. $\bar{\varepsilon} = 1$, since $\eta(1, \cdot)$ as given by Theorem 1.2.7 maps S into S, the argument of Theorem 3.1.1 shows c is a critical value of I. $\square$

Next we give two useful theorems:

Theorem 3.1.3 (Saddle Point Theorem, see [159]) *Let $E = V \oplus X$, where E is a real Banach space and $V \neq \{0\}$ is finite dimensional. Suppose $I \in C^1(E, \mathbb{R})$, satisfies the* (PS) *condition, and*

(I_3) *there is a constant $\alpha > 0$ and a bounded neighborhood D of 0 in V such that $I|_{\partial D} \geq \alpha$ and*

(I_4) *there is a constant $\beta > \alpha$ such that $I|_X \geq \beta$.*

Then I possesses a critical value $c \geq \beta$. Moreover, c can be characterized as

$$c = \inf_{h \in \Gamma} \max_{u \in \bar{D}} I\big(h(u)\big),$$

where $\Gamma = \{h \in C(\bar{D}, E)|h = \mathrm{id} \ on \ \partial D\}$.

Definition 3.1.1 (See [168]) Let $\Phi : M \times [0, \infty) \to M$ be a semi-flow on a manifold M. A family $\mathcal{F}$ of subsets of M is called (positively) Φ-invariant if $\Phi(F, t) \in \mathcal{F}$ for all $F \in \mathcal{F}, t \geq 0$.

Theorem 3.1.4 (Minimax Principle, see [168]) *Suppose M is a complete Finsler manifold of class $C^{1,1}$ and $I \in C^1(M)$ satisfies* (PS) *condition. Also assume $\mathcal{F} \subset \mathcal{P}(M)$ (where $\mathcal{P}(M) := \{U|U \subseteq M\}$) is a collection of sets which is invariant with respect to any continuous semi-flow $\Phi : M \times [0, \infty) \to M$ such that $\Phi(\cdot, 0) = \mathrm{id}$,*

$\Phi(\cdot, t)$ is a homeomorphism of M for any $t \geq 0$, and $E(\Phi(u, t))$ is non-increasing in t for any $u \in M$. Then, if

$$\beta = \inf_{F \in \mathcal{F}} \sup_{u \in F} I(u)$$

is finite, β is a critical value of I.

3.2 Linking Methods

Definition 3.2.1 Let S be a closed subset of a Banach space E, Q a sub-manifold of E with relative boundary ∂Q. We say S and ∂Q link if

(i) $S \cap \partial Q = \emptyset$, and
(ii) for any map $h \in C(E, E)$ such that $h|_{\partial Q} = \text{id}$ there holds $h(Q) \cap S \neq \emptyset$.

More generally, if S and Q are as above and if Γ is a subset of $C(E, E)$, then S and ∂Q will be said to link w.r.t. Γ, if (i) holds and if (ii) is satisfied for any $h \in \Gamma$.

Example 3.2.1 Let $E = E_1 \oplus E_2$ be decomposed into closed subspaces E_1, E_2, where $\dim E_2 < \infty$. Let $S = E_1$, $Q = B_R(0; E_2)$ with relative boundary $\partial Q = \{u \in E_2 : \|u\| = R\}$. Then S and ∂Q link.

Proof Let $\pi : E \to E_2$ be the continuous projection of E onto E_2, and let h be any continuous map such that $h|_{\partial Q} = \text{id}$. We have to show that $0 \in \pi(h(Q))$.
  For $t \in [0, 1]$, $u \in E_2$ we define

$$h_t(u) = t\pi\big(h(u)\big) + (1 - t)u.$$

Note that $h_t \in C^0(E_2; E_2)$ defines a homotopy of $h_0 = \text{id}$ with $h_1 = \pi \circ h$. Moreover, $h_t|_{\partial Q} = \text{id}$ for all t. Hence the topological degree $\deg(h_t, Q, 0)$ is well defined for all t. By homotopy invariance and normalization of the degree, we have

$$\deg(\pi \circ h, Q, 0) = \deg(\text{id}, Q, 0) = 1.$$

Hence $0 \in \pi \circ h(Q)$, as was to be shown. $\square$

Example 3.2.2 Let $E = E_1 \oplus E_2$ be decomposed into closed subspaces E_1, E_2 with $\dim E_2 < \infty$, and let $\underline{u} \in E_1$ with $\|\underline{u}\| = 1$ be given. Suppose $0 < \rho < R_1$, $0 < R_2$ and let

$$S = \big\{u \in E_1 : \|u\| = \rho\big\},$$

$$Q = \big\{s\underline{u} + u_2 : 0 \leq s \leq R_1, u_2 \in E_2, \|u_2\| \leq R_2\big\},$$

with relative boundary $\partial Q = \{s\underline{u} + u_2 \in Q : s \in \{0, R_1\} \text{ or } \|u_2\| = R_2\}$. Then S and ∂Q link.

Proof Let $\pi : E \to E_2$ be the projection onto E_2, and let $h \in C^0(E; E)$ satisfy $h|_{\partial Q} = \mathrm{id}$. We must show that there exists $u \in Q$ such that the relations $\|h(u)\| = \rho$ and $\pi(h(u)) = 0$ simultaneously hold. For $t \in [0, 1]$, $s \in \mathbb{R}$, $u_2 \in E_2$ let

$$\bar{h}_t(s, u_2) = \big(t\|h(u) - \pi(h(u))\| + (1 - t)s - \rho, \, t\pi(h(u)) + (1 - t)u_2 \big),$$

where $u = s\underline{u} + u_2$. This defines a family of maps $\bar{h}_t : \mathbb{R} \times E_2 \to \mathbb{R} \times E_2$ depending continuously on $t \in [0, 1]$. Moreover, if $u = s\underline{u} + u_2 \in \partial Q$, we have

$$\bar{h}_t(s, u_2) = \big(t\|u - u_2\| + (1 - t)s - \rho, \, u_2 \big) = (s - \rho, u_2) \neq 0$$

for all $t \in [0, 1]$. Hence, if we identify Q with a subset of $\mathbb{R} \times E_2$ via the decomposition $u = s\underline{u} + u_2$, the topological degree $\deg(\bar{h}_t, Q, 0)$ is well defined and by homotopy invariance

$$\deg(\bar{h}_1, Q, 0) = \deg(\bar{h}_0, Q, 0) = 1,$$

where $\bar{h}_0(s, u_2) = (s - \rho, u_2)$. Thus, there exists $u = s\underline{u} + u_2 \in Q$ such that $\bar{h}_1(u) = 0$, which is equivalent to

$$\pi(h(u)) = 0 \quad \text{and} \quad \|h(u)\| = \rho,$$

as desired. $\square$

Lemma 3.2.1 (Deformation Lemma [168]) *Let E be a closed subset of a Banach space. Suppose $I \in C^1(E, \mathbb{R})$ satisfies* (PS) *condition. Let $\beta \in \mathbb{R}$, $\bar{\varepsilon}$ be given and let N be any neighborhood of K_β. Then there exist a number $\varepsilon \in (0, \bar{\varepsilon})$ and a continuous 1-parameter family of homeomorphisms $\Phi(\cdot, t)$ of E, $0 \le t < \infty$, with the properties*

(1) $\Phi(u, t) = u$, *if* $t = 0$, *or* $I'(u) = 0$, *or* $|I(u) - \beta| \ge \bar{\varepsilon}$;
(2) $I(\Phi(u, t))$ *is non-increasing in t for any $u \in E$;*
(3) $\Phi(I_{\beta+\varepsilon} \setminus N, 1) \subset I_{\beta-\varepsilon}$, *and* $\Phi(I_{\beta+\varepsilon}, 1) \subset I_{\beta-\varepsilon} \cup N$.

   Moreover, $\Phi : E \times [0, \infty) \to E$ has the semi-group property; that is, $\Phi(\cdot, t) \circ \Phi(\cdot, s) = \Phi(\cdot, s + t)$ for all $s, t \ge 0$.

Theorem 3.2.1 (See [168]) *Suppose $I \in C^1(E, \mathbb{R})$ satisfies* (PS) *condition. Consider a closed subset $S \subset E$ and a sub-manifold $Q \subset E$ with relative boundary ∂Q. Suppose*

 (i) *S and ∂Q link,*
(ii) *$\alpha = \inf_{u \in S} I(u) > \sup_{u \in \partial Q} I(u) = \alpha_0$.*

   Let

$$\Gamma = \big\{ h \in C(E, E) : h|_{\partial Q} = \mathrm{id} \big\}.$$

Then the number

$$\beta = \inf_{h\in\Gamma} \sup_{u\in Q} I\big(h(u)\big)$$

defines a critical value $\beta \geq \alpha$ of I.

Proof Suppose by contradiction that $K_\beta = \emptyset$. For $\bar\varepsilon = \alpha - \alpha_0 > 0$, $N = \emptyset$, let $\varepsilon > 0$ and $\Phi : E \times [0, 1] \to E$ be the pseudo-gradient flow constructed in Lemma 3.2.1. Note that by choice of $\bar\varepsilon$ there holds $\Phi(\cdot, t)|_{\partial Q} = \mathrm{id}$ for all t. Let $h \in \Gamma$ such that $I(h(u)) < \beta + \varepsilon$ for all $u \in Q$. Define $h' = \Phi(\cdot, t) \circ h$. Then $h' \in \Gamma$ and

$$\sup_{u\in Q} I\big(h'(u)\big) < \beta - \varepsilon$$

by Lemma 3.2.1, contradicting the definition of β. $\square$

3.3 Local Linking Methods

Definition 3.3.1 (Li and Willem [126]) Let X be a Banach space with a direct sum decomposition

$$X = X^1 \oplus X^2.$$

The function $f \in C^1(X, \mathbb{R})$ has a local linking at 0, with respect to (X^1, X^2), if, for some $r > 0$,

$$f(u) \geq 0, \qquad u \in X^1, \qquad \|u\| \leq r, \tag{3.5}$$

$$f(u) \leq 0, \qquad u \in X^2, \qquad \|u\| \leq r. \tag{3.6}$$

It is clear that 0 is a critical point of f. The notion of local linking was introduced by Jiaquan Liu and Shujie Li in [125, 139] under stronger assumptions:

$$f(u) \geq c > 0, \qquad u \in X^1, \qquad \|u\| = r,$$

$$\dim X^2 < \infty.$$

The notion of local linking generalized the notions of local minimum and local maximum. When 0 is a non-degenerate critical point of a functional of class C^2 defined on a Hilbert space and $f(0) = 0$, f has a local linking at 0.

Existence results of nontrivial critical points have been established for functional which are

(a) bounded below;
(b) super-quadratic;
(c) asymptotically quadratic.

3.3.1 Deformation Lemmas

Let us recall some standard notations:

$$S_\delta := \{u \in X : \operatorname{dist}(u, S) \le \delta\},$$
$$f^c := \{u \in X : f(u) \le c\},$$
$$K_c := \{u \in X : f(u) = c, f'(u) = 0\}.$$

Lemma 3.3.1 (Li and Willem [126]) *Let f be a function of class C^1 defined on a real Banach space X. Let $S \subset X$, $\varepsilon, \delta > 0$, and $c \in \mathbb{R}$ be such that, for every $u \in f^{-1}([c - 2\varepsilon, c + 2\varepsilon]) \cap S_{2\delta}$,*

$$\|f'(u)\| \ge 4\varepsilon/\delta,$$

then there exists $\eta \in C([0, 1] \times X, X)$ such that

(i) $\eta(0, u) = u, \ \forall u \in X$,
(ii) $f(\eta(\cdot, u))$ *is nonincreasing*, $\forall u \in X$,
(iii) $f(\eta(t, u)) < c, \ \forall t \in (0, 1], \ \forall u \in f^c \cap S$,
(iv) $\eta(1, f^{c+\varepsilon} \cap S) \subset f^{c-\varepsilon}$,
(v) $\|\eta(t, u) - u\| \le \delta, \ \forall t \in [0, 1], \ \forall u \in X$.

Proof Denote

$$A := S_{2\delta} \cap f^{-1}\big([c - 2\varepsilon, c + 2\varepsilon]\big),$$
$$B := f^{-1}\big([c - \varepsilon, c + \varepsilon]\big),$$

and let $\psi : X \to [0, 1]$ be the locally Lipschitz continuous function given by

$$\psi(u) := \frac{\operatorname{dist}(u, X/A)}{\operatorname{dist}(u, X/A) + \operatorname{dist}(u, B)}.$$

Choose a pseudo-gradient vector field $v : A \to X$ for f and let $g : X \to X$ be the locally Lipschitz continuous vector field defined by

$$g(u) := \begin{cases} -\psi(u)v(u)/\|v(u)\|, & u \in A, \\ 0, & u \in X \setminus A. \end{cases}$$

The corresponding Cauchy problem has, for any $u \in X$, a unique solution $\sigma(t, u)$ defined on $\mathbb{R}$. Letting

$$\eta(t, u) = \sigma(\delta t, u),$$

it is easy to check that η satisfies the desired properties. $\square$

Let X be a Banach space with a direct sum decomposition

$$X = X^1 \oplus X^2.$$

Consider two sequences of subspaces:

$$X_0^1 \subset X_1^1 \subset \cdots \subset X^1, \qquad X_0^2 \subset X_1^2 \subset \cdots \subset X^2$$

such that

$$X^j = \overline{\bigcup_{n \in \mathbb{N}} X_n^j}, \quad j = 1, 2.$$

For every multi-index $\alpha = (\alpha_1, \alpha_2) \in \mathbb{N}^2$, we denote by X_α the space

$$X_{\alpha_1}^1 \oplus X_{\alpha_2}^2.$$

Let us recall that

$$\alpha \leq \beta \quad \Longleftrightarrow \quad \alpha_1 \leq \beta_1, \quad \alpha_2 \leq \beta_2.$$

A sequence $(\alpha_n) \in \mathbb{N}^2$ is admissible if, for every $\alpha \in \mathbb{N}^2$ there is $m \in \mathbb{N}$ such that

$$n \geq m \quad \Longrightarrow \quad \alpha_n \geq \alpha.$$

For every $f : X \to \mathbb{R}$, we denote by f_α the function f restricted to X_α.

Definition 3.3.2 Let $c \in \mathbb{R}$ and $f \in C^1(X, \mathbb{R})$. The function f satisfies the $(PS)_c^*$ condition if every sequence (u_{α_n}) such that (α_n) is admissible and

$$u_{\alpha_n} \in X_{\alpha_n}, \qquad f(u_{\alpha_n}) \to c, \qquad f_{\alpha_n}'(u_{\alpha_n}) \to 0,$$

contains a subsequence which converges to a critical point f.

Definition 3.3.3 Let $f \in C^1(X, \mathbb{R})$. The function f satisfies the $(PS)^*$ condition if every sequence (u_{α_n}) such that (α_n) is admissible and

$$u_{\alpha_n} \in X_{\alpha_n}, \qquad \sup_n f(u_{\alpha_n}) < \infty, \qquad f_{\alpha_n}'(u_{\alpha_n}) \to 0,$$

contains a subsequence which converges to a critical point of f.

Remark 3.3.1

1. In the above definitions and in the following lemmas, it is easy to replace $\mathbb{N}^2$ by any directed set.
2. The $(PS)^*$ condition implies the $(PS)_c^*$ condition for every $c \in \mathbb{R}$.
3. When $X_n^1 := X$, $X_n^2 := \{0\}$ for every $n \in \mathbb{N}$, the $(PS)_c^*$ condition is the usual Palais–Smale condition at the local c.
4. When $\dim X_n^1 < \infty$, $\dim X_n^2 = 0$ for every $n \in \mathbb{N}$, the $(PS)^*$ is the compactness condition used in [139].

5. Without loss of generality, we assume that the norm in X satisfies

$$\|u_1 + u_2\|^2 = \|u_1\|^2 + \|u_2\|^2, \quad u_j \in X_j, \ j = 1, 2.$$

Definition 3.3.4 Let A, B be closed subsets of X. By definition, $A \prec^\infty B$ if there is $\beta \in \mathbb{N}^2$ such that for every $\alpha \geq \beta$ there exists $\eta_\alpha \in C([0, 1] \times X_\alpha, X_\alpha)$ such that

(i) $\eta(0, u) = u, \forall u \in X_\alpha,$
(ii) $\eta(1, u) \in B, \forall u \in A \cap X_\alpha.$

Lemma 3.3.2 (Li and Willem [126]) *Let $f \in C^1(X, \mathbb{R})$ and $c \in \mathbb{R}$, $\rho > 0$. Let N be an open neighborhood of K_c. Assume that f satisfies* $(PS)_c^*$. *Then, for all $\varepsilon > 0$ small enough,*

$$f^{c+\varepsilon} \setminus N \prec^\infty f^{c-\varepsilon}.$$

Moreover, the corresponding deformations

$$\eta_\alpha : [0, 1] \times X_\alpha \to X_\alpha$$

satisfy

$$\left\| \eta_\alpha(t, u) - u \right\| \leq \rho, \quad \forall t \in [0, 1], \ \forall u \in X_\alpha, \tag{3.7}$$

$$f\left(\eta_\alpha(t, u) \right) < c, \quad \forall t \in (0, 1], \ \forall u \in f_\alpha^c \setminus N. \tag{3.8}$$

Proof The condition $(PS)_c^*$ implies the existence of $\gamma > 0$ and $\beta \in \mathbb{N}^2$ such that, for every $\alpha \geq \beta$ and $u \in f_\alpha^{-1}([c - 2\gamma, c + 2\gamma]) \cap (X_\alpha \setminus N)_{2\gamma}$,

$$\left\| f_\alpha'(u) \right\| \geq \gamma.$$

It suffices then to choose

$$\delta := \min\{\gamma/2, \rho, 4\}, \quad 0 < \varepsilon \leq \delta\gamma/4,$$

and to apply Lemma 3.3.1 to $S := X_\alpha \setminus N$. □

Lemma 3.3.3 (Li and Willem [126]) *Let $f \in C^1(X, \mathbb{R})$ be bounded below and let $d := \inf_X f$. If $(PS)_d^*$ condition holds then d is a critical value of f.*

Proof If d is not a critical value of f, then, by Lemma 3.3.2, there exists $\varepsilon > 0$ such that

$$f^{d+\varepsilon} \prec^\infty f^{d-\varepsilon}. \tag{3.9}$$

From the definition of d, $f_\alpha^{d+\varepsilon}$ is nonempty for α large enough. This contradicts (3.9). □

Lemma 3.3.4 (Li and Willem [126]) *Let $f \in C^1(X, \mathbb{R})$ be bounded below. If $(PS)_c^*$ holds for all $c \in \mathbb{R}$, then f is coercive.*

Proof If f is bounded below and not coercive then

$$c := \sup\{d \in \mathbb{R} : f^d \text{ is bounded}\}$$

is finite. It is easy to verify that K_c is bounded. Let N be an open bounded neighborhood of K_c. By Lemma 3.3.2, there exists $\varepsilon > 0$ such that

$$f^{c+\varepsilon} \setminus N \prec^{\infty} f^{c-\varepsilon}. \tag{3.10}$$

Moreover, we can assume that the corresponding deformation satisfy (3.7) with $\rho = 1$. It follows from the definition of c that $f^{c+\varepsilon/2} \setminus N$ is unbounded and that $f^{c-\varepsilon} \subset B(0, R)$ for some $R > 0$. It follows from (3.7) and (3.10) that, for all α large enough,

$$f^{c+\varepsilon} \setminus N \subset B(0, R + 1).$$

But then

$$f^{c+\varepsilon/2} \setminus N \subset B(0, R + 1).$$

This is a contradiction. $\qquad\qquad\qquad\qquad\qquad\qquad\qquad\qquad\qquad\qquad\qquad\square$

3.3.2 The Three Critical Points Theorem for Functionals Bounded Below

If f has a local linking at 0 and satisfies $(PS)_c^*$ or $(PS)^*$, we always assume that the same decomposition of X holds for the two properties and that $\dim X_n^j < \infty$, $j = 1, 2, n \in \mathbb{N}$.

Theorem 3.3.1 (Li and Willem [126]) *Suppose that $f \in C^1(X, \mathbb{R})$ satisfies the following assumptions*:

(A1) *f has a local linking at 0,*
(A2) *f satisfies $(PS)^*$,*
(A3) *f maps bounded sets into bounded sets,*
(A4) *f is bounded below and $d := \inf_X f < 0$.*

Then f has at least three critical points.

Proof (1) We assume that $\dim X^1$ and $\dim X^2$ are positive, since the other cases are similar. By Lemma 3.3.3, f achieves its minimum at some point v_0. Supposing

$$K := \{0, v_0\}$$

to be the critical set of f, we will be led to a contradiction. We may suppose that $r < \|v_0\|/3$ and

$$B(v_0, r) \subset f^{d/2}. \tag{3.11}$$

By assumption (A2) and Lemma 3.3.2, applied to f and to $g := -f$, there exists $\varepsilon \in (0, -d/2)$ such that

$$f^{\varepsilon} \setminus B(0, r/3) \prec^{\infty} f^{-\varepsilon}, \tag{3.12}$$

$$g^{\varepsilon} \setminus B(0, r/3) \prec^{\infty} g^{-\varepsilon}, \tag{3.13}$$

$$f^{d+\varepsilon} \setminus B(v_0, r) \prec^{\infty} f^{d-\varepsilon} = \emptyset. \tag{3.14}$$

Moreover, we can assume that the corresponding deformations exist for $\alpha \geq (m_0, m_0)$ and satisfy (3.7) with $\rho = r/2$. Assumption (A2) implies also the existence of $m_1 \geq m_0$ and $\delta > 0$ such that, for $\alpha \geq (m_1, m_1)$,

$$u \in f_{\alpha}^{-1}\big([d+\varepsilon, -\varepsilon]\big) \quad \Longrightarrow \quad \left\| f_{\alpha}'(u) \right\| \geq \delta. \tag{3.15}$$

(2) Let us write $\alpha := (n, n)$ where $n \geq m_1$ is fixed. It follows from (3.11) and (3.14) that

$$f_{\alpha}^{d+\varepsilon} \subset X_{\alpha} \cap B(v_0, r) \subset f_{\alpha}^{d/2}. \tag{3.16}$$

Using (3.12) and (3.15), it is easy to construct a deformation

$$\sigma : [0, 1] \times S_n^2 \to X_{\alpha},$$

where

$$S_n^j := \big\{ u \in X_n^j : \|u\| = r \big\}, \quad j = 1, 2,$$

such that

$$\begin{aligned} f\big(\sigma(t, u)\big) &< 0 & \forall t \in (0, 1], \ \forall u \in S_n^2, \\ f\big(\sigma(t, u)\big) &= d + \varepsilon, & \forall u \in S_n^2. \end{aligned} \tag{3.17}$$

By (3.16) there exists $\psi \in C(B_n^2, X_{\alpha})$, where

$$B_n^j := \big\{ u \in X_n^j : \|u\| \leq r \big\}, \quad j = 1, 2,$$

such that

$$\begin{aligned} \psi(u) &= \sigma(1, u), \quad \forall u \in S_n^2, \\ \psi\big(B_n^2\big) &\subset X_{\alpha} \cap B(v_0, r) \subset f_{\alpha}^{d/2}. \end{aligned} \tag{3.18}$$

Set

$$Q := [0, 1] \times B_n^2$$

and define a mapping $\Phi : \partial Q \to f_{\alpha}^0$ by

$$\Phi(t, u) = \begin{cases} u, & t = 0, \ u \in B_n^2, \\ \sigma(t, u), & 0 < t < 1, \ u \in S_n^2, \\ \psi(u), & t = 1, \ u \in B_n^2. \end{cases}$$

Lemma 3.3.4 implies the existence of $R > 0$ such that

$$f^0 \subset B(0, R).$$

Hence there is a continuous extension of Φ,

$$\tilde{\Phi} : Q \to X_\alpha,$$

such that

$$\sup_Q f(\tilde{\Phi}) \le c_0 := \sup_{B(0,R)} f. \tag{3.19}$$

By assumption (A3), c_0 is finite.

(3) Let η, depending on α, be the deformation given by (3.13). We claim that $\Phi(\partial Q)$ and $S := \eta(1, S_n^1)$ link nontrivially. We have to prove that, for any extension $\tilde{\Phi} \in C(Q, X_\alpha)$ of Φ, $\tilde{\Phi}(Q) \cap S \ne \emptyset$.

Assume, by contradiction, that

$$\eta(1, u_1) \ne \tilde{\Phi}(t, u_2) \tag{3.20}$$

for all $u_1 \in S_n^1, u_2 \in B_n^2, t \in [0, 1]$. It follows from (3.13), (3.17) and (3.7) that (3.20) holds for all $u_1 \in B_n^1, u_2 \in S_n^2, t \in [0, 1]$. By (3.18) we obtain (3.20) for $t = 1$ and for all $u_1 \in B_n^1, u_2 \in B_n^2$. Using homotopy invariance and Kronecker property of the degree, we have

$$\deg(F_0, \Omega, 0) = \deg(F_1, \Omega, 0) = 0, \tag{3.21}$$

where

$$\Omega := B_n^1 \times B_n^2,$$

$$F_t(u) := \eta(1, u_1) - \tilde{\Phi}(t, u_2).$$

We obtain from (3.7)

$$\eta(t, u_1) \ne u_2 \tag{3.22}$$

for all $u_1 \in B_n^1, u_2 \in S_n^2, t \in [0, 1]$. It follows from (3.13) that (3.22) holds for all $u_1 \in S_n^1, u_2 \in B_n^2, t \in [0, 1]$. Let us define, on $[0, 1] \times \Omega$,

$$G_t(u) := \eta(t, u_1) - u_2.$$

Using (3.21) and homotopy invariance of the degree, we have

$$0 = \deg(G_1, \Omega, 0) = \deg(G_0, \Omega, 0) = \deg\big(P_n^1 - P_n^2, \Omega, 0\big) \ne 0,$$

a contradiction.

(4) Let us define

$$c := \inf_{\tilde{\Phi} \in \Gamma} \sup_{u \in Q} f\big(\tilde{\Phi}(u)\big)$$

where

$$\Gamma := \left\{ \tilde{\Phi} \in C(Q, X_\alpha) : \tilde{\Phi}(u) = \Phi(u), \forall u \in \partial Q \right\}.$$

It follows from (3.19) and from the preceding step that

$$\varepsilon \leq c \leq c_0.$$

Assumption (A2) implies the existence of $m_2 \geq m_1$ and $\gamma > 0$ such that, for $\alpha \geq (m_2, m_2)$,

$$u \in f_\alpha^{-1}([\varepsilon, c_0]) \quad \Longrightarrow \quad \left\| f_\alpha'(u) \right\| \geq \gamma. \tag{3.23}$$

By the standard minimax argument, $c \in [\varepsilon, c_0]$ is a critical value of f_α, contrary to (3.23). $\qquad\square$

Corollary 3.3.1 *Assume that $f \in C^1(X, \mathbb{R})$ satisfies (A2) and (A3). If f has a global minimum and a local maximum, then f has a third critical point.*

3.3.3 Super-quadratic Functionals

If f is not bounded below and has a local linking at 0, we can still find a nontrivial critical point.

Theorem 3.3.2 (Li and Willem [126]) *Suppose that $f \in C^1(X, \mathbb{R})$ satisfies the following assumptions:*

(B1)   *f has a local linking at 0 and $X^1 \neq \{0\}$,*
(B2)   *f satisfies (PS)*,*
(B3)   *f maps bounded sets into bounded sets,*
(B4)   *for every $m \in \mathbb{N}$, $f(u) \to -\infty$, $|u| \to \infty$, $u \in X_m^1 \oplus X^2$.*

Then f has at least two critical points.

Proof (1) We assume that $\dim X^2$ is positive and $\dim X^1$ is infinite, since the other cases are simpler.

Supposing 0 to be the only critical point of f, we will be led to a contradiction.

By assumption (B2) and Lemma 3.3.2, applied to f and to $g := -f$, there exists $\varepsilon > 0$ such that

$$f^\varepsilon \setminus B(0, r/3) \prec^\infty f^{-\varepsilon}, \tag{3.24}$$

$$g^\varepsilon \setminus B(0, r/3) \prec^\infty g^{-\varepsilon}. \tag{3.25}$$

Moreover, we can assume that the corresponding deformations exist for $\alpha \geq (m_0, m_0)$ and satisfy (3.7) with $\rho = r/2$. Assumption (B2) implies also the existence of $m_1 \geq m_0$ and $\delta > 0$ such that, for $\alpha \geq (m_1, m_1)$,

$$u \in f_\alpha^{-1}((-\infty, -\varepsilon]) \quad \Longrightarrow \quad \left\| f_\alpha'(u) \right\| \geq \delta. \tag{3.26}$$

(2) By assumption (B4), there exists $R > 0$ such that, for $u \in X^1_{m_1+1} \oplus X^2$,

$$\|u\| \geq R/\sqrt{2} \quad \Longrightarrow \quad f(u) \leq -\varepsilon. \tag{3.27}$$

It follows from assumption (B3) that

$$\mu := \inf\{f(u) : \|u\| \leq R\} \tag{3.28}$$

is finite. Without loss of generality, we can assume that there exists $v_0 \in X^1_{m_1+1} \setminus X^1_{m_1}$ and that the norm on $X^1_{m_1+1}$ is Euclidean. Let us write $\alpha := (m_1, n)$ where $n \geq m_1 + 1$ is fixed. Using (3.24) and (3.26), it is easy to construct a deformation

$$\sigma : [0, 1] \times S^2_n \to X_\alpha$$

where

$$S^j_n := \{u \in X^j_n : \|u\| = r\}, \quad j = 1, 2,$$

such that

$$f(\sigma(t, u)) < 0, \qquad \forall t \in (0, 1], \ \forall u \in S^2_n, \tag{3.29}$$

$$f(\sigma(1, u)) = \mu - 1, \quad \forall u \in S^2_n. \tag{3.30}$$

Set

$$B^j_n := \{u \in X^j_n : \|u\| \leq r\}, \quad j = 1, 2,$$

$$Q := [0, 1] \times B^2_n,$$

and define a mapping $\Phi : \partial Q \to f^0$ by

$$\Phi(t, u) = \begin{cases} u, & t = 0, \ u \in B^2_n, \\ \sigma(2t, u), & 0 < t \leq 1/2, \ u \in S^2_n, \\ 2(1 - t)\sigma(1, t) + (2t - 1)v_0, & 1/2 < t < 1, \ u \in S^2_n, \\ v_0, & t = 1, \ u \in B^2_n, \end{cases}$$

where $\|v_0\| \geq R$. Using (3.27)–(3.30), it is easy to verify that $\Phi(\partial Q) \subset f^0$. Moreover, we have, by construction, $\Phi(\partial Q) \subset X^1_{m_1+1} \oplus X^2_n$. Hence there exists a continuous extension of Φ,

$$\tilde{\Phi} : Q \to X^1_{m_1+1} \oplus X^2_n,$$

such that

$$\sup_Q \tilde{\Phi} \leq c_0 := \sup_{X^1_{m_1+1} \oplus X^2} f. \tag{3.31}$$

By assumptions (B3) and (B4), c_0 is finite.

(3) As in the proof of the preceding theorem, it is easy to verify that $\Phi(\partial Q)$ and $s = \eta(1, S_n^1)$ link nontrivially, where η, depending on $\beta := (n, n)$, is the deformation given by (3.25).

(4) Let us define

$$c := \inf_{\tilde{\Phi} \in \Gamma} \sup_{u \in Q} f\big(\tilde{\Phi}(u)\big),$$

where

$$\Gamma := \big\{ \tilde{\Phi} \in C(Q, X_\beta) : \tilde{\Phi}(u) = \Phi(u), \forall u \in \partial Q \big\}.$$

It follows from (3.31) and from the preceding step that $\varepsilon \leq c \leq c_0$. It is then easy to obtain a contradiction as in the proof of the preceding theorem. $\square$

3.3.4 Asymptotically Quadratic Functionals

Let X be a real Hilbert space and let $f \in C^1(X, \mathbb{R})$. Throughout this section we assume that the gradient of f has the form

$$\nabla f(u) = Au + B(u),$$

where A is a bounded self-adjoint operator, 0 is not in the essential spectrum of A, and B is a nonlinear compact mapping.

We also assume that there exist an orthogonal decomposition

$$X = X^1 \oplus X^2$$

and two sequences of finite-dimensional subspaces

$$X_0^1 \subset X_1^1 \subset \cdots \subset X^1, \qquad X_0^2 \subset X_1^2 \subset \cdots \subset X^2,$$

such that

$$X^j = \overline{\bigcup_{n \in \mathbb{N}} X_n^j}, \quad j = 1, 2,$$

$$A X_n^j \subset X_n^j, \quad j = 1, 2, n \in \mathbb{N}. \tag{3.32}$$

We denote by $P_\alpha : X \to X$ the orthogonal projector onto X_α and by $M^-(L)$ the Morse index of a self-adjoint operator L.

Theorem 3.3.3 (Li and Willem [126]) *Suppose that f satisfies the following assumptions*:

(C1) *f has a local linking at 0 with respect to* (X_1, X_2),
(C2) *there exists a compact self-adjoint operator* B_∞ *such that*

$$B(u) = B_\infty u + o\big(\|u\|\big), \quad \|u\| \to \infty,$$

(C3) $A + B_\infty$ *is invertible,*
(C4) *for infinitely many multi-indices* $\alpha := (n, n)$,

$$M^-\big((A + P_\alpha B_\infty)|_{X_\alpha}\big) \neq \dim X_n^2.$$

Then f has at least two critical points.

Let us define the self-adjoint operator $L_\alpha : X_\alpha \to X_\alpha$ by

$$L_\alpha u := Au + P_\alpha B_\infty u.$$

Let us denote by X_α^- (resp., X_α^+) the negative map (resp., positive) spectral space of L_α and by P_α^- (resp., P_α^+) the orthogonal projection onto X_α^- (resp., X_α^+).

Lemma 3.3.5 (Li and Willem [126]) *Suppose that f satisfies* (C2) *and* (C3). *Then we have*

$$\nabla f_\alpha(u) = L_\alpha u + o\big(\|u\|\big), \quad \|u\| \to \infty, \; u \in X_\alpha, \tag{3.33}$$

uniformly with respect to α. *Moreover, there exist* $\gamma_0 \in \mathbb{N}^2$ *and* $c > 0$ *such that, for any* $\alpha \geq \gamma_0$, L_α *is invertible and*

$$\big\|L_\alpha^{-1}\big\| \leq c, \tag{3.34}$$

$$(L_\alpha u, u) \leq -\frac{1}{c}\|u\|^2, \quad \forall u \in X_\alpha^-. \tag{3.35}$$

Proof (1) We have, on X_α,

$$\big\|\nabla f_\alpha(u) - L_\alpha u\big\| = \big\|P_\alpha B(u) - P_\alpha B_\infty u\big\|$$
$$\leq \big\|B(u) - B_\infty(u)\big\|.$$

It suffices then to use (C2) to obtain (3.33).

(2) Since B_∞ is compact, (3.32) implies that

$$\|B_\infty - P_\alpha B_\infty\| \to 0, \quad \alpha_1 \to \infty, \; \alpha_2 \to \infty.$$

By (C3), there exist $\gamma_0 \in \mathbb{N}^2$ and $c > 0$ such that, for $\alpha \geq \gamma_0$, $A + P_\alpha B_\infty$ is invertible and

$$\big\|(A + P_\alpha B_\infty)^{-1}\big\| \leq c.$$

It is easy to verify (3.34) and (3.35). $\square$

Lemma 3.3.6 (Li and Willem [126]) *If f verifies* (C2) *and* (C3), *then every sequence* (u_{α_n}) *such that* (α_n) *is admissible and*

$$u_{\alpha_n} \in X_{\alpha_n}, \quad \nabla f_{\alpha_n}(u_{\alpha_n}) \to 0 \tag{3.36}$$

contains a sequence which converges to a critical point of f.

Proof From (3.33), (3.34) and (3.36), the sequence (u_{α_n}) is bounded in X. Going if necessary to a subsequence, we can assume that $u_{\alpha_n} \rightharpoonup u$ and $B(u_{\alpha_n}) \to v$. Using (3.36), we obtain

$$Au_{\alpha_n} = \nabla f_{\alpha_n}(u_{\alpha_n}) - P_{\alpha_n} B(u_{\alpha_n}) \to v, \quad n \to \infty.$$

Since 0 is not in the essential spectrum of A, it is easy to verify that $u_{\alpha_n} \to u$ and $\nabla f(u) = 0$. $\qquad\square$

Lemma 3.3.7 (Li and Willem [126]) *Suppose that f satisfies (C2) and (C3). Then there exist $R_0 > 0$ and $\beta_0 \geq \gamma_0$ such that, for every $\alpha \geq \beta_0$, $\frac{3}{2} L_\alpha$ is a pseudo-gradient vector field for f_α on $X_\alpha \setminus B(0, R_0)$.*

Proof Assume, by contradiction, that for every $n \in \mathbb{N}$, there exists $\alpha \geq (n, n)$ such that $\frac{3}{2} L_\alpha$ is a pseudo-gradient vector field for f_α on $X_\alpha \setminus B(0, n)$.

Then there exists $u_n \in X_{\alpha_n}$ such that $\|u_n\| \geq n$ and either

$$\frac{3}{2} \|L_{\alpha_n} u_n\| > \left\| \nabla f_{\alpha_n}(u_n) \right\|$$

or

$$\frac{3}{2} \left(\nabla f_{\alpha_n}(u_n), L_{\alpha_n} u_n \right) < \left\| \nabla f_{\alpha_n}(u_n) \right\|^2.$$

It follows from (3.33) and (3.34) that

$$c^{-1} \leq \|L_{\alpha_n} u_n\| / \|u_n\| \to 0, \quad n \to \infty,$$

a contradiction. $\qquad\square$

Proof of Theorem 3.3.3 (1) By (C4), after replacing, if necessary, f by $-f$, we can assume that for every $\alpha \in \mathbb{N}^2$,

$$\dim X_\alpha^- = M^-(L_\alpha) > \dim X_n^2.$$

Moreover, we assume that $\dim X^2$ is positive, since the other case is simpler. Supposing 0 to be the only critical point of f, we will be led to a contradiction. We may suppose that $r < R_0$. By Lemmas 3.3.2 and 3.3.6, applied to f and to $g := -f$, there exists $\varepsilon > 0$ such that

$$f^\varepsilon \setminus B(0, r/3) \prec^\infty f^\varepsilon, \tag{3.37}$$

$$g^\varepsilon \setminus B(0, r/3) \prec^\infty g^{-\varepsilon}. \tag{3.38}$$

Moreover, we can assume that the corresponding deformations exist for $\alpha \geq (m_0, m_0) \geq \beta_0$ and satisfy (3.7) with $\rho = r/2$. Lemma 3.3.6 implies also the existence of $m_1 \geq m_0$ and $\delta > 0$ such that, for $\alpha \geq (m_1, m_1)$,

$$u \in f_\alpha^{-1}\big((-\infty, -\varepsilon]\big) \quad \Longrightarrow \quad \left\| \nabla f_\alpha(u) \right\| \geq \delta. \tag{3.39}$$

(2) Let us write $\alpha := (n, n)$ where $n \geq m_1$ is fixed. Let R_0 be given by Lemma 3.3.7. By (3.33) and (3.35) there exists $R > 0$ such that, for any $u \in X_\alpha$,

$$\left\| P_\alpha^+ u \right\| \leq 2R_0, \quad \left\| P_\alpha^- u \right\| \geq R \quad \Longrightarrow \quad f_\alpha(u) \leq -\varepsilon. \tag{3.40}$$

It follows from assumption (C2) that

$$\mu := \inf\left\{ f(u) : \|u\| \leq R + 2R_0 \right\} \tag{3.41}$$

is finite. Using (3.37) and (3.39) it is easy to construct a deformation

$$\sigma_1 : [0, 1] \times S_n^2 \to X_\alpha$$

where

$$S_n^j := \left\{ u \in X_n^j : \|u\| \leq r \right\}, \quad j = 1, 2,$$

such that

$$\begin{aligned} f\big(\sigma_1(t, u)\big) &< 0, \qquad \forall t \in (0, 1], \ \forall u \in S_n^2, \\ f\big(\sigma(t, u)\big) &= \mu - 1, \quad \forall u \in S_n^2. \end{aligned} \tag{3.42}$$

Moreover, by Lemma 3.3.7, we can assume that

$$\left\| \sigma_1(t, u) \right\| \geq 2R_0 \quad \Longrightarrow \quad D_t \sigma_1(t, u) = -\frac{3}{2} L_\alpha \sigma_1(t, u). \tag{3.43}$$

Hence, we obtain

$$\left\| P_\alpha^+ \sigma_1(1, u) \right\| \leq 2R_0, \quad \forall u \in S_n^2, \tag{3.44}$$

and from (3.41) and (3.43),

$$\left\| P_\alpha^- \sigma_1(1, u) \right\| \geq R, \quad \forall u \in S_n^2. \tag{3.45}$$

Let us now define

$$\sigma_2 : [0, 1] \times S_n^2 \to X_\alpha$$

by

$$\sigma_2(t, u) := (1 - t)\sigma_1(1, u) + t R P_\alpha^- \sigma_1(1, u) / \left\| P_\alpha^- \sigma_1(1, u) \right\|.$$

It is clear, by construction, that

$$\begin{aligned} \sigma_2\big(1, S_n^2\big) &\subset S_\alpha^- := \left\{ u \in X_\alpha^- : \|v\| = R \right\}, \\ f\big(\sigma_2(t, u)\big) &< 0, \quad \forall t \in [0, 1], \ \forall u \in S_n^2. \end{aligned} \tag{3.46}$$

Let us recall that $\dim X_\alpha^- > \dim X_n^2$. By the fact that $\pi_p(S^q) \simeq 0$ for $p < q$, there exists

$$\sigma_3 : [0, 1] \times S_n^2 \to S_\alpha^-$$

such that, for any $u \in S_n^2$,

$$\sigma_3(0, u) = \sigma_2(1, u),$$

$$\sigma_3(1, u) = y$$

where $y \in S_\alpha^-$ is fixed. Combining σ_1, σ_2 and σ_3, we finally get a deformation

$$\sigma : [0, 1] \times S_n^2 \to X_\alpha$$

such that $\sigma(1, S_n^2) = y$. It is easy to verify that $\sigma([0, 1] \times S_n^2) \subset f_\alpha^0$ by using (3.40), (3.42) and (3.44)–(3.46). Set

$$B_n^j := \left\{ u \in X_n^j : \|u\| \leq r \right\}, \quad j = 1, 2,$$

$$Q := [0, 1] \times B_n^2$$

and define a mapping $\Phi : \partial Q \to f_\alpha^0$ by

$$\Phi(t, u) = \begin{cases} u, & t = 0,\, u \in B_n^2, \\ \sigma(t, u), & 0 < t < 1,\ u \in S_n^2, \\ y, & t = 1,\ u \in B_n^2. \end{cases}$$

(3) As before, it is easy to prove that $\Phi(\partial Q)$ and $S = \eta(1, S_n^1)$ link nontrivially when η, depending on α, is the deformation given by (3.38).

(4) Let us define c as before. It follows from the preceding step that $\varepsilon \leq c$. It is then easy to get a contradiction. $\qquad\qquad\square$

3.3.5 Applications to Elliptic Boundary Value Problems

Consider the problem

$$\begin{cases} -\Delta u + a(x)u = g(x, u) & \text{in } \Omega, \\ u = 0 & \text{on } \partial\Omega, \end{cases} \tag{3.47}$$

where $\Omega \subset \mathbb{R}^N$ is a bounded domain whose boundary is a smooth manifold.

We also assume

(g1) $a \in L^\infty(\Omega)$, $g \in C(\bar{\Omega} \times \mathbb{R}, \mathbb{R})$;

(g2) there are constants $a_1, a_2 \geq 0$ such that

$$\left| g(x, u) \right| \leq a_1 + a_2|u|^s$$

where $0 \leq s < (N + 2)/(N - 2)$ if $N \geq 3$.

If $N = 1$, (g2) can be dropped. If $N = 2$ we assume that

$$|g(x, u)| \leq a_1 \exp g(u)$$

where $g(u)/u^2 \to 0$, as $|u| \to \infty$.

(g3) $g(x, u) = o(|u|)$, $|u| \to 0$ uniformly on Ω;
(g4) there are constants $\mu > 2$ and $R \geq 0$ such that, for $|u| \geq R$,

$$0 < \mu G(x, u) \leq u g(x, u)$$

where

$$G(x, u) := \int_0^u g(x, s) \, ds.$$

(g5) For some $\delta > 0$, either

$$G(x, u) \geq 0 \quad \text{for } |u| \leq \delta, \ x \in \Omega,$$

or

$$G(x, u) \leq 0 \quad \text{for } |u| \leq \delta, x \in \Omega.$$

Theorem 3.3.4 (Li and Willem [126]) *Suppose that g satisfies* (g1)–(g4). *If 0 is an eigenvalue of* $-\Delta + a$ *(with Dirichlet boundary condition) assume also* (g5). *Then problem* (3.47) *has at least one nontrivial solution.*

Proof (1) We shall apply Theorem 3.3.2 to the functional

$$f(u) := \frac{1}{2} \int_\Omega |\nabla u|^2 \, dx + \frac{1}{2} \int_\Omega a(x) u^2 \, dx - \int_\Omega G(x, u) \, dx$$

defined on $X := H_0^1(\Omega)$, with the inner product $(u, v) = \int_\Omega \nabla u \cdot \nabla v \, dx$ and $\|u\| = \sqrt{(u, u)}$. We consider only the case when 0 is an eigenvalue of $-\Delta + a$ and

$$G(x, u) \leq 0 \quad \text{for } |u| \leq \delta. \tag{3.48}$$

Then other cases are similar and simpler.

Let X^2 be the (finite-dimensional) space spanned by the eigenfunctions corresponding to negative eigenvalues of $-\Delta + a$ and let X^1 be its orthogonal complement in X. Choose an Hilbertian basis $\{e_n\}_{n \geq 0}$ for X^1 and define

$$X_n^1 := \text{span}\{e_0, \ldots, e_n\}, \quad n \in \mathbb{R},$$
$$X_n^2 := X^2, \quad n \in \mathbb{R}.$$

It is well known that $f \in C^1(X, \mathbb{R})$ and maps bounded sets into bounded sets.

(2) We claim that f has a local linking at 0 with respect to (X^1, X^2). Decompose X^1 into $V + W$ when $V = \ker(-\Delta + a)$, $W = (X^2 + V)^\perp$. Also set $u = v + w$,

$u \in X^1$, $v \in V$, $w \in W$. Since V is a finite-dimensional space, there exists $C > 0$ such that

$$\|v\|_\infty \le C\|v\|_{H^1}, \quad \forall v \in V. \tag{3.49}$$

It follows from (g2) and (g3) that, for any $\varepsilon > 0$, there exists c_ε such that

$$\left| G(x, u) \right| \le \varepsilon u^2 + c_\varepsilon |u|^{s+1}. \tag{3.50}$$

We obtain, on X^2, for some $c > 0$,

$$f(u) \le \frac{1}{2} \int_\Omega \left[|\nabla u|^2 + a|u|^2 \right] dx + \varepsilon \int_\Omega u^2 \, dx + c\|u\|_{H^1}^{s+1}$$

and hence, for $r > 0$ small enough,

$$f(u) \le 0, \qquad u \in X^2, \qquad \|u\|_{H^1} \le r.$$

Let $u = v + w \in X^1$ be such that $\|u\|_{H^1} \le \delta/(2C)$ and set

$$\Omega_1 := \left\{ x \in \Omega : \left| w(x) \right| \le \delta/2 \right\},$$
$$\Omega_2 := \Omega \setminus \Omega_1.$$

On Ω_1, we have, by (3.49),

$$\left| u(x) \right| \le \left| v(x) \right| + \left| w(x) \right| \le \|v\|_\infty + \delta/2 \le \delta,$$

hence, by (3.48)

$$\int_{\Omega_1} G(x, u) \, dx \le 0.$$

On Ω_2, we have, also by (3.49),

$$\left| u(x) \right| \le \left| v(x) \right| + \left| w(x) \right| \le 2 \left| w(x) \right|$$

hence, by (3.50),

$$G(x, u) \le \varepsilon u^2 + c_\varepsilon |u|^{s+1} \le 4\varepsilon w^2 + 2^{s+1} c_\varepsilon |w|^{s+1}$$

and, for some $c > 0$, $\int_{\Omega_2} G(x, u) \, dx \le 4\varepsilon \int_\Omega w^2 \, dx + c\|w\|_{H^1}^{s+1}$. Therefore we deduce that

$$f(u) \ge \frac{1}{2} \int_\Omega \left(|\nabla w|^2 + a|w|^2 \right) dx - 4\varepsilon \int_\Omega w^2 \, dx - c\|w\|_{H^1}^{s+1} - \int_{\Omega_1} G(x, u) \, dx$$

and for $0 < r < \delta/(2C)$ small enough,

$$f(u) \ge 0, \qquad u \in X^1, \qquad \|u\|_{H^1} \le r.$$

(3) We claim that f satisfies (PS)*. Consider a sequence (u_{α_n}) such that (α_n) is admissible and

$$u_{\alpha_n} \in X_{\alpha_n}, \qquad c := \sup_n f(u_{\alpha_n}) < \infty, \qquad f'_{\alpha_n}(u_{\alpha_n}) \to 0. \qquad (3.51)$$

Integrating (g4) we obtain the existence of constants $a_3, a_4 > 0$ such that

$$G(x, u) \geq a_3 |u|^\mu - a_4. \qquad (3.52)$$

Choose $\nu \in (\mu^{-1}, 2^{-1})$.

For n large, (3.51), (3.52) and (g4) imply, with $u := u_{\alpha_n}$,

$$c + 1 + \nu \|u\|_{H^1} \geq f(u) - \nu \langle f'(u), u \rangle$$

$$= \int_\Omega \left[\left(\frac{1}{2} - \nu \right) |\nabla u|^2 + \left(\frac{1}{2} - \nu \right) au^2 + \nu g(x, u)u - G(x, u) \right] dx$$

$$\geq \left(\frac{1}{2} - \nu \right) \|u\|_{H^1}^2 - \left(\frac{1}{2} - \nu \right) \|a\|_\infty \|u\|_{L^2}^2$$

$$+ (\mu\nu - 1) \int_\Omega G(x, u)\, dx - c_1$$

$$\geq \left(\frac{1}{2} - \nu \right) \|u\|_{H^1}^2 - \left(\frac{1}{2} - \nu \right) \|a\|_\infty \|u\|_{L^2}^2$$

$$+ (\mu\nu - 1)a_3 \|u\|_{L^\mu}^\mu - c_2,$$

where c_1 and c_2 are independent of n. Since $\|u\|_{L^2} \leq c(\Omega) \|u\|_{L^\mu}$, we see that (u_{α_n}) is bounded in X. Going if necessary to a subsequence, we can assume that $u_{\alpha_n} \rightharpoonup u$ in X. Note that, with $u_n := u_{\alpha_n}$,

$$\|u_n - u\|^2 = \langle f'(u_n) - f'(u), u_n - u \rangle$$

$$+ \int_\Omega \left[-a(u_n - u)^2 + \big(g(x, u_n) - g(x, u)\big)(u_n - u) \right] dx.$$

It follows then that $u_{\alpha_n} \to u$ in X and $f'(u) = 0$.

(4) Finally, we claim that, for every $m \in \mathbb{N}$,

$$f(u) \to -\infty, \qquad |u| \to \infty, \qquad u \in X_m^1 \oplus X^2.$$

By (3.52) we have

$$f(u) \leq \int_\Omega \left[\frac{|\nabla u|^2}{2} + \frac{au^2}{2} - a_3 |u|^\mu + a_4 \right] dx$$

$$\leq \frac{1}{2} \|u\|_{H^1}^2 + \frac{1}{2} \|a\|_\infty \|u\|_{L^2}^2 + a_4 |\Omega| - a_3 \|u\|_{L^\mu}^\mu$$

and since, on the finite-dimensional space $X_m^1 \oplus X^2$, all the norms are equivalent, it is easy to conclude. $\square$

We consider now asymptotically quadratic functions. Let

$$\lambda_1 \leq \lambda_2 \leq \cdots \leq \lambda_j \leq \cdots$$

be the eigenvalues of $-\Delta + a$ and $\lambda_0 := -\infty$. We assume that

(g6) $g(x, u) = g_\infty(u) + o(|u|)$, $|u| \to \infty$, uniformly in Ω and $\lambda_k < g_\infty < \lambda_{k+1}$.

Theorem 3.3.5 (Li and Willem [126]) *Suppose that g satisfies* (g1), (g3), (g6) *and one of the following conditions*:

(a) $\lambda_j < 0 < \lambda_{j+1}$, $j \neq k$,
(b) $\lambda_j = 0 < \lambda_{j+1}$, $j \neq k$ *and for some* $\delta > 0$,

$$G(x, u) \geq 0 \quad for \ |u| \leq \delta, x \in \Omega,$$

(c) $\lambda_j < 0 = \lambda_{j+1}$, $j \neq k$ *and for some* $\delta > 0$,

$$G(x, u) \leq 0 \quad for \ |u| \leq \delta, x \in \Omega.$$

Then the problem (3.47) *has at least one nontrivial solution.*

Proof The proof which depends on Theorem 3.3.3 is similar to that of Theorem 3.3.4, it is omitted. $\square$

Finally, we give an application of Theorem 3.3.1 to the problem

$$\begin{cases} -\Delta u + a(x)u = \lambda g(u) & \text{in } \Omega, \\ u = 0 & \text{on } \partial\Omega. \end{cases} \tag{3.53}$$

We assume that $a \in L^\infty$, g is smooth,

$$g(u) = o(|u|) \quad \text{as } |u| \to 0,$$

$$\limsup_{|u| \to \infty} \frac{g(u)}{u} < 0,$$

$$G(u) > 0$$

for some $u \in \mathbb{R}$ where

$$G(u) := \int_0^u g(s)\, ds.$$

If 0 is an eigenvalue of $-\Delta + a$, we assume also that, for some $\delta > 0$, either

$$G(u) \geq 0 \quad \text{for } |u| \leq \delta$$

or

$$G(u) \le 0 \quad \text{for } |u| \le \delta.$$

Theorem 3.3.6 (Li and Willem [126]) *Under the above assumptions, for every λ sufficiently large, there are at least two nontrivial solutions of* (3.53).

Remark 3.3.2 Theorem 3.3.6 is due to Brezis and Nirenberg [35] except to the case $G(u) \le 0$ for $|u| \le \delta$.

3.3.6 Local Linking and Critical Groups

Assume that X is a Banach space, $f \in C^1(X, \mathbb{R})$ satisfies the Palais–Smale condition and has only isolated critical values, with each critical value corresponding to a finite number of critical points.

Definition 3.3.5 (A little stronger condition than Definition 3.3.1, see [137, 151]) We say that f has a local linking near the origin if X has a direct sum decomposition $X = X^1 \oplus X^2$ with dim $X^2 < \infty$, $f(0) = 0$ and for some $r > 0$

$$f(u) \le 0, \qquad u \in X^2, \qquad \|u\| \le r, \tag{3.54}$$

$$f(u) > 0, \qquad u \in X^1, \qquad 0 < \|u\| \le r. \tag{3.55}$$

It is clear that 0 is a critical point of f. It was proved in Liu [137] that

Theorem 3.3.7 (Jiaquan Liu [137]) *Let X be a Banach space, $f : X \to \mathbb{R}$ a C^1-function satisfying the* (PS) *condition, f satisfies the local linking as in Definition 3.3.5, if 0 is an isolated critical point of f, dim $X^2 = j < +\infty$, then $C_j(f, 0) \ne 0$.*

Theorem 3.3.8 (K. Perera [151]) *Suppose that there is a critical point u_0 of f, $f(u_0) = c$ with $C_j(f, u_0) \ne 0$ for some $j \ge 0$ and regular values a, b of f, $a < c < b$ such that $H_j(f_b, f_a) = 0$. Then f has a critical point u with either*

$$c < f(u) < b \quad and \quad C_{j+1}(f, u) \ne 0, \quad or$$

$$a < f(u) < c \quad and \quad C_{j-1}(f, u) \ne 0.$$

Here $f_a := \{u \in X : f(u) \le a\}$.

The following topological lemma is needed for the proof of Theorem 3.3.8.

Lemma 3.3.8 (K. Perera [151]) *If $B' \subset B \subset A \subset A'$ are topological spaces such that $H_j(A, B) \ne 0$ and $H_j(A', B') = 0$ then either*

$$H_{j+1}(A', A) \ne 0 \quad or \quad H_{j-1}(B', B) \ne 0.$$

Proof Suppose that $H_{j+1}(A', A) = 0$. Since $H_j(A', B')$ is also trivial, it follows from the following portion of the exact sequence of the triple (A', A, B') that $H_j(A, B') = 0$:

$$H_{j+1}(A', A) \xrightarrow{\partial_*} H_j(A, B') \xrightarrow{i_*} H_j(A', B').$$

Since $H_j(A, B) \neq 0$, now it follows from the following portion exact sequence of the triple (A, B, B') that $H_{j-1}(B', B) \neq 0$:

$$H_j(A, B') \xrightarrow{j_*} H_j(A, B) \xrightarrow{\partial_*} H_{j-1}(B, B'). \qquad \square$$

Proof of Theorem 3.3.8 Take ε, $0 < \varepsilon < \min\{c - a, b - c\}$ such that c is the only critical value of f in $[c - \varepsilon, c + \varepsilon]$. Then, since $C_j(f, u_0) = 0$, it follows from Chap. I, Theorem 4.2 of Chang [49] that $H_j(f_{c+\varepsilon}, f_{c-\varepsilon}) \neq 0$. Since $H_j(f_b, f_a) = 0$, by Lemma 3.3.8, either $H_{j+1}(f_b, f_{c+\varepsilon}) \neq 0$ or $H_{j-1}(f_{c-\varepsilon}, f_a) \neq 0$, and the conclusion follows from Chap. I, Theorem 4.3 and Corollary 4.1 of Chang [49]. $\qquad \square$

Hence the following corollary is immediate from Theorems 3.3.7 and 3.3.8.

Corollary 3.3.2 (K. Perera [151]) *Suppose that f has a local linking near the origin, $\dim X^2 = j$. Assume also that the regular values a, b of f, $a < 0 < b$ such that $H_j(f_b, f_a) = 0$. Then f has a critical point u with either*

$$0 < f(u) < b \quad and \quad C_{j+1}(f, u) \neq 0, \quad or$$

$$a < f(u) < 0 \quad and \quad C_{j-1}(f, u) \neq 0.$$

If X is a Hilbert space, f is C^2, and u is a critical point of f, we denote by $m(u)$ the Morse index of u and by $m^*(u) = m(u) + \dim \ker d^2 f(u)$ the large Morse index of u. We recall that if u is non-degenerate and $C_q(f, u) = 0$, then $m(u) = q$ (see Chap. I, Theorem 4.1, [49]). Let us also recall that it follows from the Shifting theorem (Chap. I, Theorem 5.4, [49]) that if u is degenerate, 0 is an isolated point of the spectrum of $d^2 f(u)$, and $C_q(f, u) \neq 0$, then $m(u) \leq q \leq m^*(u)$. Hence we have the following corollary:

Corollary 3.3.3 (K. Perera [151]) *Let X be a Hilbert space and f be C^2 in Theorem 3.3.8. Assume that for every degenerate critical point u of f, 0 is an isolated point of the spectrum of $d^2 f(u)$. Then f has a critical point u with either*

$$c < f(u) < b \quad and \quad m(u) \leq j + 1 \leq m^*(u) \quad or$$

$$a < f(u) < c \quad and \quad m(u) \leq j - 1 \leq m^*(u).$$

If X is a Hilbert space and df is Lipschitz in a neighborhood of the origin, we can relax the local linking condition as in (3.5–3.6). This follows from the following extension of the result of Liu [137]:

Theorem 3.3.9 (K. Perera [151]) *Let X be a Hilbert space and df be Lipschitz in a neighborhood of the origin. Suppose that f satisfies the local linking conditions (3.5) and (3.6), $\dim X^2 = j$. Then $C_j(f, 0) \neq 0$.*

The proof of Theorem 3.3.9 uses the following deformation lemma:

Lemma 3.3.9 *Under the assumptions of Theorem 3.3.9, there exist a closed ball B centered at the origin and a homeomorphism h of X onto X such that*

(1) *0 is the only critical point of f in $h(B)$,*
(2) *$h|_{B \cap X^2} = \mathrm{id}_{B \cap X^2}$,*
(3) *$f(u) > 0$ for $u \in h(B \cap X^1 \setminus \{0\})$.*

Proof Take open balls B', B'' centered at the origin, with $\overline{B'} \subset B''$, such that 0 is the only critical point of f in B' and df is Lipschitz in B'', and let $B \subset B'$ be a closed ball centered at the origin with radius $\leq r$ (in (3.5)–(3.6)). Since B and $(B')^c$ are disjoint closed sets there is a locally Lipschitz nonnegative function $g \leq 1$ satisfying

$$g = \begin{cases} 1 & \text{on } B \\ 0 & \text{outside } B'. \end{cases}$$

Consider the vector field

$$V(u) = g(u)\|Pu\|df(u)$$

where P is the orthogonal projection onto X^1. Clearly V is locally Lipshitz and bounded on X. Consider the flow $\eta(t) = \eta(t, u)$ defined by

$$\frac{d\eta}{dt} = V(\eta), \qquad \eta|_{t=0} = u.$$

Clearly, η is defined for $t \in [0, 1]$. Let $h = \eta(1, \cdot)$. Since $h|_{(B')^c} = \mathrm{id}_{(B')^c}$ and h is one-to-one, $h(B) \subset B'$ and (1) follows. For $u \in B \cap X^1 \setminus \{0\}$,

$$f\big(h(u)\big) = f(u) + \int_0^1 g\big(\eta(t)\big)\|P\eta(t)\|\,\big\|df\big(\eta(t)\big)\big\|^2\, dt > 0$$

since $f(u) \geq 0$ and $g(u)\|Pu\|\|df(u)\|^2 > 0$. $\qquad\square$

Proof of Theorem 3.3.9 By (1) of Lemma 3.3.9. $C_j(f, 0) = H_j(f_0 \cap h(B), f_0 \cap h(B) \setminus \{0\})$.

By the local linking condition and (2) and (3) of Lemma 3.3.9, $\partial B \cap X^2 \subset f_0 \cap h(B) \setminus \{0\} \subset h(B \setminus X^1)$ and $B \cap X^2 \subset f_0 \cap h(B)$. Since $h|_{\partial B \cap X^2} = \mathrm{id}_{\partial B \cap X^2}$, the inclusion $\partial B \cap X^2 \hookrightarrow h(B \setminus X^1)$ can also be written as the composition of the inclusion $\partial B \cap X^2 \overset{i'}{\hookrightarrow} B \setminus X^1$ and the restriction of h to $B \setminus X^1$. Hence, we have the following commutative diagram induced by inclusion and h:

$$
\begin{array}{ccccc}
H_{j-1}(B \setminus X^1) & \xleftarrow{\;i'_*\;} & H_{j-1}(\partial B \cap X^2) & \longrightarrow & H_{j-1}(B \cap X^2) \\
\Big\downarrow{\scriptstyle h_*} & & \Big\downarrow{\scriptstyle i''_*} & & \Big\downarrow \\
H_{j-1}\big(h(B \setminus X^1)\big) & \longleftarrow & H_{j-1}\big(f_0 \cap h(B) \setminus \{0\}\big) & \xrightarrow{\;i_*\;} & H_{j-1}\big(f_0 \cap h(B)\big).
\end{array}
$$

Since $\partial B \cap X^2$ is a strong deformation retract of $B \setminus X^1$ and h is a homeomorphism, i'_* and h_* are isomorphisms and hence i''_* is a monomorphism.

Since rank $H_{j-1}(B \cap X^2) <$ rank $H_{j-1}(\partial B \cap X^2)$, then it follows that i_* is not a monomorphism.

Now it follows from the following portion of the exact sequence of the pair $(f_0 \cap h(B), f_0 \cap h(B) \setminus \{0\})$ that $C_j(f, 0) = H_j(f_0 \cap h(B), f_0 \cap h(B) \setminus \{0\}) \neq 0$:

$$
C_j(f, 0) \xrightarrow{\;\partial_*\;} H_j\big(f_0 \cap h(B) \setminus \{0\}\big) \xrightarrow{\;i_*\;} H_{j-1}\big(f_0 \cap h(B)\big). \qquad \square
$$

Next we consider the following problem

$$
\begin{cases}
-\Delta u = g(u) & \text{in } \Omega, \\
u = 0 & \text{on } \partial\Omega,
\end{cases}
\tag{3.56}
$$

where Ω is a bounded domain in $\mathbb{R}^N$ with smooth boundary $\partial\Omega$ and $g \in C^1(\mathbb{R}, \mathbb{R})$ satisfying that

(g_1) $|g(u)| \leq C(1 + |u|^{p-1})$ with $2 < p < \frac{2N}{N-2}$, for some $C > 0$,
(g_2) $g(0) = 0 = g(a)$ for some $a > 0$,
(g_3) there are $\mu > 2$ and $A > 0$ such that

$$
0 < \mu G(u) \leq u g(u) \quad \text{for } u \geq A,
$$

where $G(u) = \int_0^u g(t)\,dt$.

Let $\lambda = g'(0)$ and let $0 < \lambda_1 \leq \lambda_2 \leq \lambda_3 \leq \cdots$ be the eigenvalues of $-\Delta$ with Dirichlet boundary condition.

Theorem 3.3.10 (K. Perera [151]) *Assume that g satisfies* (g_1)–(g_3) *and one of the following conditions holds*:

(a) $\lambda_j < \lambda < \lambda_{j+1}$,
(b) $\lambda_j = \lambda < \lambda_{j+1}$ *and, for some* $\delta > 0$,

$$
G(u) \geq \frac{1}{2}\lambda u^2 \quad \textit{for } |u| \leq \delta,
$$

(c) $\lambda_j < \lambda = \lambda_{j+1}$ *and, for some* $\delta > 0$,

$$G(u) \le \frac{1}{2}\lambda u^2 \quad for \ |u| \le \delta.$$

If $j \ge 3$, then problem (3.56) has at least four nontrivial solutions.

Proof Solutions of (3.56) are critical points of the C^2 functional

$$F(u) = \int_\Omega \left[\frac{1}{2}|\nabla u|^2 - G(u)\right] dx$$

defined on $X = H_0^1(\Omega)$. It is well known that F satisfies Palais–Smale condition.

By a standard argument involving a cut-off technique and the strong maximum principle, F has a local minimizer u_0 with $0 < u_0 < a$,

$$\operatorname{rank} C_q(F, u_0) = \delta_{q0}.$$

Since $\lim_{t\to\infty} F(\pm t\phi_1) = -\infty$, where $\phi_1 > 0$ is the first Dirichlet eigenfunction of $-\Delta$, then F also has two mountain pass points $u_1^\pm$ with $u_1^- < u_0 < u_1^+$,

$$\operatorname{rank} C_q\left(F, u_1^\pm\right) = \delta_{q1}$$

(see the proof of Theorem B in [51]). Let X^2 be a j-dimensional space spanned by the eigenfunctions corresponding to $\lambda_1, \lambda_2, \ldots, \lambda_j$ and let X^1 be its orthogonal complement in X. Then f has a local linking near the origin with respect to the decomposition $X = X^1 \oplus X^2$ (see the proof of Theorem 3.3.4) and hence

$$C_j(f, 0) \ne 0.$$

Also, for $\alpha < 0$ and $|\alpha|$ sufficiently large,

$$H_q(X, f_\alpha) = 0, \quad \forall q \in \mathbb{Z}$$

(see Lemma 3.2 of Wang [186]). Therefore by Theorem 3.3.8, f has a nontrivial critical point u_j with either

$$C_{j+1}(f, u_j) \ne 0 \quad \text{or} \quad C_{j-1}(f, u_j) \ne 0.$$

Since $j \ge 3$, a comparison of the critical groups shows that $u_0, u_1^\pm, u_j$ are distinct nontrivial critical points of f. $\square$

Now we apply such an abstract theorem to the problem (3.47) where $a \in L^\infty(\Omega)$ and $g \in C^1(\bar{\Omega} \times \mathbb{R}, \mathbb{R})$ satisfies

(g_1') $|g(x, u)| \le C(1 + |u|^{p-1})$ with $2 < p < \frac{2N}{N-2}$, for some $C > 0$,
(g_2') $g(x, 0) = g_u(x, 0) = 0$,
(g_3') $\lim_{u\to-\infty} \frac{g(x,u)}{u} < \lambda_1$, uniformly in $\bar{\Omega}$,

(g_4') $\limsup_{u\to-\infty}(G(x,u) - \frac{1}{2}ug(x,u)) < +\infty$, uniformly in $\bar{\Omega}$,
(g_5') there are $\mu > 2$ and $A > 0$ such that

$$0 < \mu G(x,u) \le ug(x,u) \quad \text{for } u \ge A,$$

where $G(x,u) = \int_0^u g(x,t)\,dt$.

Here $\lambda_1 < \lambda_2 \le \lambda_3 \le \cdots$ denotes the eigenvalues of $-\Delta + a$ with Dirichlet boundary condition.

Theorem 3.3.11 (K. Perera [151]) *Assume that g satisfies* (g_1')–(g_5') *and one of the following conditions holds*:

(a) $\lambda_j < 0 < \lambda_{j+1}$,
(b) $\lambda_j = 0 < \lambda_{j+1}$ *and, for some $\delta > 0$,*

$$G(x,u) \ge 0 \quad \text{for } |u| \le \delta,$$

(c) $\lambda_j < 0 = \lambda_{j+1}$ *and, for some $\delta > 0$,*

$$G(x,u) \le 0 \quad \text{for } |u| \le \delta.$$

If $j \ge 3$, then (3.47) *has at least three nontrivial solutions.*

We seek critical points of

$$F(u) = \int_\Omega \frac{1}{2}\big(|\nabla u|^2 + a(x)u^2\big)\,dx - \int_\Omega G(x,u)\,dx$$

on $X = H_0^1(\Omega)$.

Lemma 3.3.10 (K. Perera [151]) *If g satisfies* (g_1'), (g_3')–(g_5'), *then for $\alpha < 0$ and $|\alpha|$ sufficiently large,*

$$H_q(X, F_\alpha) = 0 \quad \forall q \in \mathbb{Z}.$$

Proof Let $\tilde{X} = C_0^1(\bar{\Omega})$ and $\tilde{F} = F|_{\tilde{X}}$. By elliptic regularity, F and $\tilde{F}$ have the same critical set. If F does not have any critical values in (α, α'), then F_α (respectively $\tilde{F}_\alpha$) is a strong deformation retract of $\{u \in X : F(u) < \alpha'\}$ (respectively $\{u \in \tilde{X} : \tilde{F}(u) < \alpha'\}$) (see Chap. I, Theorem 3.2 and Chap. III, Theorem 1.1 of [49]). Since $\tilde{X}$ is dense in X, by a theorem of Palais [149],

$$H_q\big(X, \{F < \alpha'\}\big) \cong H_q\big(\tilde{X}, \{\tilde{F} < \alpha'\}\big).$$

Therefore it suffices to prove that, for $\alpha < 0$ and $|\alpha|$ large,

$$H_q(\tilde{X}, \tilde{F}_\alpha) = 0 \quad \forall q \in \mathbb{Z}.$$

Let $S^\infty = \{u \in \tilde{X} : \|u\|_X = 1\}$ be the unit sphere in $\tilde{X}$ and let $S^\infty_+ = \{u \in S^\infty : u > 0$ somewhere$\}$, which is a relatively open subset of S^∞, contractible to $\{\phi_1\}$ via $(t, u) \mapsto \frac{(1-t)u+t\phi_1}{\|(1-t)u+t\phi_1\|}$, $t \in [0, 1]$. We shall show that $\tilde{F}_\alpha$ is homotopy equivalent to S^∞_+ for $\alpha < 0$ and $|\alpha|$ large.

By (g'_3) and (g'_5),

$$-C\left(1 + u^2\right) \le G(x, u) \le \frac{1}{2}\lambda_1 u^2 + C \quad \text{for } u \le A,$$

$$G(x, u) \ge Cu^\mu \quad \text{for } u \ge A,$$

where C denotes (possibly different) positive constants. Thus for $u \in S^\infty_+$,

$$\tilde{F}(tu) = \frac{1}{2}\left(1 + \int_\Omega au^2\right)t^2 - \int_\Omega G(x, tu)\,dx$$

$$\le C\left(1 + t^2 - t^\mu \int_{tu \ge A} u^\mu\right)$$

and it follows that

$$\lim_{t \to \infty} \tilde{F}(tu) = -\infty.$$

On the other hand, in $N = \{u \in \tilde{X} : u \le 0$ everywhere$\}$, the nonpositive cone in $\tilde{X}$,

$$\tilde{F}(u) \ge \frac{1}{2}\int_\Omega \left(|\nabla u|^2 + a(x)u^2 - \lambda_1 u^2\right) - C \ge -C.$$

By (g'_4) and (g'_5),

$$\gamma := \sup_{\bar{\Omega} \times \mathbb{R}}\left(G(x, u) - \frac{1}{2}ug(x, u)\right) < +\infty.$$

Thus for $u \in S^\infty_+$ and $t > 0$,

$$\frac{d}{dt}\tilde{F}(tu) = \left(1 + \int_\Omega au^2\right)t - \int_\Omega ug(x, tu)$$

$$= \frac{2}{t}\left\{\tilde{F}(tu) + \int_\Omega\left[G(x, tu) - \frac{1}{2}tug(x, tu)\right]\right\}$$

$$\le \frac{2}{t}\left\{\tilde{F}(tu) + \gamma|\Omega|\right\} < 0$$

if $\tilde{F}(tu) < -\gamma|\Omega|$.

Fix $\alpha < \min\{\inf_N \tilde{F}, -\gamma|\Omega|, \inf_{\|u\|<1} \tilde{F}\}$. Then it follows that for each u_+^∞ there exists a unique $T(u) \geq 1$ such that

$$\tilde{F}(tu) \begin{cases} > \alpha, & 0 \leq t < T(u), \\ = \alpha, & t = T(u), \\ < \alpha, & t > T(u), \end{cases}$$

and

$$\tilde{F}_\alpha = \{tu : u \in S_+^\infty, t \geq T(u)\}.$$

By the implicit function theorem, $T \in C(S_+^\infty, [1, \infty))$. Hence

$$\eta(s, tu) = \begin{cases} (1-s)tu + sT(u)u & \text{if } 1 \leq t < T(u), \\ tu & \text{if } t \geq T(u) \end{cases}$$

defines a strong deformation retraction of $\{tu : u \in S_+^\infty, t \geq 1\} \simeq S_+^\infty$ onto $\tilde{F}_\alpha$.   □

Proof of Theorem 3.3.11 Since $F(-t\phi_1) < 0$ for $t > 0$ sufficiently small, by standard arguments, F has a local minimizer u_0 with $u_0 < 0$,

$$\operatorname{rank} C_q(F, u_0) = \delta_{q0}.$$

Since $\lim_{t \to \infty} F(t\phi_1) = -\infty$, then F also has a mountain pass point u_1,

$$\operatorname{rank} C_q(F, u_1) = \delta_{q1}.$$

As in the proof of Theorem 3.3.10,

$$C_j(F, 0) \neq 0,$$

so using Lemma 3.3.10, F also has a nontrivial critical point u_j with either

$$C_{j+1}(F, u_j) \neq 0 \quad \text{or} \quad C_{j-1}(F, u_j) \neq 0.$$

Since $j \geq 3$, u_0, u_1, u_j are distinct nontrivial solutions of (3.47).   □

Finally we give an application of Theorem 3.3.8 to the problem

$$\begin{cases} -\Delta u + a(x)u = \lambda g(u) & \text{in } \Omega, \\ u = 0 & \text{on } \partial\Omega, \end{cases} \tag{3.57}$$

where $a \in L^\infty(\Omega)$ and $g \in C^1(\mathbb{R}, \mathbb{R})$ satisfies

(g_1'') $\limsup_{|u| \to \infty} \frac{g(u)}{u} < 0$,
(g_2'') $g(0) = g'(0) = 0$.

Theorem 3.3.12 (K. Perera [151]) *Assume that g satisfies* (g_1''), (g_2'') *and one of the following conditions holds*:

(a) $\lambda_j < 0 < \lambda_{j+1}$,
(b) $\lambda_j = 0 < \lambda_{j+1}$ *and for some* $\delta > 0$,

$$G(x,u) \geq 0 \quad \text{for } |u| \leq \delta,$$

(c) $\lambda_j < 0 = \lambda_{j+1}$ *and for some* $\delta > 0$,

$$G(x,u) \leq 0 \quad \text{for } |u| \leq \delta.$$

If $j \geq 3$, *then problem (3.57) has at least four nontrivial solutions for* λ *sufficiently large*.

Proof Since, for λ sufficiently large, there is an a-priori estimate for the solutions of problem (3.57) by the maximum principle, we may also assume that $g(u) = bu$ with $b < 0$ for $|u|$ large. Then the functional

$$F(u) = \int_\Omega \left[\frac{1}{2} \left(|\nabla u|^2 + a(x)u^2 \right) - \lambda G(x,u) \right] dx$$

is well defined on $X = H_0^1(\Omega)$ and bounded below and satisfies Palais–Smale condition for λ large.

Since $F(\pm t\phi_1) < 0$ for $t > 0$ sufficiently small, F has two local minimizers $u_0^\pm$ with $u_0^- < 0 < u_0^+$ and

$$\operatorname{rank} C_q \left(F, u_0^\pm \right) = \delta_{q0}.$$

Then F also has a mountain pass point u_1 with

$$\operatorname{rank} C_q(F, u_1) = \delta_{q1}.$$

As before, we have

$$C_j(F, 0) \neq 0,$$

and for $\alpha < \inf F$,

$$\operatorname{rank} H_q(X, F_\alpha) = \delta_{q0},$$

so F has a fourth critical point u_j with either

$$C_{j+1}(F, u_j) \neq 0 \quad \text{or} \quad C_{j-1}(F, u_j) \neq 0. \qquad \square$$

Example 3.3.1 $g(u) = \pm |u|u - u^3$ satisfies Theorem 3.3.12.

Chapter 4
Bifurcation and Critical Point

4.1 Introduction

We consider the following semilinear elliptic equation:

$$\begin{cases} -\Delta u = \lambda f(u) & \text{in } \Omega, \\ u|_{\partial \Omega} = 0, \end{cases} \tag{4.1}$$

where $-\Delta$ is the Laplacian, Ω is a smooth bounded domain in $\mathbb{R}^n$, $n \geq 1$, λ is a positive parameter. Let λ_k be the kth eigenvalue of

$$\begin{cases} -\Delta \phi = \lambda \phi & \text{in } \Omega, \\ \phi|_{\partial \Omega} = 0. \end{cases} \tag{4.2}$$

Let ϕ_k be the corresponding eigenfunction with the property that $\int_\Omega \phi_k^2 = 1$ for $k = 1, 2, \ldots$. It is well known that $\lambda_1 > 0$, λ_1 is simple, $\phi_1 > 0$. Here we assume that all the eigenvalues λ_k are simple, i.e., $0 < \lambda_1 < \lambda_2 < \cdots < \lambda_k < \cdots$. Moreover, we assume

(f_1) $f \in C^1(\mathbb{R}, \mathbb{R})$, $f(0) = 0$, $f'(0) > 0$;
(f_2) $\lim_{|u| \to +\infty} \frac{f(u)}{u} = f'(\infty) > 0$;
(f_3) $f(u)/u$ is increasing in $(0, \infty)$ and is decreasing in $(-\infty, 0)$;
(f_4) $f(u)/u$ is decreasing in $(0, \infty)$ and is increasing in $(-\infty, 0)$.

Remark 4.1.1 (f_3) and (f_4) are contradictory, they are used alternatively in the following.

Nontrivial solutions of (4.1) correspond to nontrivial critical points of the following functional on $H := H_0^1(\Omega)$ with the inner product $(u, v) = \int_\Omega \nabla u \cdot \nabla v \, dx$ and $\|u\| = \sqrt{(u, u)}$:

$$J_\lambda(u) = \frac{1}{2} \int_\Omega |\nabla u|^2 \, dx - \lambda \int_\Omega F(u) \, dx, \tag{4.3}$$

Z. Zhang, *Variational, Topological, and Partial Order Methods with Their Applications*, Developments in Mathematics 29,
DOI 10.1007/978-3-642-30709-6_4, © Springer-Verlag Berlin Heidelberg 2013

where $F(u) = \int_0^u f(s)\,ds$.

Let u be an isolated critical point of the functional J, and let $c = J(u)$. We recall

$$C_q(J, u) = H_q\big(J^c \cap V_u, \big(J^c \backslash \{u\} \cap V_u\big), G\big)$$

the qth critical group, with coefficient group G of J at u, $q = 0, 1, 2, \ldots$, where $J^c = \{z \in H : J(z) \leq c\}$, V_u is a neighborhood of u such that the critical set K satisfies $K \cap (J^c \cap V_u) = \{u\}$, and $H_*(X, Y; G)$ stands for the singular relative homology groups with the abelian coefficient group G.

Let $W(x) \in L^\infty(\Omega)$, consider an eigenvalue problem:

$$\begin{cases} -\Delta\phi = W(x)\phi + \mu_i(W)\phi & \text{in } \Omega, \\ \phi|_{\partial\Omega} = 0. \end{cases}$$

It is well known that, for $i = 1, 2, \ldots$,

$$\mu_i(W) = \min_i \max_i \frac{\int_\Omega (|\nabla z|^2 - W(x)z^2)\,dx}{\int_\Omega z^2\,dx},$$

where $\max_i$ is over all $z \,(\neq 0) \in T_i$, and $\min_i$ is over all linear subspaces T_i of H of dimension i. For $W_1, W_2 \in L^\infty(\Omega)$ satisfying $W_2(x) \geq W_1(x)$ almost everywhere, $\mu_i(W_2) \leq \mu_i(W_1)$. If in addition $\text{mes}\{x \in \Omega : W_2(x) > W_1(x)\} > 0$, then $\mu_i(W_2) < \mu_i(W_1)$.

Let

$$\lambda_k^0 = \frac{\lambda_k}{f'(0)}, \qquad \lambda_k^\infty = \frac{\lambda_k}{f'(\infty)}, \qquad \lambda_k^m = \frac{\lambda_k}{\inf_{u \in \mathbb{R} \backslash \{0\}} f'(u)} \quad \text{and}$$

$$\lambda_k^M = \frac{\lambda_k}{\sup_{u \in \mathbb{R} \backslash \{0\}} f'(u)}. \tag{4.4}$$

When (f_3) is satisfied, we have $f'(u) \geq f(u)/u$ for $u \in \mathbb{R} \backslash \{0\}$, and for $k \in N$, f satisfying $(f_1), (f_2), (f_3), \lambda_k^M \leq \lambda_k^\infty < \lambda_k^0$. If in addition, $uf''(u) \geq 0$, then $\lambda_k^M = \lambda_k^\infty$. For $k \in N$, we define open intervals

$$I_k = \big(\lambda_k^\infty, \lambda_k^0\big), \qquad \tilde{I}_k = \big(\lambda_k^M, \lambda_k^0\big).$$

We define the Morse index $M(u)$ of a solution u to (4.1) to be the number of negative eigenvalues of the following problem:

$$\begin{cases} -\Delta\phi = \lambda f'(u)\phi + \mu\phi & \text{in } \Omega, \\ \phi = 0 & \text{on } \partial\Omega. \end{cases} \tag{4.5}$$

If u is a solution to (4.1), and 0 is not an eigenvalue of (4.5), then u is a nondegenerate solution, otherwise it is degenerate. In [162], the author gets the following main results (for simplicity, it is assumed that $\lambda_k^M = \lambda_k^\infty$, i.e., $I_k = \tilde{I}_k$):

(A) If $\lambda \notin \bigcup_{j \in N} I_j$, then (4.1) has only the trivial solution $u = 0$;

(B) $\lambda \in I_k \setminus \bigcup_{j \neq k} I_j$, then (4.1) has exactly two nontrivial solutions which are non-degenerate and with Morse index $M(u) = k$;

(C) If $\lambda \in (I_k \cap I_{k+1}) \setminus \bigcup_{j \neq k, k+1} I_j$, then there exists $\epsilon > 0$ such that (4.1) has exactly four nontrivial solutions for $\lambda \in (\lambda_{k+1}^\infty, \lambda_{k+1}^\infty + \epsilon) \cup (\lambda_k^0 - \epsilon, \lambda_k^0)$ (which is near the boundary of $I_k \cap I_{k+1}$), and all of them are non-degenerate with two of them having Morse index $M(u) = k$, the other two $M(u) = k + 1$.

4.2 Main Results with Parameter

When f satisfies (f_1), (f_2), (f_3), for $u \in \mathbb{R} \setminus \{0\}$, we have

$$f'(0) < \frac{f(u)}{u} < f'(u) \leq \sup_{u \in \mathbb{R}} f'(u), \qquad \frac{f(u)}{u} < f'(\infty). \tag{4.6}$$

If u is a nontrivial solution of (4.1), then

$$\mu_k\left(\lambda f'(0)\right) > \mu_k\left(\frac{\lambda f(u)}{u}\right) > \mu_k\left(\lambda f'(u)\right) \geq \mu_k\left(\lambda \sup_{u \in \mathbb{R}} f'(u)\right),$$

$$\mu_k\left(\frac{\lambda f(u)}{u}\right) > \mu_k\left(\lambda f'(\infty)\right).$$

Our main results are:

Theorem 4.2.1 (Zhitao Zhang [201]) *Suppose that f satisfies (f_1), (f_2), (f_3), $\lambda_k^0 \leq \lambda_{k+1}^M$, for $k = 1, 2, \ldots, N$; Moreover, suppose that $uf''(u) \geq 0$ in a neighborhood of 0. Then for $k = 1, 2, \ldots, N$ (see Fig. 4.1)*

(a) *The branch of the solutions $(\lambda \in (I_k \cup \{\lambda_k^0\}) \setminus \bigcup_{j \neq k} \tilde{I}_j)$ to (4.1) has the same critical groups $C_q^\pm(J_\lambda, u^\pm(\lambda, \cdot)) = \delta_{qk} G$, and the critical values are positive; let $\Sigma_k^\pm = \{(\lambda, u_k^\pm(\lambda, \cdot)) : \lambda \in I_k\}$, then Σ_k^+ and Σ_k^- join at $(\lambda_k^0, 0)$, and*

$$\lim_{\lambda \to (\lambda_k^\infty)^+} \left\| u_k^\pm(\lambda, \cdot) \right\| = \infty.$$

(b) *For $\lambda \in I_k$, $k \geq 2$, the nontrivial solutions are sign-changing; for $\lambda \in I_1$, the solutions are positive or negative.*

(c) *For $s_1 < s_2$, $s_i \in I_1$, $i = 1, 2$, the positive solutions $u_{s_1} \not\leq u_{s_2}$, and the negative solutions $u_{s_1} \not\geq u_{s_2}$. ($u_1 \leq u_2 \Leftrightarrow u_1(x) \leq u_2(x), \forall x \in \Omega$.)*

Proof By Theorem 1.3 of [163], (4.1) has only trivial solution if $\lambda \notin \bigcup_{k \geq 1} I_k$, and has exactly two nontrivial solutions $u_k^\pm(\lambda, \cdot)$ if $\lambda \in I_k$. Moreover, $u_k^\pm(\lambda, \cdot)$ are non-degenerate and $M(u_k^\pm(\lambda, \cdot)) = k$; for $\lambda \in (\lambda_k^0, \lambda_k^\infty)$, all nontrivial solutions of (1) lie on two smooth curves $\Sigma_k^\pm$, Σ_k^+ and Σ_k^- join at $(\lambda_k^0, 0)$, and

Fig. 4.1 The structure of the solutions

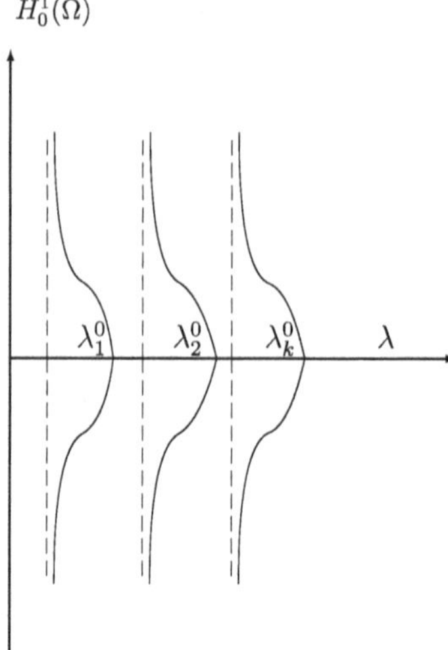

$$\lim_{\lambda \to (\lambda_k^\infty)^+} \left\| u_k^\pm(\lambda, \cdot) \right\| = \infty.$$

(a) So on the branch $\Sigma_k^\pm \cup \{(\lambda_k^0, 0)\}$, for each nontrivial solution $u^\pm(\lambda, \cdot)$, we have the critical groups

$$C_q^\pm\left(J_\lambda, u^\pm(\lambda, \cdot)\right) = \delta_{qk} G.$$

Next for the point $(\lambda_k^0, 0)$ on the branch $\Sigma_k^\pm \cup \{(\lambda_k^0, 0)\}$, we use the Shifting-theorem (see Theorem 5.4 of [49], p. 50) to prove

$$C_q(J_{\lambda_k^0}, 0) = \delta_{qk} G. \tag{4.7}$$

First we know $(\lambda_k^0, 0)$ is degenerate and $M(u_k(\lambda_k^0, 0)) = k - 1$ since $-\Delta \phi_k = \lambda_k^0 f'(0)\phi_k$.

Let

$$Z = \ker\{-\Delta - \lambda_k\} = \{t\phi_k : t \in R\}, \qquad \tilde{J}_{\lambda_k^0}(u) := J_{\lambda_k^0}|_Z$$

(since λ_k is simple, $\dim Z = 1$). So

$$\tilde{J}_{\lambda_k^0}(t\phi_k) = \frac{1}{2} \int_\Omega \left|\nabla(t\phi_k)\right|^2 dx - \lambda_k^0 \int_\Omega F(t\phi_k)\, dx,$$

$$\frac{d\tilde{J}_{\lambda_k^0}(t\phi_k)}{dt} = t\lambda_k - \lambda_k^0 \int_\Omega f(t\phi_k)\phi_k\, dx,$$

$$\frac{d^2 \tilde{J}_{\lambda_k^0}(t\phi_k)}{dt^2} = \lambda_k - \lambda_k^0 \int_\Omega f'(t\phi_k)\phi_k^2\, dx.$$

Therefore,

$$\frac{d\tilde{J}_{\lambda_k^0}(t\phi_k)}{dt}\bigg|_{t=0} = t\lambda_k - \lambda_k^0 \int_\Omega f(t\phi_k)\phi_k\,dx\bigg|_{t=0} = 0,$$

$$\frac{d^2\tilde{J}_{\lambda_k^0}(t\phi_k)}{dt^2}\bigg|_{t=0} = \lambda_k - \lambda_k^0 \int_\Omega f'(t\phi_k)\phi_k^2\,dx\bigg|_{t=0} = 0.$$

Noticing that $uf''(u) \geq 0$ in a neighborhood of 0, $\forall t \neq 0$ sufficiently small, we have

$$\frac{d^2\tilde{J}_{\lambda_k^0}(t\phi_k)}{dt^2} = \lambda_k - \lambda_k^0 \int_\Omega f'(t\phi_k)\phi_k^2\,dx < \lambda_k - \lambda_k^0 \int_\Omega f'(0)\phi_k^2\,dx = 0. \tag{4.8}$$

Hence, we know that 0 is a local maximum of $\tilde{J}_{\lambda_k^0}$, and by the Shifting-theorem we get

$$C_q(J_{\lambda_k^0}, 0) = C_{q-(k-1)}(\tilde{J}_{\lambda_k^0}, 0) = \delta_{q,(k-1)+1}G = \delta_{qk}G.$$

Now we consider the critical values: Suppose u is a nontrivial solution of (4.1), $-\Delta u = \lambda f(u)$, $u|_{\partial\Omega} = 0$, then

$$J_\lambda(u) = \frac{1}{2}\int_\Omega |\nabla u|^2\,dx - \lambda \int_\Omega F(u)\,dx = \lambda \int_\Omega \left(\frac{1}{2}f(u)u - F(u)\right).$$

Since $f'(s) > \frac{f(s)}{s}$, for $s > 0$ we have $f'(s)s > f(s)$, and $\int_0^s f'(s)s\,ds > F(s)$, then $f(s)s > 2F(s)$; for $s < 0$ we have $f'(s)s < f(s)$ and $\int_s^0 f'(s)s\,ds < \int_s^0 f(s)\,ds$, then $f(s)s > 2F(s)$ too. Therefore, $J_\lambda(u) > 0$.

(b) Suppose u is a nontrivial solution of (4.1), then $\int_\Omega(-\Delta u)\phi_1 = \int_\Omega \lambda f(u)\phi_1$, and $\int_\Omega \lambda_1\phi_1 u = \int_\Omega \lambda f(u)\phi_1$. Thus by meas$(\Omega_0) = 0$ (where $\Omega_0 = \{x \in \Omega : u(x) = 0\}$, by $f(0) = 0$ and the strong maximum principle), we have

$$\int_{\Omega\backslash\Omega_0} \left(\lambda_1 - \lambda\frac{f(u)}{u}\right)u\phi_1 = 0. \tag{4.9}$$

Noticing (4.6), we have $\lambda_1 - \lambda\frac{f(u)}{u} < \lambda_1 - \lambda f'(0) \leq 0$ when $\lambda \geq \frac{\lambda_1}{f'(0)}$. Therefore, by (4.9) we see that u is sign-changing.

When $\lambda \in I_1$, it is obvious that there are only one positive and one negative solution for (4.1).

(c) Let $C_0(\Omega) = \{u \in C(\bar{\Omega})|u(x) = 0, \forall x \in \partial\Omega\}$, $P := \{u \in C_0(\Omega)|u(x) \geq 0\}$, and define operators $A_s(u) := (-\Delta)^{-1}sf(u)$, $\forall u \in C_0(\Omega)$. Then P is a cone with nonempty interior points, A_s is compact and increasing for each $s > 0$. The solution u_s of

$$-\Delta u = sf(u), \quad u > 0 \quad \text{in } \Omega, \qquad u|_\Omega = 0 \tag{4.10}$$

is equivalent to the fixed point of A_s in P.

Suppose u_{s_1}, u_{s_2} ($s_1 < s_2$, $s_i \in I_1$) are the solutions of (4.10), then u_{s_2} is the super-solution of (4.10) with $s = s_1$, thus by Lemma 4.1 of [100] (p. 294) the fixed point index

$$i\left(A_{s_1}, (0, u_{s_2}), P\right) = 1, \tag{4.11}$$

where $(0, u_{s_2}) := \{u \in C_0(\Omega) | 0 < u(x) < u_{s_2}(x), \forall x \in \Omega\}$.

For A_s, $s \in I_1$, we prove that its Fréchet differential operator at ∞ along P is $(-\Delta)^{-1} s f'(\infty)$.

$\forall \varepsilon > 0, \exists R > 0$ such that

$$\left| \frac{f(t)}{t} - f'(\infty) \right| < \varepsilon, \quad \forall t, \ |t| > R.$$

Thus $\forall u \in C_0(\Omega)$ with $\|u\|_{C_0(\Omega)} \geq R$, we denote $\Omega_0 = \{x \in \Omega : |u(x)| \leq R\}$, $\Omega_1 = \Omega \backslash \Omega_0$, then

$$\left\| (-\Delta)^{-1} f(u) - s f'(\infty)(-\Delta)^{-1} u \right\|_{C_0(\Omega)}$$

$$\leq \left\| (-\Delta)^{-1} \right\| \cdot \left\| f(u) - s f'(\infty) u \right\|_{C_0(\Omega)}$$

$$= \left\| (-\Delta)^{-1} \right\| \cdot \max_{x \in \bar{\Omega}} \left| f\left(u(x)\right) - s f'(\infty) u(x) \right|$$

$$\leq \left\| (-\Delta)^{-1} \right\| \cdot \left(\max_{x \in \bar{\Omega}_0} \left| f\left(u(x)\right) - s f'(\infty) u(x) \right| \right.$$

$$\left. + \max_{x \in \bar{\Omega}_1} \left| f\left(u(x)\right) - s f'(\infty) u(x) \right| \right)$$

$$\leq C + \varepsilon \left\| (-\Delta)^{-1} \right\| \cdot \|u\|_{C_0(\Omega)}, \tag{4.12}$$

where $C > 0$ is a constant independent on u. Therefore, we have

$$\limsup_{u \in P, \|u\|_{C_0(\Omega)} \to \infty} \left\| (-\Delta)^{-1} \left[f(u) - s f'(\infty) u \right] \right\|_{C_0(\Omega)} / \|u\|_{C_0(\Omega)} \leq \varepsilon \left\| (-\Delta)^{-1} \right\|.$$

Since ε is arbitrary, we get

$$\lim_{u \in P, \|u\|_{C_0(\Omega)} \to \infty} \left\| (-\Delta)^{-1} \left[f(u) - s f'(\infty) u \right] \right\|_{C_0(\Omega)} / \|u\|_{C_0(\Omega)} = 0.$$

Hence, A_s has its Fréchet differential operator at ∞ along P and

$$A_s'(\infty) = (-\Delta)^{-1} s f'(\infty).$$

Since $s f'(\infty) \in (\lambda_1, \lambda_2)$, we know that 1 is not an eigenvalue of $A_s'(\infty)$, and $A_s'(\infty)$ has $s f'(\infty)/\lambda_1$ as the first eigenvalue which is greater than 1 with positive eigenvalue ϕ_1. By Lemma 4.10 of [100], p. 328, we know that $\exists \sigma > 0$ such that

$$i(A_s, P_{r_s}, P) = 0 \quad (\forall \sigma \leq r_s < \infty). \tag{4.13}$$

By the additivity of the fixed point index, we know

$$i\left(A_{s_1}, P_{r_{s_1}} \setminus [0, u_{s_2}], P\right) = -1,$$

which implies that there is a fixed point $\bar{u}_{s_1}$ of A_{s_1} in $P_{r_{s_1}} \setminus [0, u_{s_2}]$, noticing that the uniqueness of the positive solution for each $s \in I_1$, we get $u_{s_1} = \bar{u}_{s_1}$.

So $u_{s_1} \not\le u_{s_2}$.

Similarly we have the results for the negative solutions. $\qquad\qquad\square$

Remark 4.2.1 $\lambda_k^0 \le \lambda_{k+1}^M$ is equivalent to

$$\lambda_{k+1}/\lambda_k \ge \sup_u f'(u)/f'(0).$$

Since $\lambda_{k+1}/\lambda_k \to 1$ as $k \to \infty$, then the λ-ranges of solution curves must have overlaps for large k.

Theorem 4.2.2 (Zhitao Zhang [201]) *Suppose that f satisfies (f_1), (f_2), (f_4), $\lambda_k^m \le \lambda_{k+1}^0$, for $k = 1, 2, \ldots, N$. Moreover, suppose that $uf''(u) \le 0$ in a neighborhood of 0. Then for $k = 1, 2, \ldots, N$ (see Fig. 4.2)*

(a) *The branch of the solutions $(\lambda \in (I_k \cup \{\lambda_k^0\}) \setminus \bigcup_{j \ne k} \tilde{I}_j)$ to (4.1) has the same critical groups $C_q^\pm(J_\lambda, u(\lambda, \cdot)) = \delta_{q(k-1)}G$, and the critical values are negative, where $I_k = (\frac{\lambda_k}{f'(0)}, \frac{\lambda_k}{f'(\infty)})$, $\Sigma_k^\pm = \{(\lambda, u_k^\pm(\lambda, \cdot)) : \lambda \in I_k\}$, Σ_k^+ and Σ_k^- join at $(\lambda_k^0, 0)$, and*

$$\lim_{\lambda \to (\lambda_k^\infty)^+} \left\| u_k^\pm(\lambda, \cdot) \right\| = \infty.$$

(b) *For $\lambda \in (\lambda_k/f'(0), \lambda_k/f'(\infty))$, $k \ge 2$, the nontrivial solutions are sign-changing; for $\lambda \in I_1$, the solutions are positive or negative.*

(c) *For $s_1 < s_2$, $s_i \in I_1$, $i = 1, 2$, the positive solutions $u_{s_1} \le u_{s_2}$, and the negative solutions $u_{s_1} \ge u_{s_2}$.*

Proof (a) When f satisfies (f_1), (f_2), (f_4), we have for $u \ne 0$,

$$f'(0) > \frac{f(u)}{u} > f'(u) \ge \inf_{u \in R} f'(u), \qquad \frac{f(u)}{u} > f'(\infty).$$

Similar to the proof of Theorem 4.2.1, noticing that $uf''(u) \le 0$ in a neighborhood of 0, now $\forall t \ne 0$ sufficiently small, (4.8) becomes

$$\frac{d^2 \tilde{J}_{\lambda_k^0}(t\phi_k)}{dt^2} = \lambda_k - \lambda_k^0 \int_\Omega f'(t\phi_k)\phi_k^2 \, dx > \lambda_k - \lambda_k^0 \int_\Omega f'(0)\phi_k^2 \, dx = 0. \qquad (4.14)$$

Hence, we know that 0 is a local minimum of $\tilde{J}_{\lambda_k^0}$, and by the Shifting-theorem we get

Fig. 4.2 The structure of the solutions

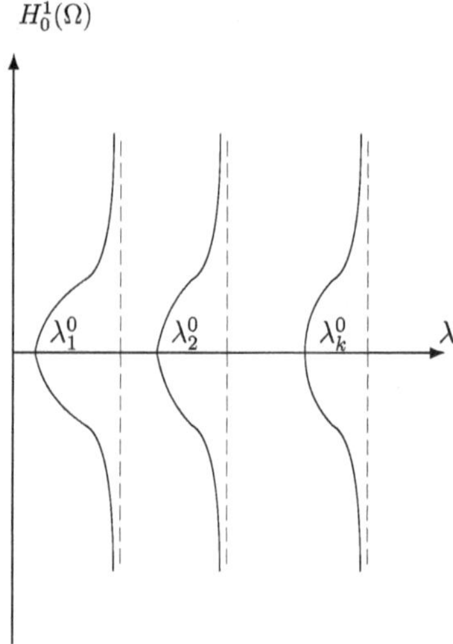

$$C_q(J_{\lambda_k^0}, 0) = C_{q-(k-1)}(\tilde{J}_{\lambda_k^0}, 0) = \delta_{q(k-1)}G.$$

And similarly we prove that the critical values are negative.

(b) For (4.9), now we have $\lambda_1 - \lambda\frac{f(u)}{u} < \lambda_1 - \lambda f'(\infty) \le 0$ when $\lambda \ge \frac{\lambda_1}{f'(\infty)}$. Therefore, by (4.9) we get u is sign-changing.

(c) Suppose u_{s_1}, u_{s_2} ($s_1 < s_2$, $s_i \in I_1$) are the solutions of (4.10), then u_{s_2} is the super-solution of (4.10) with $s = s_1$. Noticing that $s_1 f'(0) > \lambda_1$, we see that $\varepsilon\phi_1$ ($\varepsilon > 0$ sufficiently small) is a sub-solution of (4.10) with $s = s_1$. So by the strongly increasing property of $(-\Delta)^{-1}$, we know the unique positive solution $u_{s_1} \in (\varepsilon\phi_1, u_{s_2})$, where $(\varepsilon\phi_1, u_{s_2}) := \{u \in C_0(\Omega) | \varepsilon\phi_1(x) < u(x) < u_{s_2}(x), \forall x \in \Omega\}$.

Now we consider super-linear problems. $\square$

Theorem 4.2.3 (Zhitao Zhang [201]) *Suppose that f satisfies $(f_1), (f_2), (f_3)$, and $f'(\infty) = +\infty$. Moreover, suppose that $u f''(u) \ge 0$ in a neighborhood of 0; suppose that $|f'(t)| \le C(1 + |t|^{\alpha-1})$, $\alpha \in (1, \frac{n+2}{n-2})$, $n \ge 3$, (if $n \le 2$, α has no restriction) and $\exists \theta > 2, M > 0$ such that $\theta F(t) \le t f(t)$ for $|t| \ge M$. Then:*

(a) *$\forall \lambda \in (0, \frac{\lambda_1}{f'(0)})$, (4.1) has at least one positive, one negative, one sign-changing solution.*

(b) *(4.1) has one-signed nontrivial solutions $\Leftrightarrow \lambda \in (0, \frac{\lambda_1}{f'(0)})$.*

(c) *$\forall \lambda \in (0, \infty)$, (4.1) has at least one sign-changing solution.*

(d) *$\forall k \ge 2, \exists \varepsilon > 0$, such that $\forall \lambda \in (\frac{\lambda_k}{f'(0)} - \varepsilon, \frac{\lambda_k}{f'(0)})$, (4.1) has at least three sign-changing solutions.*

(e) *u is a nontrivial critical point of J_λ implies that $J_\lambda(u) > 0$.*

Proof (a) We can use similar (but simpler) proof of Theorem 1 of [71] (see Theorem 8.2.2) to prove it. (Here the condition is strong enough.)

(b) By the proof of (b) of Theorem 4.2.1, and (a), we get it.

(c) We consider the critical points of the functionals $J_\lambda(u)$ of (4.3). Under the conditions of this theorem, we know $J_\lambda(u)$ satisfies the (PS) condition for each $\lambda \in (0, \infty)$.

First we compute the critical groups of 0.

As $\lambda \in (\frac{\lambda_k}{f'(0)}, \frac{\lambda_{k+1}}{f'(0)})$, $k = 1, 2, \ldots$, then 0 is a non-degenerate critical point of J_λ, the Morse index of 0 is k, so the critical group is $C_q(J_\lambda, 0) = \delta_{qk} G$.

As $\lambda \in (0, \frac{\lambda_1}{f'(0)})$, the critical group $C_q(J_\lambda, 0) = \delta_{q0} G$.

As $\lambda = \frac{\lambda_k}{f'(0)}$, $k = 1, 2, \ldots$, by the proof (a) of Theorem 4.2.1, we find that the critical group is $C_q(J_\lambda, 0) = \delta_{qk} G$.

Following the proof of Wang [186], since $f(t)t \geq 0$ and $F(t) > 0$ for any $|t| > M$, we have

$$\frac{f(t)}{F(t)} \geq \frac{\theta}{t}, \quad \forall t \geq M, \qquad \frac{f(t)}{F(t)} \leq \frac{\theta}{t}, \quad \forall t \leq -M.$$

Hence

$$F(t) \geq C|t|^\theta, \quad \text{as } |t| \geq M.$$

Thus $\forall u \in S^\infty$, the unit sphere in H,

$$J_\lambda(tu) \to -\infty \quad \text{as } t \to +\infty. \tag{4.15}$$

We now prove that $\forall a < 0$, $J_\lambda(tu) \leq a$ implies that $\frac{dJ_\lambda(tu)}{dt} < 0$.

If $J_\lambda(tu) = \frac{t^2}{2} - \lambda \int_\Omega F(tu(x))\,dx \leq a < 0$, then noticing that (f_3) implies that $sf(s) > 2F(s)$, $\forall s \neq 0$, we have

$$\frac{dJ_\lambda(tu)}{dt} = (dJ_\lambda(tu), u)$$

$$= t - \lambda \int_\Omega f(tu(x))u(x)\,dx$$

$$= \frac{2}{t}\left\{\frac{t^2}{2} - \frac{\lambda}{2}\int_\Omega tu(x)f(tu(x))\right\}$$

$$\leq \frac{2}{t}\left\{\frac{t^2}{2} - \frac{2\lambda}{2}\int_\Omega F(tu(x))\,dx\right\} < 0. \tag{4.16}$$

The implicit function theorem is employed to obtain a unique $T(u) \in C(S^\infty, \mathbb{R})$ such that

$$J_\lambda(T(u)u) = a, \quad \forall u \in S^\infty.$$

Since J_λ is continuous in 0, there is a open neighborhood V of 0, such that

$$J_\lambda(u) > \frac{a}{2}, \quad \forall u \in V.$$

Thus $\|T(u)\|$ possesses a positive lower bound $\delta > 0$.

Finally, let us define a deformation retract $\eta : [0, 1] \times (H \setminus B_\varepsilon(0)) \to H \setminus B_\varepsilon(0)$, where $B_\varepsilon(0)$ is the ε-ball with center 0, by

$$\eta(s, u) = (1 - s)u + s T(u)u, \quad \forall u \in H \setminus B_\varepsilon(0).$$

This proves that $H \setminus B_\varepsilon(0) \simeq J_\lambda^a$, i.e., $J_\lambda^a \simeq S^\infty$, where $J_\lambda^a = \{u \in H \mid J_\lambda(u) \le a\}$. Thus $H_q(H, J_\lambda^a) \simeq H_q(H, S^\infty) \simeq 0$, the Betti numbers

$$\beta_q = 0, \quad \forall q = 0, 1, 2, \ldots. \tag{4.17}$$

Now $\forall \lambda \in [\frac{\lambda_k}{f'(0)}, \frac{\lambda_{k+1}}{f'(0)})$, $k = 1, 2, \ldots$, by the Morse inequality (Theorem 1.2.10), we know that J_λ has at least one nontrivial critical point, i.e., (4.1) has at least one nontrivial solution, by (b) we know that it is sign-changing. Thus noticing (a) we find that $\forall \lambda \in (0, \infty)$, (4.1) has at least one sign-changing solution.

(d) By the bifurcation theory, we know that $(\frac{\lambda_k}{f'(0)}, 0)$, $k = 1, 2, \ldots$ is a bifurcation point. Similarly to the proof of Lemma 2.6 of [163], we know $\lambda(s) < \frac{\lambda_k}{f'(0)}$ for $0 < |s| < \delta$, i.e. the curve turns to the left of $\frac{\lambda_k}{f'(0)}$.

So there is $\varepsilon > 0$ such that there is a curve $\Sigma_k^\pm$, and $\forall \lambda \in (\frac{\lambda_k}{f'(0)} - \varepsilon, \frac{\lambda_k}{f'(0)})$, we know $u_1(\lambda, \cdot), u_2(\lambda, \cdot) \in \Sigma_k^\pm$.

Next we prove they are non-degenerate as $\varepsilon > 0$ sufficiently small.

Since $\|u_i(\lambda, \cdot)\|$, $i = 1, 2$ are sufficiently small as $\varepsilon > 0$ sufficiently small, by the regularity of the elliptic operator we know that $u_i(\lambda, \cdot) \in C^2(\Omega)$, and $\|u_i(\lambda, \cdot)\|_{C_0^1(\Omega)}$ are sufficiently small such that $\frac{\lambda_k f'(u_i(\lambda, \cdot))}{f'(0)} < \lambda_{k+1}$, then $\frac{\lambda f(u_i(\lambda, \cdot))}{u_i(\lambda, \cdot)} < \lambda f'(u_i(\lambda, \cdot)) < \lambda_{k+1}$, $\forall \lambda \in (\frac{\lambda_k}{f'(0)} - \varepsilon, \frac{\lambda_k}{f'(0)})$. Now consider (4.5) with $u = u_i(\lambda, \cdot)$, we get

$$\mu_{k-1}(\lambda f'(0)) < 0, \qquad \mu_k(\lambda f'(0)) > 0;$$

$$\mu_{k-1}\left(\lambda \frac{f(u_i)}{u_i}\right) < 0, \qquad \mu_{k+1}\left(\lambda \frac{f(u_i)}{u_i}\right) > 0, \qquad \mu_k\left(\lambda \frac{f(u_i)}{u_i}\right) = 0.$$

Thus $\mu_k(\lambda f'(u_i)) < 0$, $\mu_{k+1}(\lambda f'(u_i)) > 0$. So $u_i(\lambda, \cdot)$, $\forall \lambda \in (\frac{\lambda_k}{f'(0)} - \varepsilon, \frac{\lambda_k}{f'(0)})$ are non-degenerate, the critical groups are

$$C_q\left(J_\lambda, u_i(\lambda, \cdot)\right) = \delta_{qk} G.$$

By the proof of (c), we know $C_q(J_\lambda, 0) = \delta_{q(k-1)} G$. Suppose there were no more critical points of J_λ. The Morse type numbers over (H, J_λ^a) would be

$$M_{k-1} = 1, \qquad M_k = 2, \qquad M_q = 0, \quad q \ne k - 1, k,$$

but the Betti numbers

$$\beta_q = 0, \quad \forall q = 0, 1, 2, \ldots.$$

By the Morse inequality, this is a contradiction. Thus there are at least 3 sign-changing solutions for $\lambda \in (\frac{\lambda_k}{f'(0)} - \varepsilon, \frac{\lambda_k}{f'(0)})$.

(e) By the proof (a) of Theorem 4.2.1, we get it. $\square$

Remark 4.2.2 If f satisfies the assumptions of Theorem 4.2.3, and f is odd, then (4.1) has infinitely many solutions (see Theorem 9.38 of [159]); and when $\lambda \geq \frac{\lambda_1}{f'(0)}$, (4.1) has infinitely many sign-changing solutions.

4.3 Equations Without the Parameter

$$\begin{cases} -\Delta u = f(u) & \text{in } \Omega, \\ u|_{\partial\Omega} = 0. \end{cases} \tag{4.18}$$

Theorem 4.3.1 (Zhitao Zhang [201]) *Suppose f satisfies $(f_1), (f_2), (f_3)$, and $\lambda_{k-1} < f'(0) \leq \lambda_k \leq f'(\infty) < \sup_{u \in \mathbb{R}} f'(u) < \lambda_{k+1}$, as $k \geq 2$; $0 < f'(0) \leq \lambda_1 \leq f'(\infty) < \sup_{u \in \mathbb{R}} f'(u) < \lambda_2$, as $k = 1$. (It is clear that the equalities do not hold simultaneously.) Then*

(a) *If $f'(0) \neq \lambda_k, f'(\infty) \neq \lambda_k$, then (4.18) has exactly two nontrivial non-degenerate solutions with same critical groups $C_q(J, u_i) = \delta_{qk}G, i = 1, 2$, and the critical values are positive.*

   If $f'(0) = \lambda_k, uf''(u) \geq 0$ in a neighborhood of 0, then (4.18) has only the trivial solution and $C_q(J, 0) = \delta_{qk}G$, where

$$J(u) = \frac{1}{2}\int_\Omega |\nabla u|^2\,dx - \int_\Omega F(u)\,dx. \tag{4.19}$$

   If $f'(\infty) = \lambda_k, \lambda_{k-1} < f'(0) < \lambda_k$, then (4.18) has only the trivial solution and $C_q(J, 0) = \delta_{q(k-1)}G$. (In fact, since $\lambda_{k-1} < f'(0) < \frac{f(u)}{u} < f'(\infty) = \lambda_k$, $\mu_{k-1}(\frac{f(u)}{u}) < 0$, $\mu_k(\frac{f(u)}{u}) > 0$, we get $-\Delta u = (f(u)/u)u$ in Ω, $u|_{\partial\Omega} = 0$ has no solution.)

(b) *For $k \geq 2$, the two nontrivial solutions are sign-changing; for $k = 1$, (4.18) has exactly one positive and one negative solution.*

Theorem 4.3.2 (Zhitao Zhang [201]) *Suppose f satisfies $(f_1), (f_2), (f_4)$, $\lambda_{k-1} < f'(\infty) \leq \lambda_k \leq f'(0) < \lambda_{k+1}$, as $k \geq 2$, $0 < f'(\infty) \leq \lambda_1 \leq f'(0) < \sup_{u \in \mathbb{R}} f'(u) < \lambda_2$, as $k = 1$. (It is clear that the equalities do not hold simultaneously.) Then*

(a) *If $f'(\infty) \neq \lambda_k, f'(0) \neq \lambda_k$, then (4.18) has exactly two nontrivial non-degenerate solutions with the same critical groups $C_q(J, u_i) = \delta_{q(k-1)}G, i = 1, 2$, and the critical values are negative. (Now $C_q(J, 0) = \delta_{qk}G$.)*

   If $f'(0) = \lambda_k, f'(\infty) < \lambda_k, uf''(u) \leq 0$ in a neighborhood of 0, then (4.18) has only the trivial solution and $C_q(J, 0) = \delta_{q(k-1)}G$.

   If $f'(\infty) = \lambda_k, \lambda_k < f'(0) < \lambda_{k+1}$, then (4.18) has only the trivial solution and $C_q(J, 0) = \delta_{qk}G$.

(b) *For $k \geq 2$, the nontrivial solutions are sign-changing; for $k = 1$, the solutions are positive or negative.*

Theorem 4.3.3 (Zhitao Zhang [201]) *Suppose that f satisfies (f_1), (f_2), (f_3), $f'(\infty) = +\infty$. Moreover, suppose that $uf''(u) \geq 0$ in a neighborhood of 0 if $f'(0) = \lambda_k$; suppose that $\exists \theta > 2, M > 0$ such that $\theta F(t) \leq tf(t)$ for $|t| \geq M$; $|f'(t)| \leq C(1 + |t|^{\alpha-1})$, $\alpha \in (1, \frac{n+2}{n-2})$, $n \geq 3$ (if $n \leq 2$, α has no restriction). Then*

(a) *$f'(0) < \lambda_1$, (4.18) has at least one positive, one negative, one sign-changing solution,*
(b) *(4.18) has no one-signed nontrivial solution if $f'(0) \geq \lambda_1$,*
(c) *if $f'(0) < +\infty$, (4.18) has at least one sign-changing solution,*
(d) *$\forall k \geq 2, \exists \varepsilon > 0$ such that $f'(0) \in (\lambda_k - \varepsilon, \lambda_k)$ implies that (4.18) has at least three sign-changing solutions $(-\Delta u = f'(0)\frac{f(u)}{f'(0)})$,*
(e) *u is a nontrivial critical point of J implies that the critical value $J(u) > 0$.*

Remark 4.3.1 If f satisfies the assumptions of Theorem 4.3.3, and f is odd, then (4.18) has infinitely many solutions. If $f'(0) \geq \lambda_1$, (4.18) has infinitely many sign-changing solutions. (By the Morse inequality and the proof of Theorem 4.2.3.)

Chapter 5
Solutions of a Class of Monge–Ampère Equations

5.1 Introduction

The Monge–Ampère equations are a type of important fully nonlinear elliptic equations; that is, nonlinear elliptic equations that are not quasilinear [95, 182, 187]. We consider the boundary value problems for a class of Monge–Ampère equations:

$$\begin{cases} \det D^2 u = e^{-u} & \text{in } \Omega, \\ u = 0 & \text{on } \partial\Omega, \end{cases} \tag{5.1}$$

where Ω is a bounded convex domain in $\mathbb{R}^n$ $(n \geq 1)$ with smooth boundary, and the matrix $D^2 u = (u_{ij}) = (\frac{\partial^2 u}{\partial x_i \partial x_j})$, $i, j = 1, 2, \ldots, n$ is the Hessian of u; in the following we will simply denote the first derivative by u_i $(i = 1, 2, \ldots, n)$, second derivative by u_{ij} $(i, j = 1, 2, \ldots, n)$ and so on; we also use (u_{ij}) instead of $D^2 u$ sometimes. We only consider convex solutions of (5.1) in order to ensure the ellipticity of the equation. In fact, any convex solution u of (5.1) is smooth, negative and strictly convex on $\overline{\Omega}$ (we can get $C^3(\overline{\Omega})$ estimate by Theorem 17.23 in [95], then by a standard bootstrap argument and Schauder estimate we can prove higher order estimates).

This equation is an analogue of the important complex Monge–Ampère equation arising from the Kähler–Einstein metric in the case of positive first Chern class in geometry. The equation written in the local coordinates has the form

$$\frac{\det(g_{i\bar{j}} + \frac{\partial^2 u}{\partial z_i \partial \bar{z}_j})}{\det(g_{i\bar{j}})} = e^{f-u},$$

where $\sum_{1 \leq i, j \leq n} g_{i\bar{j}} \, dz^i \, d\bar{z}^j$ is the Kähler metric in the class of the first Chern class, and f is a known function. The equation has been studied by many mathematicians and many problems remain still open cf. [179] and references therein.

We denote $\lambda\Omega := \{\lambda x : x \in \Omega\}$, $\lambda > 0$; noticing that (5.1) is invariant under translations, we assume without loss of generality that $0 \in \Omega$.

Z. Zhang, *Variational, Topological, and Partial Order Methods with Their Applications*, Developments in Mathematics 29,
DOI 10.1007/978-3-642-30709-6_5, © Springer-Verlag Berlin Heidelberg 2013

We also consider a problem with a parameter $t \geq 0$ in Sect. 5.4:

$$\begin{cases} \det u_{ij} = e^{-tu} & \text{in } \Omega, \\ u = 0 & \text{on } \partial\Omega. \end{cases} \tag{5.2}$$

This is equivalent to (5.1) in $t^{\frac{1}{2}}\Omega$ for $t > 0$ through a scaling.

Our main result is

Theorem 5.1.1 (Zhang and Wang [207]) *Given a smooth bounded convex domain* Ω, *there exists a critical value* $T^* > 0$ *such that*

1. *for* $t \in (0, T^*)$, *there exist at least two solutions of* (5.2);
2. *for* $t = T^*$, *there exists a unique solution of* (5.2);
3. *for* $t > T^*$, *there exists no solution of* (5.2).

Moreover, we get some other results, like the local structure of the branch near the first degenerate point.

Concerning (5.2), Theorem 5.1.1 implies:

Theorem 5.1.2 (Zhang and Wang [207]) *Given a smooth bounded convex domain* Ω, *there exists a critical value* $\lambda^* > 0$ *such that*

1. *for* $\lambda \in (0, \lambda^*)$, *there exist at least two solutions of* (5.1) *in* $\lambda\Omega$;
2. *there exists a unique solution of* (5.1) *in* $\lambda^*\Omega$;
3. *for* $\lambda > \lambda^*$, *there exists no solution of* (5.1) *in* $\lambda\Omega$.

Sometimes we would like to write (5.1) in the form

$$\log \det D^2 u = -u.$$

This form is more natural, because the function $\log \det$ defined on the space of positive definite symmetric matrices has many interesting properties, like concavity (see Appendix 5.5).

For a more general right hand side term our method still goes through, we can have a result analogues to Theorem 5.1.1 for the problem

$$\begin{cases} \log \det D^2 u = -tk(u) & \text{in } \Omega, \\ u = 0 & \text{on } \partial\Omega, \end{cases} \tag{5.3}$$

where $t \geq 0$ is a parameter and $k(\cdot) : (-\infty, 0] \to (-\infty, 0]$ is a C^2 function satisfying some conditions (see Remark 5.4.2).

In Sect. 5.2 using the argument of moving plane, we prove that any solution of (5.1) on the ball is radially symmetric. This method has been used by several authors in slightly different settings (see [84] and [188]).

In Sect. 5.3 we can reduce the equation on the ball to an ODE, and prove there exists a critical radius such that if the radius of a ball is smaller than this critical value there exists a solution, and vice versa. Using the comparison between domains we prove that this phenomenon occurs for every domain. We calculate the one dimensional case explicitly, which also indicates some kind of bifurcation phenomenon may exist (for related results about bifurcation, see [60, 157, 201]).

In Sect. 5.4, for the fixed domain, by using Lyapunov–Schmidt Reduction method we get the local structure of the solutions near a degenerate point, and prove existence of at least two solutions for a certain range of parameters. Finally we study the global structure of the branch emerging from $t = 0$ by the Leray–Schauder degree theory, a priori estimates and bifurcation theory. From all of these results we prove Theorem 5.1.1 at last. In Appendix 5.5 we collect some results concerning matrices.

5.2 Moving Plane Argument

In this section we prove a symmetry result for a $C^3(\overline{\Omega})$ solution u of (5.1) using the moving planc mcthod (of coursc from thc regularity theory we can relax the regularity assumption). With the property of being C^3 continuous up to the boundary, there exist two positive constants λ and Λ such that

$$\lambda|\xi|^2 \le g^{ij}\xi_i\xi_j \le \Lambda|\xi|^2 \quad \forall \xi \in \mathbb{R}^n, \ \xi \ne 0.$$

Here (g^{ij}) is the inverse matrix of the Hessian of u.

Given three positive constants λ, Λ and C_0, for any smooth domain Γ in $\mathbb{R}^n$ of sufficiently small measure, here we give a generalization of the weak maximum principle for the elliptic operator $Lf := h^{ij}\frac{\partial^2 f}{\partial x_i \partial x_j} + f$ (here and in the sequel we will use the summation convention that repeated indices are summed from 1 to n) if $h^{ij}(x)$ satisfy $\lambda|\xi|^2 \le h^{ij}(x)\xi_i\xi_j \le \Lambda|\xi|^2$, $\forall x \in \Gamma$, $\forall \xi \in \mathbb{R}^n$, $\xi \ne 0$, and $h^{ij}(x)$ are uniformly bounded in $C^1(\overline{\Gamma})$ by the constant C_0.

Lemma 5.2.1 (Zhang and Wang [207]) *There exists a positive constant δ, which only depends on λ, Λ, and C_0, such that for any smooth domain Γ in $\mathbb{R}^n$ of measure smaller than δ, if $Lf = 0$ in Γ, and $f \ge 0$ on $\partial\Gamma$, then $f \ge 0$ in Γ.*

Proof Use $f^- = \max\{-f, 0\}$ as a test function and integrate by parts to get

$$\int_\Gamma (f^-)^2 = \int_\Gamma h^{ij} f_{ij} f^-$$

$$= \int_\Gamma \frac{\partial}{\partial x_j}\left(h^{ij} f_i f^-\right) - f_i \frac{\partial}{\partial x_j}\left(h^{ij} f^-\right)$$

$$= \int_\Gamma f_i^- \frac{\partial}{\partial x_j}\left(h^{ij} f^-\right)$$

$$= \int_\Gamma h^{ij} f_i^- f_j^- + \frac{\partial}{\partial x_j}\left(h^{ij}\right) f^- f_i^-.$$

Because h^{ij} are uniformly bounded in $C^1(\Gamma)$ by a constant which is independent of the domain Γ, there is a constant C_1 such that $|\frac{\partial}{\partial x_j}(h^{ij})| := |\sum_j \frac{\partial h^{ij}}{\partial x_j}| \le C_1$, so using the uniform ellipticity and Cauchy inequality we get

$$\int_\Gamma (f^-)^2 \ge \int_\Gamma \lambda |\nabla f^-|^2 - \int_\Gamma \frac{\lambda}{2}|\nabla f^-|^2 - \int_\Gamma \frac{C^2}{2\lambda}|f^-|^2. \tag{5.4}$$

So we have

$$\int_\Gamma (f^-)^2 \ge C \int_\Gamma |\nabla f^-|^2, \tag{5.5}$$

here C only depends on h^{ij}.

If $n > 2$ we can use Sobolev embedding theorem to obtain

$$\int_\Gamma |\nabla f^-|^2 \ge C \left(\int_\Gamma (f^-)^{\frac{2n}{n-2}} \right)^{\frac{n-2}{n}}. \tag{5.6}$$

Then from Hölder inequality we get

$$(\text{meas}(\Gamma))^{\frac{2}{n}} \left(\int_\Gamma (f^-)^{\frac{2n}{n-2}} \right)^{\frac{n-2}{n}} \ge \int_\Gamma (f^-)^2. \tag{5.7}$$

Combining (5.5)–(5.7), we get

$$(\text{meas}(\Gamma))^{\frac{2}{n}} \left(\int_\Gamma (f^-)^{\frac{2n}{n-2}} \right)^{\frac{n-2}{n}} \ge C \left(\int_\Gamma (f^-)^{\frac{2n}{n-2}} \right)^{\frac{n-2}{n}}. \tag{5.8}$$

Note that the constant in (5.8) is independent on Γ, so (5.8) implies $f^- = 0$ in the case of meas(Γ) is small enough, that is, $f \ge 0$ in Γ.

If $n = 2$, (5.6)–(5.8) do not make sense. However, we can use the Hölder Inequality to obtain a similar result. Choose some $\alpha \in (1, 2)$, and we have similarly to (5.6)–(5.8)

$$\int_\Gamma |\nabla f^-|^2 \ge C \left(\int_\Gamma |\nabla f^-|^\alpha \right)^{\frac{2}{\alpha}}$$
$$\ge C \left(\int_\Gamma (f^-)^{\frac{2\alpha}{2-\alpha}} \right)^{\frac{2-\alpha}{\alpha}},$$

$$(\text{meas}(\Gamma))^{\frac{2\alpha-2}{\alpha}} \left(\int_\Gamma (f^-)^{\frac{2\alpha}{2-\alpha}} \right)^{\frac{2-\alpha}{\alpha}} \ge \int_\Gamma (f^-)^2,$$

$$(\text{meas}(\Gamma))^{\frac{2\alpha-2}{\alpha}} \left(\int_\Gamma (f^-)^{\frac{2\alpha}{2-\alpha}} \right)^{\frac{2-\alpha}{\alpha}} \ge C \left(\int_\Gamma (f^-)^{\frac{2\alpha}{2-\alpha}} \right)^{\frac{2-\alpha}{\alpha}},$$

where C is a constant depending only on Ω. Then we get the same result too. The proof for the case $n = 1$ is exactly the same as above $n = 2$. $\square$

Theorem 5.2.1 (Zhang and Wang [207]) *Let Ω be symmetric with respect to a hyperplane, then any solution $u \in C^3(\overline{\Omega})$ of (5.1) is symmetric with respect to the hyperplane too.*

Proof Without loss of generality, we can assume the hyperplane is the coordinate plane $x_1 = 0$. We denote $\Omega_t := \Omega \cap \{x_1 \le t\}$ $(t \le 0)$, and $u_t(x_1, x_2, \ldots, x_n) := u(2t - x_1, x_2, \ldots, x_n)$ in Ω_t. Then we have

$$D^2 u_t(x_1, x_2, \ldots, x_n) = P D^2 u(2t - x_1, x_2, \ldots, x_n) P^T, \tag{5.9}$$

where the diagonal matrix $P = \mathrm{diag}\{-1, 1, \ldots, 1\}$ and P^T is the transpose of P. Because $\det P = -1$, we have

$$\det D^2 u_t(x_1, x_2, \ldots, x_n) = \det D^2 u(2t - x_1, x_2, \ldots, x_n)$$
$$= e^{-u(2t - x_1, x_2, \ldots, x_n)}$$
$$= e^{-u_t(x_1, x_2, \ldots, x_n)},$$

so u_t still satisfies (5.1) in Ω_t.

Now

$$-u + u_t = \log \det(u_{ij}) - \log \det\big((u_t)_{ij}\big)$$
$$= \int_0^1 \frac{d}{d\tau} \log \det\big(\tau u_{ij} + (1 - \tau)(u_t)_{ij}\big) d\tau$$
$$= \left[\int_0^1 g_\tau^{ij} d\tau\right](u - u_t)_{ij},$$

Here (g_τ^{ij}) is the inverse matrix of $(\tau u_{ij} + (1 - \tau)(u_t)_{ij})$.

Let $w_t = u - u_t$, which satisfies the equation $-[\int_0^1 g_\tau^{ij} d\tau](w_t)_{ij} = w_t$. We have $u - u_t$ on $\partial\Omega_t \cap \{x_1 = t\}$, and for $t < 0$ we have $u = 0$ and $u_t < 0$ on $\partial\Omega_t \cap \partial\Omega$ because the reflection of this part lies in the interior of Ω. Here we use the fact that $u < 0$ in the interior of Ω because u is convex with vanishing boundary value, this fact will be used a lot in the sequel. Thus $w_t \ge 0$ on $\partial\Omega_t$. Because $\int_0^1 g_\tau^{ij} d\tau$ are uniformly elliptic with the same constant λ and Λ as for g^{ij} above and uniformly bounded in C^1 with the constant which only depends on u, by Lemma 5.2.1 we conclude that if t is so close to $\min\{x_1 | x \in \Omega\}$ that $\mathrm{meas}(\Omega_t)$ is small enough, then $w_t \ge 0$ in Ω_t. Now $-[\int_0^1 g_\tau^{ij} d\tau](w_t)_{ij} \ge 0$, by the strong maximum principle (see Theorem 3.5 of [95]), we find that either $w_t > 0$ strictly in the interior point of Ω_t or $w_t \equiv 0$ in Ω_t.

Now we can move the plane towards right. Define

$$T = \sup\{t < 0 \mid w_t \ge 0 \text{ in } \Omega_t\}. \tag{5.10}$$

If $T < 0$, because $w_T > 0$ on $\partial\Omega_T \cap \partial\Omega$, we have $w_T > 0$ strictly in the interior point of Ω_T. Thus there exists a compact set $K \subset \operatorname{int}\Omega_T$ such that $\operatorname{meas}(\Omega_T \setminus K)$ is small and there exists a positive constant ϵ such that $w_T > \epsilon$ in K. Noticing w_t is continuous with respect to t, there exists a positive constant σ with $T + \sigma < 0$ such that $w_t > 0$ in K, $\forall t \in (T, T + \sigma)$. Of course, we also have $w_t \geq 0$ on $\partial(\Omega_t \setminus K)$ for $t \in (T, T + \sigma)$(we only need to check that $w_t \geq 0$ on ∂K, which is guaranteed by the fact $w_t \geq 0$ in K). Now if σ is so small that $\operatorname{meas}(\Omega_t \setminus K)$ is small enough, we can use Lemma 5.2.1 again to conclude that $w_t \geq 0$ in $\Omega_t \setminus K$. So we find that $w_t \geq 0$ in Ω_t, $\forall t \in (T, T + \sigma)$, which contradicts (5.10).

Therefore we have $T \geq 0$, in particular $w_0 \geq 0$ in Ω_0, which implies as $x_1 < 0$

$$u(x_1, x_2, \ldots, x_n) \geq u(-x_1, x_2, \ldots, x_n).$$

Now we can move the plane from the right towards the left and get the reverse inequality. So we have

$$u(x_1, x_2, \ldots, x_n) = u(-x_1, x_2, \ldots, x_n), \tag{5.11}$$

which means u is symmetric with respect to the hyperplane $x_1 = 0$. $\square$

From this theorem we can easily get a corollary:

Corollary 5.2.1 (Zhang and Wang [207]) *If Ω is a ball, then any solution of (5.1) is radially symmetric.*

Remark 5.2.1 From the proof we can see the theorem still holds in the case of (5.3)

$$\det D^2 u = e^{-k(u)}, \tag{5.12}$$

when $k(u)$ is a Lipschitz continuous function in its domain $[\inf u, 0]$.

Remark 5.2.2 Solutions of the related equation in the entire space $\mathbb{R}^n$ $(n \geq 2)$ may be not radially symmetric. For example, the following problem has a non-radially symmetric solution:

$$\begin{cases} \det D^2 u = e^{-u} & \text{in } \mathbb{R}^n, \\ u \geq 0 & \text{in } \mathbb{R}^n, \\ u(0) = 0. \end{cases} \tag{5.13}$$

Next we use the solution $f(t)$ of this equation in one dimension to construct a non-radially symmetric solution of (5.13) as $n \geq 2$. We know

$$\begin{cases} f'' = e^{-f} & \text{in } \mathbb{R}, \\ f \geq 0 & \text{in } \mathbb{R}, \\ f(0) = 0 \end{cases} \tag{5.14}$$

has an unique solution $f(t) = 2\log(1 + e^{\sqrt{2}t}) - \sqrt{2}t - \log 4$ for $t > 0$, and for $t < 0$ we have $f(t) = f(-t)$ in $\mathbb{R}$, which is asymptotically linear. We define

$$u(x_1, x_2, \ldots, x_n) = f(x_1) + f(x_2) + \cdots + f(x_n), \tag{5.15}$$

which is a solution of (5.13) and not radially symmetric. In fact because u is convex, $\frac{u(tx)}{t}$ $(t > 0)$ is increasing in t, and we have $v(x) = \lim_{t \to +\infty} \frac{u(tx)}{t} = C(|x_1| + |x_2| + \cdots + |x_n|)$ for some positive constant C.

5.3 Existence and Non-existence Results

From Sect. 5.2 we know that any solution of (5.1) in the ball $B_R(0) \subset \mathbb{R}^n$ is radially symmetric. So we may write $u(x) = u(r)$, here $r = |x|$. Moreover, 0 is the minimal point of u and u is increasing in $[0, R]$. Now we have

$$u_i = u'(r)\frac{x_i}{r}, \tag{5.16}$$

$$u_{ij} = u''(r)\frac{x_i x_j}{r^2} + u'(r)\left(\frac{\delta_{ij}}{r} - \frac{x_i x_j}{r^3}\right), \tag{5.17}$$

where δ_{ij} is the Kronecker δ. By an elementary calculation we get

$$\det(u_{ij}) = \left(\frac{u'(r)}{r}\right)^{n-1} u''(r). \tag{5.18}$$

So (5.1) becomes

$$\left(\frac{u'(r)}{r}\right)^{n-1} u''(r) = e^{-u}. \tag{5.19}$$

Next, we try to find a solution in $[0, R]$ which is strictly convex and satisfies $u(R) = 0$. Assume $u(0) = -C$ for some positive constant C. We write the equation as

$$\frac{d}{dr}\left(u'(r)\right)^n = nr^{n-1}e^{-u}. \tag{5.20}$$

By integration from 0 to r ($r \in [0, R]$), we get (noticing $u'(0) = 0$)

$$\left(u'(r)\right)^n = n\int_0^r s^{n-1}e^{-u(s)}\,ds. \tag{5.21}$$

Since u is increasing, e^{-u} is decreasing. Thus we have

$$\left(u'(r)\right)^n \geq r^n e^{-u(r)}, \tag{5.22}$$

that is

$$u'(r) \geq r e^{\frac{-u(r)}{n}}, \tag{5.23}$$

$$\frac{d}{dr} e^{\frac{u(r)}{n}} \geq \frac{r}{n}.$$

(5.24)

By integration we get

$$e^{\frac{u(r)}{n}} \geq \frac{r^2}{2n} + e^{\frac{-C}{n}}.$$

(5.25)

In particular, since $u(R) = 0$, as $r = R$ we have

$$1 \geq \frac{R^2}{2n} + e^{\frac{-C}{n}}.$$

(5.26)

Thus $R \leq (2n)^{\frac{1}{2}}$, which means in particular in balls with radius large enough (5.1) has no solution.

On the other hand, if R is small we can use $u(x) = \frac{n}{R^2}(|x|^2 - |R|^2)$ as a sub-solution in the ball B_R, and since 0 is always a sup-solution, we can construct a solution from this sub-solution. In fact, we have $u_{ij} = \frac{2n}{R^2}\delta_{ij}$. So if $R \leq (\frac{2n}{e})^{\frac{1}{2}}$, then we have

$$\det(u_{ij}) = \left(\frac{2n}{R^2}\right)^n$$

$$\geq e^n$$

$$\geq e^{-u}.$$

We can use sup-solution and sub-solution method to show the existence of a solution by iteration (for the proof, see that of Lemma 5.3.1 below). So in balls with small radius there exists a solution.

In conclusion we have

Theorem 5.3.1 (Zhang and Wang [207]) *There is no solution of (5.1) for $\Omega = B_R(0)$ with $R > 0$ large enough, and for sufficiently small $R > 0$ there is a solution of (5.1).*

Now we use sub-solution and sup-solution method to construct a solution by iteration in an arbitrary domain. Notice 0 is always a sup-solution, so we just need the existence of a negative sub-solution. This is standard and well known, we include it here just for completeness.

Lemma 5.3.1 (Zhang and Wang [207]) *If we have a strictly convex function $f \in C^3(\overline{\Omega})$, such that $\det(f_{ij}) \geq e^{-f}$ in Ω and $f \leq 0$ on $\partial\Omega$, then (5.1) has a solution u in Ω.*

Proof Set $u^0 = f$ and define the iteration as

$$\begin{cases} \det(u_{ij}^{k+1}) = e^{-u^k} & \text{in } \Omega, \\ u^{k+1} = 0 & \text{on } \partial\Omega. \end{cases}$$

(5.27)

By Theorem 17.23 of [95], we know that u^{k+1} exists with the $C^3(\overline{\Omega})$ norm controlled by the $C^2(\overline{\Omega})$ norm of u^k.

Noticing

$$\begin{cases} \det(u^1_{ij}) \leq \det(u^0_{ij}) & \text{in } \Omega, \\ u^1 \geq u^0 & \text{on } \partial\Omega, \end{cases} \tag{5.28}$$

we have $u^1 \geq u^0$ in Ω by the comparison principle (Theorem 17.1 of [95]). Then $\det(u^1_{ij}) = e^{-u^0} \geq e^{-u^1}$. By induction we have $u^{k+1} \geq u^k$ and $\det(u^{k+1}_{ij}) \geq e^{-u^{k+1}}$ for any k.

From the higher order estimate of Monge–Ampère equation (Theorem 17.26 of [95]) we know that the $C^3(\overline{\Omega})$ norm of u^k can be controlled by the $C^3(\overline{\Omega})$ norm of f, so the sequence u^k is compact in $C^2(\overline{\Omega})$, and $u^k(x)$ also converge increasingly to some $u(x)$, $\forall x \in \Omega$, which is a convex function with vanishing boundary value. Combining these facts we know u^k converge to u in $C^2(\overline{\Omega})$. By taking the limit in (5.27), we know that u is a solution of (5.1). $\qquad\square$

Next we prove a lemma concerning the comparison between domains:

Lemma 5.3.2 (Zhang and Wang [207]) *Given two bounded convex domains Ω_1 and Ω_2 such that $\Omega_1 \subset \Omega_2$. If we have a solution u of (5.1) in Ω_2, then there exists a solution v of (5.1) in Ω_1, or equivalently if there is no solution of (5.1) in Ω_1, then there is no solution of (5.1) in Ω_2.*

Proof Just take the restriction of u in Ω_1 as a sub-solution, then we can use Lemma 5.3.1. $\qquad\square$

Given a bounded convex domain Ω, a result of F. John (see [34] or [185]) says that there exists an ellipsoid P such that $P \subset \Omega \subset nP$. P can be transformed into a ball by a matrix A with $\det A = 1$, which leaves the equation invariant. Now our first main result is clear:

Theorem 5.3.2 (Zhang and Wang [207]) *Given a bounded convex domain Ω, there exists a positive constant λ^* such that if $\lambda < \lambda^*$ there exists a solution of (5.1) in $\lambda\Omega$; and if $\lambda > \lambda^*$ there exists no solution of (5.1) in $\lambda\Omega$. Moreover, we have the estimation $c(n)(\mathrm{meas}(\Omega))^{-\frac{1}{n}} \leq \lambda^* \leq C(n)(\mathrm{meas}(\Omega))^{-\frac{1}{n}}$ for some universal constants $c(n)$ and $C(n)$.*

Proof First from Lemma 5.3.2 we have if in $\lambda\Omega$ there exists a solution then for any $\lambda' \in (0, \lambda)$ there exists a solution in $\lambda'\Omega$; and if in $\lambda\Omega$ there exists no solution then for any $\lambda' > \lambda$ there exists no solution in $\lambda'\Omega$. So we can define

$$\lambda^* = \sup\{\lambda > 0 \mid (1) \text{ has a solution in } \lambda\Omega\}. \tag{5.29}$$

In fact, from our previous discussion in this section we know that in small balls the equation has a solution and in large balls the equation has no solution, that is, our claim is true for unit ball with the critical radius $\lambda^*(B_1) \in [(\frac{2n}{e})^{\frac{1}{2}}, (2n)^{\frac{1}{2}}]$. From the result of F. John (see [34] or [185]), without loss of generality we can assume that $B_R(0) \subset \Omega \subset nB_R(0)$, where $B_R(0)$ is the ball with radius R. From the comparison of volume we have $C_1(n)(\mathrm{meas}(\Omega))^{\frac{1}{n}} \le R \le C_2(n)(\mathrm{meas}(\Omega))^{\frac{1}{n}}$ for some universal constants $C_1(n)$ and $C_2(n)$. Now using Lemma 5.3.2 again, if λ is so large that λR is grater than $\lambda^*(B_1)$, then there is no solution in $\lambda\Omega$; and if λ is so small that $n\lambda R$ is less than $\lambda^*(B_1)$, then there is a solution in $\lambda\Omega$. Hence λ^* for Ω is positive and finite. Our estimation of $\lambda^*(\Omega)$ can be easily checked by

$$\begin{cases} \lambda^*(\Omega)R \le \lambda^*(B_1), \\ n\lambda^*(\Omega)R \ge \lambda^*(B_1). \end{cases} \tag{5.30}$$

$\square$

We include here one example in $\mathbb{R}$ to indicate how the solution varies with respect to the size of the domain. Note that any solution u of (5.1) in $\lambda\Omega$ can be scaled to $u^\lambda(x) = \lambda^{-2}u(\lambda x)$ defined in Ω, which satisfies

$$\det(u_{ij}^\lambda)(x) = \det(u_{ij})(\lambda x)$$
$$= e^{-u(\lambda x)}$$
$$= e^{-\lambda^2 u^\lambda(x)}.$$

So we can consider an equivalent problem in a fixed domain Ω with a parameter in the equation.

Example 5.3.3 $u'' = e^{-tu}$ in the interval $[-1, 1]$ with vanishing boundary value.

First we have a constant $C \ge 2$ such that

$$t(u')^2 + 2e^{-tu} = C. \tag{5.31}$$

So if $x > 0$

$$t^{\frac{1}{2}}u' = (C - 2e^{-tu})^{\frac{1}{2}}, \tag{5.32}$$

if $x < 0$ we have a negative sign before the right-hand side.

Take $f = (C - 2e^{-tu})^{\frac{1}{2}}$, then

$$f' = \frac{1}{2}(C - 2e^{-tu})^{-\frac{1}{2}} 2te^{-tu}u'$$
$$= \pm t^{\frac{1}{2}}e^{-tu}$$
$$= \pm \frac{t^{\frac{1}{2}}}{2}(C - f^2).$$

So for $x < 0$ we have

$$\frac{f'}{C - f^2} = \frac{-t^{\frac{1}{2}}}{2}. \tag{5.33}$$

We can integrate to obtain for $x < 0$

$$f(x) = C^{\frac{1}{2}} \frac{e^{-(Ct)^{\frac{1}{2}}x} - 1}{e^{-(Ct)^{\frac{1}{2}}x} + 1}. \tag{5.34}$$

For $x > 0$ we have $f(x) = f(-x)$. From the boundary value $f(0) = 0$ and $f(-1) = (C - 2)^{\frac{1}{2}}$, we have a constraint on C and t:

$$t^{\frac{1}{2}} = \frac{\log(C - 1 + (C^2 - 2C)^{\frac{1}{2}})}{C^{\frac{1}{2}}}. \tag{5.35}$$

If we use C as a parameter ($C \geq 2$), the right-hand side has an upper bound (in fact it is increasing if C is less than some value and decreasing to 0 if C is greater than the value). So we have this result: there exists a t_0 such that for $t > t_0$ there is no C satisfying (5.35), therefore no solution to the original equation, and for $t < t_0$ there are exactly two C satisfying (5.35) and therefore there exist two solutions to the original equation.

5.4 Bifurcation and the Equation with a Parameter

In this section we study the equation with a parameter in a fixed domain:

$$\begin{cases} \det u_{ij} = e^{-tu} & \text{in } \Omega, \\ u = 0 & \text{on } \partial\Omega. \end{cases} \tag{5.36}$$

We know from the previous section for $t > 0$, (5.36) is equivalent to (5.1) in the domain $t^{\frac{1}{2}}\Omega$.

In this section we first extend a branch of solutions emanating from $t = 0$, and prove some properties of this branch, then we find that this branch degenerates at a point $t = T$. Moreover, we study the local structure of the branch near the degenerate point (Theorem 5.4.1). At last, we study the global structure of this connected component emanating from $t = 0$; using an a priori estimate and Leray–Schauder degree theory we can prove Theorem 5.1.1.

We first use the Implicit Function Theorem to find a branch (emanating from $t = 0$) of solutions of (5.36). We need two function spaces: X is the space of functions in $C^{k,\alpha}(\overline{\Omega})$ with vanishing boundary value and Y is the space of functions in $C^{k-2,\alpha}(\overline{\Omega})$ with vanishing boundary value, here k is an integer greater than 2 and $\alpha \in (0, 1)$. For a function $u \in U := \{u \in X, \text{ there exists a positive constant } \epsilon(u)$ such that $D^2u - \epsilon\,\mathrm{Id}$ is positive definite in Ω, where Id is the identity matrix$\}$ (note that U is an open set of X, in fact, the elements of U are strictly convex functions), we define a map $F : \mathbb{R}^1 \times U \to Y$:

$$F(t, u) = \log \det(u_{ij}) + tu. \tag{5.37}$$

We have the formula for the derivative:

$$D_u F(t, u)v = g^{ij} v_{ij} + tv, \tag{5.38}$$

here (g^{ij}) is the inverse matrix of (u_{ij}). In order to check if $D_u F(t, u)$ is surjective we need to estimate the first eigenvalue of the elliptic operator $Lf := -g^{ij} f_{ij}$, denoted by λ_1. Note that by a result of Bakelman [22] we can get

$$\int_\Omega g^{ij} f_{ij} f \det(u_{ij}) \, dx = -\int_\Omega g^{ij} f_i f_j \det(u_{ij}) \, dx, \tag{5.39}$$

for $f|_{\partial\Omega} = 0$. The calculation is simple if u is strictly convex, we present it here for readers' convenience.

Lemma 5.4.1 (Zhang and Wang [207]) *For a strictly convex function u and two $C^1(\overline{\Omega})$ functions f and h with vanishing boundary values, we have*

$$\int_\Omega g^{ij} f h_{ij} \det(u_{ij}) \, dx = -\int_\Omega g^{ij} f_i h_j \det(u_{ij}) \, dx, \tag{5.40}$$

here (g^{ij}) is the inverse matrix of (u_{ij}).

Proof It is just an integration by parts, but some terms can be canceled.

$$\int_\Omega g^{ij} f_i h_j \det(u_{ij}) \, dx$$

$$= \int_\Omega \frac{\partial}{\partial x_i} \left(g^{ij} f h_j \det(u_{ij}) \right) dx - \int_\Omega g^{ij} f h_{ij} \det(u_{ij}) \, dx$$

$$- \int_\Omega \frac{\partial}{\partial x_i} \left(g^{ij} \right) f h_j \det(u_{ij}) \, dx - \int_\Omega g^{ij} f h_j \frac{\partial}{\partial x_i} \det(u_{ij}) \, dx.$$

On the right hand side, the first term is the divergence of a vector field which vanishes on the boundary. For the last two terms we have

$$\frac{\partial}{\partial x_i} g^{ij} = -g^{iq} g^{pj} u_{pqi}, \tag{5.41}$$

and

$$\frac{\partial}{\partial x_i} \det(u_{ij}) = g^{kl} u_{kli} \det(u_{ij}). \tag{5.42}$$

Now we have

$$g^{iq} g^{pj} u_{pqi} = g^{ij} g^{kl} u_{kli},\tag{5.43}$$

which can be seen by changing i, p, q into k, i, l, respectively, in the left-hand side of (5.43), because these are used to be summed. So the last two terms cancel each other and only the second term which we want is left. $\qquad\square$

Concerning the spectrum of the elliptic operator $Lf := -g^{ij}\frac{\partial^2 f}{\partial x_i \partial x_j}$ with Dirichlet boundary condition (here (g^{ij}) is the inverse matrix of (u_{ij}) and $u \in U$), we have a result analogues to the Laplace operator:

Lemma 5.4.2 (Zhang and Wang [207]) *The spectrum of the elliptic operator L is real, and the first eigenvalue $\lambda_1 > 0$ has a positive eigenfunction. Moreover, λ_1 is simple.*

Proof We need two spaces: X is the completion of the $C_0^\infty(\Omega)$ function with the norm $\|f\|^2 = \int_\Omega g^{ij} f_i f_j \det(u_{ij})\, dx$ and Y is the completion of the $C_0^\infty(\Omega)$ function with the norm $\|f\|^2 = \int_\Omega f^2 \det(u_{ij})\, dx$ (in fact because u is in U and strictly convex these two spaces are $H_0^1(\Omega)$ and $L^2(\Omega)$, respectively, with an equivalent norm). X can be embedded into Y compactly. So the inverse of L is a self-adjoint, compact, positive definite operator on Y.

The first eigenvalue can be characterized by

$$\lambda_1 = \inf_{f \in X, f \neq 0} \frac{\int_\Omega g^{ij} f_i f_j \det(u_{ij})\, dx}{\int_\Omega f^2 \det(u_{ij})\, dx}.\tag{5.44}$$

There exists a non-negative minimizer f, which satisfies the equation $Lf = \lambda_1 f$. Then from the strong maximum principle (see Theorem 3.5 of [95]) we know f is positive in the interior of Ω. The simpleness of the first eigenvalue is the same as the Laplace case. $\qquad\square$

It is well known that for $t = 0$ there exists a unique smooth convex solution u_0 of (5.36). Of course, the first eigenvalue $\lambda_{1,0}$ is positive. Then we know from the Implicit Function Theorem that there exist a constant $T_0 > 0$ and a C^1 map $u : [0, T_0) \to U$ such that $F(t, u_t) = 0$. In the sequel we denote the first eigenvalue associated with u_t by $\lambda_{1,t}$.

Here we need a lemma:

Lemma 5.4.3 (Zhang and Wang [207]) *The first eigenvalue λ_1 of L is continuous in u with respect to the $C^2(\overline{\Omega})$ norm.*

Proof We need to prove that if u_k converges to u in $C^2(\overline{\Omega})$ norm, then the first eigenvalue $\lambda_{1,k}$ of the elliptic operator $L_k f := -g_k^{ij} f_{ij}$ converges to the first eigenvalue λ_1 of the elliptic operator $Lf := -g^{ij} f_{ij}$, where g_k^{ij} is the inverse matrix of

$D^2 u_k$ and g^{ij} is the inverse matrix of $D^2 u$. Denote the first (positive) eigenfunction of L by f which satisfies $\int_\Omega f^2 \det(u_{ij})\,dx = 1$, and the first (positive) eigenfunction of L_k by f_k which satisfies $\int_\Omega f_k^2 \det(u_k)_{ij}\,dx = 1$.

First we have

$$
\lambda_1 = \frac{\int_\Omega g^{ij} f_i f_j \det(u_{ij})\,dx}{\int_\Omega f^2 \det(u_{ij})\,dx}
$$

$$
= \lim_{k\to+\infty} \frac{\int_\Omega g_k^{ij} f_i f_j \det((u_k)_{ij})\,dx}{\int_\Omega f^2 \det((u_k)_{ij})\,dx}
$$

$$
\geq \varlimsup_{k\to+\infty} \lambda_{1,k}.
$$

If there exists a subsequence of $\lambda_{1,k}$, still denoted by $\lambda_{1,k}$, which converges to $\lambda_1 - \delta$ for some positive constant δ, then there exists a subsequence of f_k, still denoted by f_k, converging weakly in the Sobolev space $H_0^1(\Omega)$ and strongly in $L^2(\Omega)$. Denote the limit function as h, then by taking the limit we have

$$
\int_\Omega h^2 \det(u_{ij})\,dx = 1, \tag{5.45}
$$

and

$$
\int_\Omega g^{ij} h_i h_j \det(u_{ij})\,dx \leq \lambda_1 - \delta, \tag{5.46}
$$

which contradicts the definition of λ_1. $\square$

Next we estimate the first eigenvalue $\lambda_{1,t}$ associated with u_t above, to get the maximum interval for t such that the branch exists. We show that the branch extends until a point at which the curve is not differentiable in t.

Proposition 5.4.1 (Zhang and Wang [207]) *Given the C^1 map $u : [0, T_0) \to U$ starting from u_0 such that $F(t, u_t) = 0$ above, we have $\lambda_{1,t} > t$.*

Proof First from the discussion above we have $\lambda_{1,0} > 0$. Assume there exists $t_0 \in (0, T_0)$ such that $\lambda_{1,t_0} = t_0$ with a first positive eigenfunction as f, that is,

$$
-g_{t_0}^{ij} f_{ij} = t_0 f. \tag{5.47}
$$

Without loss of generality we can assume for all $t < t_0$ we have $\lambda_{1,t} > t$. Because u_t is differentiable in t, we differentiate the equation $F(t, u_t) = 0$ in $t \in (0, T_0)$:

$$
-g_t^{ij}\left(\frac{\partial u_t}{\partial t}\right)_{ij} = t\frac{\partial u_t}{\partial t} + u_t, \tag{5.48}
$$

where g_t^{ij} is the inverse matrix of $D^2 u_t$.

Now we prove that the non-negative part $(\frac{\partial u_t}{\partial t})^+ \equiv 0$, where $(\frac{\partial u_t}{\partial t})^+ :=$ $\max\{0, \frac{\partial u_t}{\partial t}\}$.

In fact, if $(\frac{\partial u_t}{\partial t})^+ \not\equiv 0$, using $(\frac{\partial u_t}{\partial t})^+$ as a test function, then we obtain by integration by parts (note that u_t is convex and equal to 0 on the boundary, so $u_t < 0$ in Ω and $\frac{\partial u_t}{\partial t} \equiv 0$ on $\partial\Omega$, and here $t < t_0$, so by our assumption $t < \lambda_{1,t}$)

$$\int_\Omega g_t^{ij}\left(\frac{\partial u_t}{\partial t}\right)_i \left(\frac{\partial u_t}{\partial t}\right)_j^+ \det\big((u_t)_{ij}\big)\, dx$$

$$= -\int_\Omega g_t^{ij}\left(\frac{\partial u_t}{\partial t}\right)_{ij} \left(\frac{\partial u_t}{\partial t}\right)^+ \det\big((u_t)_{ij}\big)\, dx$$

$$= t\int_\Omega \left[\left(\frac{\partial u_t}{\partial t}\right)^+\right]^2 \det\big((u_t)_{ij}\big)\, dx + \int_\Omega \left(\frac{\partial u_t}{\partial t}\right)^+ u_t \det\big((u_t)_{ij}\big)\, dx$$

$$\leq t\int_\Omega \left[\left(\frac{\partial u_t}{\partial t}\right)^+\right]^2 \det\big((u_t)_{ij}\big)\, dx$$

$$< \lambda_{1,t}\int_\Omega \left[\left(\frac{\partial u_t}{\partial t}\right)^+\right]^2 \det\big((u_t)_{ij}\big)\, dx,$$

but by the definition of $\lambda_{1,t}$, it is impossible. So $(\frac{\partial u_t}{\partial t})^+ \equiv 0$. Thus we have $\frac{\partial u_t}{\partial t} \leq 0$. Then from (5.48) we have

$$-g_t^{ij}\left(\frac{\partial u_t}{\partial t}\right)_{ij} < 0 \quad \text{in } \Omega, \tag{5.49}$$

so we have $\frac{\partial u_t}{\partial t} < 0$ in Ω by the strong maximum principle (see Theorem 3.5 of [95]).

Now because $\frac{\partial u_t}{\partial t}$ is continuous in t, we have $\frac{\partial u_t}{\partial t}(t_0) \leq 0$. Then by the Hopf Lemma we have $\frac{\partial u_t}{\partial t}(t_0) < 0$ in Ω and $\frac{\partial u_t}{\partial t}(t_0)$ has no vanishing gradient on the boundary. Denote $v = -\frac{\partial u_t}{\partial t}(t_0)$, then by (5.48) we have

$$-g_{t_0}^{ij} v_{ij} > t_0 v. \tag{5.50}$$

Because $f = 0$ on $\partial\Omega$ and v has no vanishing gradient on the boundary, for any point $x \in \partial\Omega$, we can define $\frac{f(x)}{v(x)} := \frac{\partial f}{\partial v}/\frac{\partial v}{\partial v}$, where v denotes the exterior unit normal of $\partial\Omega$ at x. We find that $C = \sup_{x\in\Omega} \frac{f(x)}{v(x)}$ is a positive finite number, which is attained in $\overline{\Omega}$. Now combining (5.47) and (5.50) we get

$$-g_{t_0}^{ij}(f - Cv)_{ij} < t_0(f - Cv)$$

$$\leq 0.$$

By the definition of C, $f - Cv$ either has 0 as a maximum which is attained in the interior of Ω or has vanishing gradient at some point of $\partial\Omega$. So by the Hopf

Lemma (see Lemma 3.4 of [95]) we find that $f - Cv$ is a constant, that is, 0. Now substitute $\frac{\partial u_t}{\partial t}(t_0) = -C^{-1} f$ into (5.48), we get $u_{t_0} \equiv 0$, which contradicts $u_t < 0$ for $t \in [0, T_0)$ in Ω. So our assertion is proved. $\qquad\square$

Remark 5.4.1 We have another simple proof of this proposition, but the method in the proof above is more valuable because it can give us more information. In fact, First multiplying (5.48) by the positive first eigenfunction f and integrating by parts we get

$$\int_\Omega \left(t\frac{\partial u_t}{\partial t} + u_t \right) f \det\big((u_t)_{ij}\big) \, dx = -\int_\Omega g_t^{ij} \left(\frac{\partial u_t}{\partial t} \right)_{ij} f \det\big((u_t)_{ij}\big) \, dx$$

$$= -\int_\Omega g_t^{ij} f_{ij} \frac{\partial u_t}{\partial t} \det\big((u_t)_{ij}\big) \, dx$$

$$= \lambda_{1,t} \int_\Omega f \frac{\partial u_t}{\partial t} \det\big((u_t)_{ij}\big) \, dx.$$

So if $\lambda_{1,t} = t$, we must have $\int_\Omega u_t f \det((u_t)_{ij}) \, dx = 0$, which is impossible because in Ω we have $u_t < 0$ and $f > 0$.

From the proof (see the statement below (5.49)) we also know u_t is decreasing in t. Now with this estimate we can extend the C^1 map u_t to be defined on a maximal interval $[0, T), T > 0$. By Theorem 5.3.2 and rescaling we know that (5.36) has no solution for $t > (\lambda^*)^2$ where λ^* is the critical value as in Theorem 5.3.2, so $0 < T < +\infty$. We conclude that either

(i) u_t converges to $-\infty$ as t approaches T, or
(ii) u_t converges decreasingly to some convex function u_T as t approaches T (we can prove u_t converges to u_T in $C^2(\overline{\Omega})$, using the higher order estimate, see the proof of Lemma 5.3.1), which is a solution of (5.36) at $t = T$, but u_t is not left differentiable in t at T, that is, the solution u_T is degenerate.

In fact the case (i) cannot happen on any smooth convex domain Ω, because we have an a priori estimate:

Lemma 5.4.4 (Zhang and Wang [207]) *Given a positive constant $t_0 > 0$, any solution of (5.36) with $t > t_0$, must satisfy $\sup_\Omega |u| \le C(t_0)$ for some constant $C(t_0)$ which depends on t_0 and Ω only.*

Proof Denote $M := \sup_\Omega |u|$. From the arithmetic-geometric mean value inequality (see Theorem 6.6.9, p. 154, [190]) we have

$$\frac{1}{n}\Delta u \ge \big(\det(u_{ij})\big)^{\frac{1}{n}}$$

$$= e^{-\frac{tu}{n}}.$$

Moreover, there exists a Dirichlet Green function $G(x, y) > 0$ in Ω such that by the Green formula we have

$$u(x) = -\int_\Omega G(x, y)\Delta u(y)\, dy$$

$$\leq -n\int_\Omega G(x, y)e^{-\frac{tu(y)}{n}}\, dy.$$

Noticing that there exists a unique positive function $\phi(x)$ satisfying $-\Delta\phi = 1$ in Ω with vanishing boundary value, we get (integrating the inequality above in Ω)

$$\mathrm{meas}(\Omega)M \geq -\int_\Omega u(x)\, dx$$

$$\geq n\int_\Omega\int_\Omega G(x, y)e^{-\frac{tu(y)}{n}}\, dy\, dx$$

$$= n\int_\Omega e^{-\frac{tu(y)}{n}}\int_\Omega G(x, y)\, dx\, dy$$

$$= n\int_\Omega e^{-\frac{tu(y)}{n}}\phi(y)\, dy.$$

Now assume x_0 is the minimal point of u, that is, $u(x_0) = -M$. Without loss of generality, we can assume $x_0 = 0$ by translation. We define a function ψ to be a cone over Ω, that is, $\psi(0) = -M$, $\psi = 0$ on $\partial\Omega$ and $\psi(x) = -(1 - t)M$ where t is characterized uniquely by $\frac{x}{t} \in \partial\Omega$. Because u is convex with vanishing boundary value, we have $\psi(x) \geq u(x)$.

Now $A := \{x : \psi(x) \leq -\frac{M}{2}\} = \frac{1}{2}\Omega$. Take a small positive constant ϵ such that with $\Omega_\epsilon := \{x : \phi(x) \leq \epsilon\}$, we have $\mathrm{meas}(\Omega_\epsilon) \leq 4^{-n}\,\mathrm{meas}(\Omega)$ and

$$\int_\Omega e^{-\frac{tu(y)}{n}}\phi(y)\, dy \geq \int_{(\Omega\cap A)\backslash\Omega_\epsilon} e^{-\frac{tu(y)}{n}}\phi(y)\, dy$$

$$\geq \epsilon e^{\frac{tM}{2n}}\,\mathrm{meas}\big((\Omega\cap A)\backslash\Omega_\epsilon\big)$$

$$\geq \epsilon e^{\frac{tM}{2n}}\big[\mathrm{meas}(\Omega) - \big(1 - 2^{-n}\big)\mathrm{meas}(\Omega) - \mathrm{meas}(\Omega_\epsilon)\big]$$

$$\geq Ce^{\frac{tM}{2n}},$$

where C is a constant only depending on Ω. Now combining the two inequalities above we get

$$M \geq Ce^{\frac{tM}{2n}}$$

$$\geq Ce^{\frac{t_0 M}{2n}}.$$

This implies a uniform bound $C(t_0)$ such that $M \leq C(t_0)$, here $C(t_0)$ depends on t_0 and Ω only. $\square$

This lemma also guarantees that the branch extending from $t = 0$ always stays in U and the branch has some compactness property.

We can use the estimation of the first eigenvalue $\lambda_{1,t} > t$ to get some more results:

Proposition 5.4.2 (Zhang–Wang [207]) *If there exists another solution v of* (5.36) *for $t \in [0, T)$, then $v < u_t$ (here u_t is the solution in the branch extended from $t = 0$). Moreover, the first eigenvalue of the operator $Lf := -g^{ij} f_{ij}$ (here g^{ij} is the inverse matrix of $D^2 v$) is smaller than t.*

Proof From the concavity of the $\log \det$ function (see Proposition 5.5.3 of the appendix), we have

$$\log \det\big(\big[\tau v + (1-\tau)u_t\big]_{ij}\big) \geq \tau \log \det(v_{ij}) + (1-\tau) \log \det\big((u_t)_{ij}\big)$$

$$= -\tau t v - (1-\tau) t u_t,$$

for any $\tau \in [0, 1]$ at any fixed point $x \in \Omega$, which becomes an equality as $\tau = 0$ or $\tau = 1$. The left-hand side is a concave function $h(\tau)$, $\tau \in [0, 1]$. So we can compare the derivative in τ of both sides at $\tau = 0$ and obtain

$$-g_t^{ij}(v - u_t)_{ij} \leq tv - tu_t, \tag{5.51}$$

here (g_t^{ij}) is the inverse matrix of $((u_t)_{ij})$. Comparing the derivative in τ of both sides at $\tau = 1$ we obtain

$$-g^{ij}(v - u_t)_{ij} \geq tv - tu_t, \tag{5.52}$$

here (g^{ij}) is the inverse matrix of (v_{ij}). Here inequalities (5.51) or (5.52) become equalities if and only if the concave function $h''(\tau) \equiv 0$, that is, the matrix $((v - u_t)_{ij})(x) = 0$ (see Proposition 5.5.3 of the appendix).

Noticing $\lambda_{1,t} > t$, we can proceed as in the proof of Proposition 4.4 (using an integration by parts) to prove that $v \leq u_t$.

We assume by contradiction that $(v - u_t)^+ \not\equiv 0$, then we have

$$\int_\Omega g_t^{ij}(v - u_t)_i (v - u_t)_j^+ \det\big((u_t)_{ij}\big)\, dx$$

$$= -\int_\Omega g_t^{ij}(v - u_t)_{ij}(v - u_t)^+ \det\big((u_t)_{ij}\big)\, dx$$

$$\leq t \int_\Omega \big[(v - u_t)^+\big]^2 \det\big((u_t)_{ij}\big)\, dx$$

$$< \lambda_{1,t} \int_\Omega \big[(v - u_t)^+\big]^2 \det\big((u_t)_{ij}\big)\, dx,$$

which contradicts the definition of $\lambda_{1,t}$, so $(v - u_t)^+ \equiv 0$, that is, $v \leq u_t$. Now we have

$$\det\big((u_t)_{ij}\big) = e^{-tu_t}$$

$$\leq e^{-tv}$$

$$= \det(v_{ij}).$$

Write this in another form so that we can use the strong maximum principle (see Theorem 3.5 of [95]):

$$0 \leq \log\det(v_{ij}) - \log\det\big((u_t)_{ij}\big)$$

$$= \int_0^1 \frac{d}{d\tau} \log\det\big(\tau v_{ij} + (1 - \tau)(u_t)_{ij}\big)\,d\tau$$

$$= \left[\int_0^1 g_\tau^{ij}\,d\tau\right](v - u_t)_{ij},$$

here (g_τ^{ij}) is the inverse matrix of $(\tau v_{ij} + (1 - \tau)(u_t)_{ij})$. Now we can conclude that either $v \equiv u_t$ or $v < u_t$ in Ω by the strong maximum principle (see Theorem 3.5 of [95]).

In fact by the Hopf Lemma (see Lemma 3.4 of [95]) we have $v - u_t < 0$ in Ω and $v - u_t$ has no vanishing gradient on the boundary. Denote $\varphi := -v + u_t$, then by (5.52) we have

$$-g^{ij}\varphi_{ij} \leq t\varphi. \tag{5.53}$$

Now assume the first eigenvalue of the operator L: $\lambda_1 \geq t$ with a positive eigenfunction f. In fact we cannot have $\lambda_1 > t$, because otherwise we can proceed as the proof above to find that $v \geq u_t$ (note that above we just use the first eigenvalue to prove that $u_t \geq v$). So with our assumption we must have $\lambda_1 = t$. Because $\varphi = 0$ on $\partial\Omega$ and f has no vanishing gradient on the boundary, for any point $x \in \partial\Omega$, we can define $\frac{\varphi(x)}{f(x)} := \frac{\partial\varphi}{\partial\nu} / \frac{\partial f}{\partial\nu}$, where ν denotes the exterior unit normal of $\partial\Omega$ at x. We find that $C = \sup_{x\in\Omega} \frac{\varphi(x)}{f(x)}$ is a positive finite number, which is attained in $\overline{\Omega}$. Now from (5.53) and $-g^{ij} f_{ij} = tf$ we have

$$-g^{ij}(\varphi - Cf)_{ij} \leq t\varphi - tCf$$

$$\leq 0.$$

By the definition of C, $\varphi - Cf$ either has 0 as a maximum which is attained in the interior of Ω or has vanishing gradient at some point of $\partial\Omega$. So by the Hopf Lemma (see Lemma 3.4 of [95]) we see that $\varphi - Cf$ is a constant, that is, 0. Now (5.53), hence (5.52), becomes an equality for any $x \in \Omega$. So we must have $D^2 v = D^2 u_t$, which implies $v \equiv u_t$ by the boundary condition. This is a contradiction and our assertion is proved. $\qquad\square$

The method in the proof of the first part of Proposition 5.4.2 can also be used to prove the following result:

Proposition 5.4.3 (Zhang and Wang [207]) *If there is a solution v of (5.36) for $t > T$, then we have $v < u_s$ for any $s \in [0, T)$.*

Proof For any $s < T$, from the concavity of the log det function (see Appendix 5.5), we have

$$\log \det\big(\big[\tau v + (1 - \tau)u_s\big]_{ij}\big) \geq \tau \log \det(v_{ij}) + (1 - \tau) \log \det\big((u_s)_{ij}\big)$$

$$= -\tau t v - (1 - \tau)s u_s,$$

for any $\tau \in [0, 1]$, which becomes an equality as $\tau = 0$ or $\tau = 1$. So we can compare the derivative in τ of both sides at $\tau = 0$ and obtain

$$-g_s^{ij}(v - u_s)_{ij} \leq t v - s u_s$$

$$\leq s(v - u_s).$$

The last inequality is because $t > s$ and $v \leq 0$. Here g_s^{ij} is the inverse matrix of $((u_s)_{ij})$.

Because $\lambda_{1,s} > s$, we can proceed as in the proof of Proposition 4.7 (using an integration by parts) to prove that $v \leq u_s$.

We assume by contradiction that $(v - u_s)^+ \not\equiv 0$, then we have

$$\int_\Omega g_s^{ij}(v - u_s)_i(v - u_s)_j^+ \det\big((u_s)_{ij}\big)\, dx$$

$$= -\int_\Omega g_s^{ij}(v - u_s)_{ij}(v - u_s)^+ \det\big((u_s)_{ij}\big)\, dx$$

$$\leq s \int_\Omega \big[(v - u_s)^+\big]^2 \det\big((u_s)_{ij}\big)\, dx$$

$$< \lambda_{1,s} \int_\Omega \big[(v - u_s)^+\big]^2 \det\big((u_s)_{ij}\big)\, dx,$$

which contradicts the definition of $\lambda_{1,s}$, so $(v - u_s)^+ \equiv 0$, that is, $v \leq u_s$. Thus

$$\det\big((u_s)_{ij}\big) = e^{-s u_s}$$

$$\leq e^{-t v}$$

$$= \det(v_{ij}).$$

We can proceed as the proof of Proposition 5.4.2 to conclude that $v < u_s$ in Ω.   □

Next we study the structure of the branch near the degenerate point. We have a smooth and strictly convex function u_T satisfying the following equation (for convenience, in the following we use u instead of u_T):

$$\begin{cases} \det(u_{ij}) = e^{-Tu} & \text{in } \Omega, \\ u = 0 & \text{on } \partial\Omega, \\ \lambda_1 = T, \end{cases} \tag{5.54}$$

where λ_1 is the first eigenvalue of the operator $Lf := -g^{ij} f_{ij}$, g^{ij} is the inverse matrix of $D^2 u$. We also have a positive eigenfunction f corresponding to λ_1.

We obtain the following result about the local structure of the degenerate point of (5.36).

Theorem 5.4.1 (Zhang and Wang [207]) *For u satisfying* (5.54), *we have a unique family $u_s = u + sf + o(s)$ near u, satisfying*

$$\det\big((u_s)_{ij}\big) = e^{-(T+r(s))u(s)}, \tag{5.55}$$

where $r(s)$ is a continuously differentiable function defined in a small open neighborhood of 0 in $\mathbb{R}^1$. Moreover, $r(0) = 0$, $r(s) \leq 0$, u_s is increasing in s. That implies near T and u for $t < T$ there are two solutions of (5.36) *and no solutions for $t > T$.*

Proof In this case the Implicit Function theorem is invalid. We need use the Lyapunov–Schmidt reduction method (for the general introduction please see [22]). X can be split as a direct sum span$\{f\} \oplus W$, where $W = \{v : v \in X, \int_\Omega fv \det(u_{ij})\, dx = 0\}$. Y has a similar decomposition $Y = \text{span}\{f\} \oplus Z$, where $Z = \{v : v \in Y, \int_\Omega fv \det(u_{ij})\, dx = 0\}$.

We write the map F of (5.37) near u for $(r, s, w) \in \mathbb{R}^1 \times \mathbb{R}^1 \times W$ as

$$F(r, s, w) = \log \det(u + sf + w)_{ij} + (T + r)(u + sf + w), \tag{5.56}$$

this map is well defined in a small neighborhood of $(0, 0, 0)$. Now the equation $F(r, s, w) = 0$ can be written as

$$\begin{cases} P_1\big[\log \det(u + sf + w)_{ij} + (T + r)(u + sf + w)\big] = 0, \\ P_2\big[\log \det(u + sf + w)_{ij} + (T + r)(u + sf + w)\big] = 0, \end{cases} \tag{5.57}$$

here P_1 is the projection from Y to span$\{f\}$ and P_2 is the projection from Y to Z. The derivative

$$D_w(P_2 F)(0, 0, 0)v = g^{ij} v_{ij} + Tv, \quad \forall v \in W \tag{5.58}$$

is a linear operator from W to Z with a bounded inverse operator.

Then by the Implicit Function Theorem the second equation of (5.54) can be solved, that is, there exists a continuously differentiable map w from a neighborhood of $(0, 0)$ in $\mathbb{R}^2$ to a neighborhood of 0 in W such that $w(0, 0) = 0$ and

$$F\big(r, s, w(r, s)\big) = \lambda(r, s)f, \tag{5.59}$$

for some continuously differentiable function $\lambda(r, s)$ defined in a neighborhood of $(0, 0)$ in $\mathbb{R}^2$. For w in this neighborhood of 0 in W, $F(r, s, w) = 0$ is equivalent to $w = w(r, s)$ for some (r, s) and $\lambda(r, s) = 0$.

Now let us look at the structure of $\lambda(r, s) = 0$. First, differentiating (5.59) with r at $(0, 0)$ we obtain

$$\frac{\partial \lambda(r, s)}{\partial r} f = \frac{\partial F}{\partial r} + D_w F\left(\frac{\partial w}{\partial r}\right)$$

$$= u + g^{ij}\left(\frac{\partial w}{\partial r}\right)_{ij} + T\frac{\partial w}{\partial r}.$$

Multiply both sides by f and integrate by parts to get at $(0, 0)$ (note $g^{ij} f_{ij} + Tf = 0$)

$$\frac{\partial \lambda}{\partial r}(0, 0) \int_\Omega f^2 \det(u_{ij})\, dx$$

$$= \int_\Omega fu \det(u_{ij})\, dx + \int_\Omega \left[g^{ij}\left(\frac{\partial w}{\partial r}\right)_{ij} + T\frac{\partial w}{\partial r}\right] f \det(u_{ij})\, dx$$

$$= \int_\Omega fu \det(u_{ij})\, dx + \int_\Omega [g^{ij} f_{ij} + Tf]\frac{\partial w}{\partial r} \det(u_{ij})\, dx$$

$$= \int_\Omega fu \det(u_{ij})\, dx$$

$$< 0.$$

So at $(0, 0)$ we get

$$\frac{\partial \lambda}{\partial r} < 0. \tag{5.60}$$

Similarly we obtain the formula for $\frac{\partial \lambda(r,s)}{\partial s}$:

$$\frac{\partial \lambda}{\partial s}(0, 0) f = \frac{\partial F}{\partial s} + D_w F\left(\frac{\partial w}{\partial s}\right)$$

$$= g^{ij} f_{ij} + Tf + g^{ij}\left(\frac{\partial w}{\partial s}\right)_{ij} + T\frac{\partial w}{\partial s}$$

$$= g^{ij}\left(\frac{\partial w}{\partial s}\right)_{ij} + T\frac{\partial w}{\partial s}.$$

Multiply both sides by f and integrate by parts to get at $(0, 0)$ (note $g^{ij} f_{ij} + Tf = 0$)

$$\frac{\partial \lambda(r, s)}{\partial s} \int_\Omega f^2 \det(u_{ij})\, dx = \int_\Omega \left[g^{ij}\left(\frac{\partial w}{\partial s}\right)_{ij} + T\frac{\partial w}{\partial s}\right] f \det(u_{ij})\, dx$$

$$= \int_\Omega [g^{ij} f_{ij} + Tf]\frac{\partial w}{\partial s} \det(u_{ij})\, dx$$

$$= 0.$$

So at $(0, 0)$ we have

$$\frac{\partial \lambda}{\partial s} = 0. \tag{5.61}$$

Furthermore, this implies at the origin

$$g^{ij}\left(\frac{\partial w}{\partial s}\right)_{ij} + T\frac{\partial w}{\partial s} = 0. \tag{5.62}$$

Since T is the first eigenvalue and the first eigenfunction is unique modulo a constant, we have a constant c such that $\frac{\partial w}{\partial s}(0, 0) = cf$. However, by the definition of W we also have $\int_\Omega w(r, s) f \det(u_{ij})\, dx = 0$, which implies $\int_\Omega \frac{\partial w}{\partial s}(0, 0) f \det(u_{ij})\, dx = 0$. So $c = 0$ and $\frac{\partial w}{\partial s}(0, 0) \equiv 0$.

Now we know that there exist a small open neighborhood of 0 in $\mathbb{R}^1$ and an unique continuously differentiable function $r(s)$ defined in it such that $\lambda(r(s), s) = 0$ and $r(0) = 0$.

Differentiating $\lambda(r(s), s) = 0$ with s, we get

$$\frac{\partial \lambda}{\partial r}\frac{\partial r}{\partial s} + \frac{\partial \lambda}{\partial s} = 0, \tag{5.63}$$

which implies

$$\frac{\partial r}{\partial s}(0) = 0.$$

Differentiate (5.63) with s again

$$\frac{\partial^2 \lambda}{\partial^2 r}\left(\frac{\partial r}{\partial s}\right)^2 + 2\frac{\partial^2 \lambda}{\partial r \partial s}\frac{\partial r}{\partial s} + \frac{\partial \lambda}{\partial r}\frac{\partial^2 r}{\partial^2 s} + \frac{\partial^2 \lambda}{\partial^2 s} = 0. \tag{5.64}$$

Taking values at $(0, 0)$, we have

$$\frac{\partial \lambda}{\partial r}(0, 0)\frac{\partial^2 r}{\partial^2 s}(0) + \frac{\partial^2 \lambda}{\partial^2 s}(0, 0) = 0. \tag{5.65}$$

Next we want to calculate $\frac{\partial^2 \lambda}{\partial^2 s}(0, 0)$. We have by differentiating (5.59) in s twice

$$\frac{\partial^2 \lambda}{\partial^2 s} f = \frac{\partial^2 F}{\partial^2 s} + D_{ww}F\left(\frac{\partial w}{\partial s}, \frac{\partial w}{\partial s}\right) + D_w F\left(\frac{\partial^2 w}{\partial^2 s}\right) + \left(\frac{\partial}{\partial s}D_w F\right)\left(\frac{\partial w}{\partial s}\right)$$
$$+ D_w\left(\frac{\partial}{\partial s}F\right)\left(\frac{\partial w}{\partial s}\right). \tag{5.66}$$

The first term can be calculated by taking the second derivative along the line $(0, s, 0)$:

$$\frac{\partial^2 F}{\partial^2 s}(0,0,0) = \frac{d^2}{d^2 s}\big[\log\det(u+sf)_{ij} + T(u+sf)\big]\Big|_{s=0}$$

$$= -g^{iq}g^{pj}f_{pq}f_{ij}$$

$$< 0.$$

The second term, the fourth term and the last term are 0 at the origin because $\frac{\partial w}{\partial s}(0,0) \equiv 0$. The third term is

$$D_w F\left(\frac{\partial^2 w}{\partial^2 s}\right)(0,0) = g^{ij}\left(\frac{\partial^2 w}{\partial^2 s}\right)_{ij} + T\left(\frac{\partial^2 w}{\partial^2 s}\right). \tag{5.67}$$

Now we can multiply (5.66) with f and integrate (note that by (5.67) and integration by parts the third term has no contribution) to obtain $\frac{\partial^2 \lambda}{\partial^2 s}(0,0) < 0$. So from (5.65) we have $\frac{\partial^2 r}{\partial^2 s}(0) < 0$, which in particular implies $r(s) < 0$ in a neighborhood of 0.

Moreover, if we define $u_s = u + sf + w(s, r(s))$, we have

$$\det(u_s)_{ij} = e^{-(T+r(s))u_s}. \tag{5.68}$$

But

$$\frac{\partial}{\partial s}w\big(s, r(s)\big)\Big|_{s=0} = \frac{\partial w}{\partial s}(0,0) + \frac{\partial w}{\partial r}(0,0)\frac{\partial r}{\partial s}(0)$$

$$= 0,$$

and we have $f > 0$, so at least for s small $u + sf + w(s, r(s)) = u + s(f + \frac{w(s,r(s))}{s})$ is increasing with respect to s. So near u there exist two family solutions of equation (5.36) for those $t < T$ and no solution for $t > T$, that is, the branch u_t turns to the left at $t = T$. $\qquad\square$

Now we study the global structure of the branch, using the Leray–Schauder degree theory (see [50]). Define the space $E := C_0^2(\overline{\Omega})$ (that is, C^2 function on $\overline{\Omega}$ with vanishing boundary value) and a map $K : E \to E$ which is uniquely defined by $\log\det D^2 K(f) = f$. We know that for each $f \in E$ there exists a unique $K(f) \in C^3(\overline{\Omega})$, which depends on f continuously in E. Moreover, we have $\sup_{x\in\Omega}(|K(f)| + |DK(f)| + |D^2 K(f)| + |D^3 K(f)|)(x) \le C$, where C is a constant depending on $|f|_{C^2}$ and Ω (see Theorem 9.2, Theorem 17.21, Theorem 17.20, Theorem 17.26 in [95]). So $K : E \to E$ is a continuous compact map. With these definitions, we can write equation (5.36) in the following form (note if we define $f := \log\det D^2 u$, then $u = K(f)$):

$$T_t(f) = 0, \tag{5.69}$$

where $T_t(f) = f + tK(f)$.

Define $\Sigma := \{(f, t) \in E \times [0, +\infty) : T_t f = 0\}$, we know that $(0, 0) \in \Sigma$ (note that such t is unique for f if it exists). Since K is a continuous compact map, we know that Σ is a closed locally compact set. In fact, for any bounded set $B \subset E \times [0, +\infty)$, $B \cap \Sigma$ is a compact set.

Define Σ' to be the connected component of Σ containing $(0, 0)$ and by the following Theorem 3.5.3 (Leray–Schauder) of [50] we have

Lemma 5.4.5 (Zhang and Wang [207]) *Σ' is an unbounded set in $E \times \mathbb{R}$.*

For convenience, we give Theorem 3.5.3 here.

Theorem 3.5.3 of [50]: Let X be a real Banach space, $T : X \times \mathbb{R} \to X$ be a compact map satisfying $T(x, 0) = \theta$, and $f(x, \lambda) = x - T(x, \lambda)$. Let $S = \{(x, \lambda) \in X \times \mathbb{R} | f(x, \lambda) = \theta\}$, and let ζ be the component of S passing through $(\theta, 0)$. If $\zeta^{\pm} = \zeta \cap (X \times \mathbb{R}_{\pm})$, then both ζ^{+} and ζ^{-} are unbounded.

By Theorem 5.3.2 we can define $T^* := \sup\{t > 0 : \exists (f, t) \in \Sigma', T_t f = 0\}$, which is a finite positive number. (In fact $T^* = (\lambda^*)^2$ by the following Theorem 5.4.2, where λ^* is the critical value in Theorem 5.3.2.) The following Theorem 5.4.2 also implies that the branch Σ' extends to the maximum t such that (5.69) (or (5.36)) has a solution. We know that from the local compactness and closed property of Σ there exists a solution of (5.69) at T^*.

Theorem 5.4.2 (Zhang and Wang [207]) *For $t > T^*$, (5.69) (or (5.36)) has no solution.*

Proof Assume there exists a solution v of (5.36) for some $t_0 > T^*$.

Define $\Sigma'' := \{(f, t) : (f, t) \in \Sigma' \text{ and } K(f) \geq v\}$, here $K(f) = u$ is the solution of (5.36) by the definition of $T_t(f)$.

We want to prove that Σ'' is nonempty, open and closed relatively to Σ', so we can get $\Sigma'' = \Sigma'$. If this is proved, then there exists a sequence (f_k, t_k) in Σ' such that $K(f_k)$ diverges to $-\infty$ in $E = C_0^2(\overline{\Omega})$, we have $\inf_\Omega K(f_k)$ diverges to $-\infty$ too, because otherwise we will have a uniform bound for $K(f_k)$ in $C_0^2(\overline{\Omega})$ norm by an a priori estimate. Thus, we must have $\inf_{x \in \Omega} v(x) = -\infty$, which is a contradiction.

First, by Proposition 5.4.3 or direct comparison we have $(0, 0) \in \Sigma''$, so Σ'' is nonempty.

The closeness of Σ'' is obvious by the continuity of the operator K.

Next if $(f_0, s_0) \in \Sigma''$ for some $s_0 \leq T^*$, let $K(f_0) = w$, we get $\det D^2 w = e^{-s_0 w}$ and $w \geq v$, then we have

$$\det D^2 w = e^{-s_0 w}$$

$$\leq e^{-t_0 v}$$

$$= \det D^2 v,$$

by $0 \geq w \geq v$ and $t_0 \geq s_0$. Then by the strong maximum principle, unless $w \equiv v$ (which is impossible here because $t_0 > s_0$), we must have $w > v$ strictly in Ω, and

$\frac{\partial w}{\partial n} < \frac{\partial v}{\partial n}$ on $\partial\Omega$, where n is the outer normal vector of $\partial\Omega$ (see the proof of Proposition 4.7).

We can choose a small open neighborhood $B \subset E \times \mathbb{R}$ of (f_0, s_0) such that for any $(f, s) \in B \cap \Sigma'$ we have $|s - s_0| < \epsilon$, $\sup_{x\in\overline{\Omega}}(|w - u| + |D(w - u)| + |D^2(w - u)|)(x) < \epsilon$ for $u = K(f)$ and for some $\epsilon > 0$ by the continuity of K. Moreover, we have $u \geq v$ in $\Omega_\epsilon := \{x : w(x) \geq v(x) + \varepsilon\}$. While for ϵ small enough, in $\Omega \setminus \Omega_\epsilon$ we have

$$
\begin{aligned}
u(x) &\geq u(\pi x) - \frac{\partial u}{\partial n}(\pi x)|x - \pi x| \\
&\geq -\left(\frac{\partial w}{\partial n}(\pi x) + \epsilon\right)|x - \pi x| \\
&\geq -(1 - \epsilon)\frac{\partial v}{\partial n}(\pi x)|x - \pi x| \\
&\geq v(x),
\end{aligned}
$$

here πx is the projection of x onto $\partial\Omega$, which is a smooth map near $\partial\Omega$. We also use the fact that u, v and w are convex function with vanishing boundary value. So Σ'' is open relatively to Σ'. $\qquad\square$

Now we know by Lemma 5.4.5 and Theorem 5.4.2 that Σ' starts from $t = 0$ and reaches T^*, and at last diverges to infinity as and only as t approaches 0 (by Lemma 4.6, we know that if a sequence $(f_k, t_k) \in \Sigma'$ satisfying $T_{t_k} f_k = 0$ tends to infinity in $E \times \mathbb{R}$ (by an a priori estimate, it also tends to infinity in $C(\bar{\Omega}) \times \mathbb{R}$), then t_k tends to 0). So for $0 < t < T^*$, now we prove the existence of at least two solutions for (5.69) (or (5.36)):

Theorem 5.4.3 (Zhang and Wang [207]) *For $t \in (0, T^*)$, (5.69) (or (5.36)) has at least two solutions.*

Proof Let

$$
T_1 := \sup\{t_1 > 0 : \forall 0 < t \leq t_1, \exists f_1 \neq f_2 \text{ such that } (f_1, t) \in \Sigma', (f_2, t) \in \Sigma'\}.
$$

We know that $T_1 \geq T > 0$, where T is the first degenerate point of the branch emanating from $t = 0$ (introduced before Lemma 5.4.4). We prove that $T_1 = T^*$.

Assuming that $T_1 < T^*$, take a $T_2 \in [T_1, T^*)$ such that there exists a unique f such that $t(f) = T_2, (f, t(f)) \in \Sigma'$. We define

$$
\Sigma_1 := \{(f, t(f)) \in \Sigma' : t(f) \leq T_2\}
$$

Since $t(f)$ depends continuously on f, Σ_1 is a closed subset of Σ'. Now by the following Lemma 5.4.6 we find that Σ_1 is connected. We can repeat the proof of Theorem 5.4.2 to prove that there exists no solution of (5.36) for $t > T_2$, which is a contradiction with Σ' reaches $T^* > T_2$, so our claim follows. $\qquad\square$

Lemma 5.4.6 (Zhang–Wang [207]) Σ_1 *is connected.*

Proof If Σ_1 is not connected, there exist two disjoint closed subset K_1 K_2 of Σ' such that $\Sigma_1 = K_1 \cup K_2$

Because for T_2 there exists a unique f such that $t(f) = T_2, (f, T_2) \in \Sigma'$, and without loss of generality, we can assume $(f, T_2) \in K_1$, then $\forall(f', t(f')) \in K_2$ we have $t(f') < T_2$ strictly. By Lemma 5.4.4 we know that $\{(f', t(f')) \in K_2 : t(f') \geq \frac{T_1}{2}\}$ is a compact set. So by the compactness, we find that there exists $\delta > 0$ such that $t(f') \leq T_2 - \delta, \forall(f', t(f')) \in K_2$.

Now the set $K_3 := \{(f, t(f)) \in \Sigma' : t(f) \geq T_2\}$ is also a compact set. From the above discussion we know that $K_3 \cap K_2 = \emptyset$. So we have $\Sigma' = (K_1 \cup K_3) \cup K_2$, and $(K_1 \cup K_3)$ and K_2 are disjoint closed sets. This contradicts with the connectedness of Σ'. $\qquad\square$

Next we want to investigate the exact number of the solutions at $t = T^*$. First, we give a lemma, which is kind of the inverse of Proposition 5.4.2.

Lemma 5.4.7 (Zhang–Wang [207]) *Given $t > 0$, if there exist two solutions u and v of (5.36) with $u \geq v$, then the first eigenvalue of the operator $Lf := -g^{ij} f_{ij}$ satisfies $\lambda_1 > t$, where (g^{ij}) is the inverse matrix of (u_{ij}).*

Proof Using the comparison principle as before, we know that $u > v$ strictly in the interior of Ω and $\frac{\partial u}{\partial \nu} < \frac{\partial v}{\partial \nu}$ on $\partial\Omega$.

From the concavity of the log det function (see Proposition 5.5.3), we have

$$\log \det\left(\left[\tau v + (1 - \tau)u\right]_{ij}\right) \geq \tau \log \det(v_{ij}) + (1 - \tau) \log \det(u_{ij})$$

$$= -\tau t v - (1 - \tau)tu,$$

for any $\tau \in [0, 1]$, which becomes an equality as $\tau = 0$ or $\tau = 1$. So we can compare the derivative in τ of both sides at $\tau = 0$ and obtain

$$g^{ij}(v - u)_{ij} \geq -tv + tu, \tag{5.70}$$

where (g^{ij}) is the inverse matrix of (u_{ij}).

Denote $h = u - v$, which is a positive function and satisfies

$$-g^{ij} h_{ij} \geq th. \tag{5.71}$$

We also know that the first eigenvalue λ_1 of the operator $Lf := -g^{ij} f_{ij}$ has a positive eigenfunction f, that is,

$$-g^{ij} f_{ij} = \lambda_1 f. \tag{5.72}$$

Now we use Bakelman's formula (5.40) to integrate by parts

$$t \int_\Omega hf \det(u_{ij})\, dx \leq - \int_\Omega g^{ij} h_{ij} f \det(u_{ij})\, dx$$

$$= - \int_\Omega g^{ij} f_{ij} h \det(u_{ij}) \, dx$$

$$= \lambda_1 \int_\Omega h f \det(u_{ij}) \, dx.$$

Because the integrand is positive, we get $\lambda_1 \geq t$.

Moreover, if $\lambda_1 = t$, the inequality above becomes an equality, so from (5.71) we must have

$$-g^{ij} h_{ij} = th. \tag{5.73}$$

Noticing that

$$K(\tau) := \log \det\big([\tau v + (1 - \tau)u]_{ij}\big) - \tau \log \det(v_{ij}) - (1 - \tau) \log \det(u_{ij})$$

is concave for $\tau \in [0, 1]$ too by Proposition 5.5.3 in the appendix and its proof, and moreover that $K(0) = K(1) = 0$ and that (5.73) means that $K'(0) = 0$, we find

$$\log \det\big([\tau v + (1 - \tau)u]_{ij}\big) = \tau \log \det(v_{ij}) + (1 - \tau) \log \det(u_{ij}). \tag{5.74}$$

By the strict concavity of the function $\log \det$ (see Proposition 5.5.3), this means we must have $D^2 u \equiv D^2 v$ in Ω. Thus $u \equiv v$, which contradicts our assumption, so we must have $\lambda_1 > t$. $\qquad\square$

Theorem 5.4.4 (Zhang and Wang [207]) *For $t = T^*$, (5.36) has exactly one solution. Moreover, we have $T^* = T$, where T is the first degenerate point of the branch emanating from $t = 0$, introduced just before Lemma 5.4.4.*

Proof Concerning the case $t = T$, the existence of a solution u_T is discussed before Lemma 5.4.4. Moreover, from Proposition 5.4.2, we know that this u_T is the maximal solution, that is, if there exists another solution v such that

$$\det(v_{ij}) = e^{-Tv}, \qquad v|_{\partial\Omega} = 0, \tag{5.75}$$

then $u_T \geq v$. Noticing that the first eigenvalue λ_T associated with u_T equals T, we must have $u_T \equiv v$, in view of Lemma 5.4.7. That is, the solution (5.36) at $t = T$ is unique.

Of course $T \leq T^*$, then combining Theorem 5.4.3 and the above discussion, we get $T^* = T$. $\qquad\square$

Now Theorem 5.1.1 is proved by the above Theorems 5.4.2, 5.4.3, and 5.4.4, and (5.2) is equivalent to (5.36) through a scaling (see the discussion before the Example 5.3.3).

Remark 5.4.2 We assume

(k1) $k(0) = 0$;

(k2) $k'(u) \geq c$ for some positive constant c, in particular, $k(u) \leq cu$ for $u < 0$ and
k is increasing.

Here, the condition (k2) is essential in our proof, but the condition (k1) can always be satisfied by adding a constant and multiplying u by an appropriate constant. However, Propositions 5.4.2, 5.4.3 and Theorem 5.4.1 (which is not needed in proving the main Theorem 5.1.1) need some more conditions:

(k3) k is concave, or $k''(u) \leq 0$.

Moreover, Lemma 5.4.4 is true for (5.3), and the proof of Theorems 5.4.2 and 5.4.3 can be easily modified in the case of (5.3). For example, in Theorem 5.3.1, (5.24) can be modified into

$$\frac{d}{dr} e^{\frac{k(u(r))}{n}} \geq \frac{1}{n} e^{\frac{k(u(r))}{n}} k'\big(u(r)\big) u'(r)$$

$$\geq \frac{c}{n} r,$$

where the constant c is as in the condition (k2). The calculation below (5.24) is almost the same.

Remark 5.4.3 For those $t \in (0, T)$ (here T is the first degenerate point of the branch emanating from $t = 0$, introduced above Lemma 5.4.4; from the discussion above, we have $T = T^*$), noticing Lemma 5.4.4, there exists another method to prove the existence of the second solution by mountain pass lemma (see [182]).

First, we know (see Bakelman [22]) critical points of the functional defined on the space of smooth convex functions:

$$I(u) := - \int_\Omega \left(\frac{1}{n+1} u \det D^2 u + \frac{1}{t} e^{-tu} \right) dx \tag{5.76}$$

are weak solutions of (5.36).

For each $t \in (0, T)$ we have constructed a solution u_t. Moreover, we have $\lambda_{1,t} > t$. This implies u_t is a local minimizer of I. We also have for fixed $u(x) < 0$

$$I(\tau u) = - \int_\Omega \left(\tau^{n+1} \frac{1}{n+1} u \det D^2 u + \frac{1}{t} e^{-\tau t u} \right) dx,$$

which diverges to $-\infty$ as $\tau \to +\infty$. So there exists a mountain pass structure and we can use the logarithmic gradient heat flow to prove the existence of a mountain pass type critical point v_t (cf. [182]).

Note that by Proposition 5.4.2, v_t must have $\lambda_1 < t$, but its Morse index is 1, so we must have $\lambda_2 \geq t$.

5.5 Appendix

In this appendix we collect some formulas for matrix analysis we used.

Proposition 5.5.1 *Let $F(A) := A^{-1}$ be defined on the invertible matrix space, then its (Fréchet) differential is $DF(A)B = -A^{-1}BA^{-1}$ for any matrix B.*

Proof Because $F(A)A = \mathrm{Id}$, where Id is the identity matrix, take differential we obtain $(DF(A)B)A + F(A)B = 0$. Rewrite this and we get the formula. □

Proposition 5.5.2 *Let $G(A) = \log \det A$ be defined on the positive definite symmetric matrix space, then its (Fréchet) differential is $DG(A)B = A^{ij}B_{ij}$ (repeated indices are summed) for any symmetric matrix B, where (A^{ij}) is the inverse matrix of A.*

Proof We need to calculate $\frac{d}{dt}G(A(t))|_{t=0}$ where $A(t)$ is a curve near A with $A(0) = A$ and $\frac{d}{dt}A(t)|_{t=0} = B$.

Because A is a positive definite symmetric matrix, there exists a nonsingular matrix P with $\det P > 0$ such that $A = PP^T$, where P^T is the transpose of P. We have

$$\log \det A(t) = \log \det\left(P^{-1}A(t)\left(P^T\right)^{-1}\right) + \log \det P + \log \det P^T. \tag{5.77}$$

Now we have

$$P^{-1}A(t)\left(P^T\right)^{-1} = \mathrm{Id} + tP^{-1}B\left(P^T\right)^{-1} + o(t), \tag{5.78}$$

so

$$\det\left(P^{-1}A(t)\left(P^T\right)^{-1}\right) = 1 + t\,\mathrm{Tr}\left(P^{-1}B\left(P^T\right)^{-1}\right) + o(t), \tag{5.79}$$

here Tr is the trace. This implies

$$\log \det A(t) = \log \det P + \log \det P^T + t\,\mathrm{Tr}\left(P^{-1}B\left(P^T\right)^{-1}\right) + o(t). \tag{5.80}$$

So

$$\frac{d}{dt}\log \det A(t)|_{t=0} = \mathrm{Tr}\left(P^{-1}B\left(P^T\right)^{-1}\right). \tag{5.81}$$

Since $\mathrm{Tr}(Q_1 Q_2) = \mathrm{Tr}(Q_2 Q_1)$ for any matrices Q_1, Q_2, we have

$$\frac{d}{dt}\log \det A(t)|_{t=0} = \mathrm{Tr}\left(B\left(P^T\right)^{-1}P^{-1}\right)$$

$$= \mathrm{Tr}\left(BA^{-1}\right).$$

If we write this using coefficients of the matrices it is in the form of the proposition. □

Corollary 5.5.1 *Let $H(A) = \det A$ be defined on the positive definite symmetric matrix space, then its (Fréchet) differential is $DH(A)(B) = A^{ij}B_{ij}\det A$ for any symmetric matrix B, where (A^{ij}) is the inverse matrix of A and repeated index are summed.*

Proof Because we have $H(A) = e^{G(A)}$, $DH(A) = e^{G(A)} DG(A)$. $\qquad\square$

Proposition 5.5.3 *Let $F(A) = \log \det A$ be defined on the positive definite symmetric matrix space, then it is strictly concave.*

Proof We need to prove for two positive definite symmetric matrix A and B, we have $f(t) = \log \det(tA + (1-t)B)$ is concave for $t \in [0, 1]$, or $f''(t) \leq 0$. First we have by Proposition 5.5.2

$$f'(t) = A_t^{ij}(A - B)_{ij}, \tag{5.82}$$

where A_t^{ij} is the inverse matrix of $tA + (1-t)B$. Then we have by Proposition 5.5.1

$$f''(t) = -A_t^{iq}(A - B)_{pq} A_t^{pj}(A - B)_{ij}, \tag{5.83}$$

which is non-positive for each t which can be seen by diagonalizing A_t (in fact negative unless $A = B$). $\qquad\square$

Chapter 6
Topological Methods and Applications

6.1 Superlinear System of Integral Equations and Applications

6.1.1 Introduction

There are a lot of results about Hammerstein integral equations (see [111]), but superlinear problems are difficult, there are only a few results about it. In [99, 169] the authors study superlinear integral equations; in [140] the authors study a system of superlinear integral equations, and some existence theorems are obtained. Using different methods from [140], we obtain new results about existence of solutions, and apply them to two-point boundary problems of system of equations.

6.1.2 Existence of Non-trivial Solutions

Consider the following superlinear system of integral equations (6.1):

$$\begin{cases} \varphi_1(x) = \displaystyle\int_G k_1(x,y) f_1\big(y, \varphi_1(y), \varphi_2(y)\big)\, dy, \\[2mm] \varphi_2(x) = \displaystyle\int_G k_2(x,y) f_2\big(y, \varphi_1(y), \varphi_2(y)\big)\, dy. \end{cases} \tag{6.1}$$

Suppose that $G \subset \mathbb{R}^n$ is a bounded closed set. Let

$$A_i(\varphi_1, \varphi_2) = \int_G k_i(x,y) f_i\big(y, \varphi_1(y), \varphi_2(y)\big)\, dy, \quad i = 1, 2, \tag{6.2}$$

$$A(\varphi_1, \varphi_2) = \big(A_1(\varphi_1, \varphi_2), A_2(\varphi_1, \varphi_2)\big). \tag{6.3}$$

For convenience, we first give some conditions ($i = 1, 2$).

(H_1) $k_i(x,y) : G \times G \to \mathbb{R}$ are nonnegative and continuous, $f_i(x,u,v) : G \times \mathbb{R} \times \mathbb{R} \to \mathbb{R}$ are continuous.

(H_2) $\exists l_i > 0$ such that $f_i(x,u,v) \geq -l_i$, $\forall x \in G$, $u, v \in \mathbb{R}$.

Z. Zhang, *Variational, Topological, and Partial Order Methods with Their Applications*, Developments in Mathematics 29,
DOI 10.1007/978-3-642-30709-6_6, © Springer-Verlag Berlin Heidelberg 2013

(H_3) $\exists a_{12}(x) > 0,\ a_{21}(x) > 0,\ b_i(x) \geq 0,\ h_i(x) \geq 0$ such that

$$f_1(x, u, v) \geq a_{12}(x)v - b_1(x), \qquad \forall x \in G,\ u \geq 0,\ v \geq 0, \tag{6.4}$$

$$f_2(x, u, v) \geq a_{21}(x)u - b_2(x), \qquad \forall x \in G,\ u \geq 0,\ v \geq 0, \tag{6.5}$$

$$\left| f_i(x, u, v) \right| \leq h_i(x)[|u| + |v|], \qquad \forall x \in G,\ 0 \leq |u| + |v| \leq r_0, \tag{6.6}$$

where r_0 is a sufficiently small positive constant.

(H_4) $\exists a_{11}(x) > 0,\ a_{22}(x) > 0,\ b_i(x) \geq 0$ such that

$$f_1(x, u, v) \geq a_{11}(x)u - b_1(x), \qquad \forall x \in G,\ u \geq 0,\ v \geq 0, \tag{6.7}$$

$$f_2(x, u, v) \geq a_{22}(x)v - b_2(x), \qquad \forall x \in G,\ u \geq 0,\ v \geq 0. \tag{6.8}$$

We know $C(G)$ is a Banach space with norm $\|\varphi\| = \max_{x \in G} |\varphi(x)|$, $C(G) \times C(G)$ is a Banach space with norm

$$\|\varphi\|_1 = \|\varphi_1\| + \|\varphi_2\|, \quad \forall \varphi = \begin{pmatrix} \varphi_1 \\ \varphi_2 \end{pmatrix} \in C(G) \times C(G).$$

Now we consider the following completely continuous positive linear operator $K :$ $C(G) \times C(G) \to C(G) \times C(G)$, which satisfies $K(P \times P) \subset P \times P$, where $P = \{\varphi \in C(G) \mid \varphi(x) \geq 0\}$ is a cone of $C(G)$, and $P \times P$ is a cone of $C(G) \times C(G)$. We have

$$K \begin{pmatrix} \varphi_1 \\ \varphi_2 \end{pmatrix}(x) = \begin{pmatrix} \int_G k_1(x, y)[a_{11}(y)\varphi_1(y) + a_{12}(y)\varphi_2(y)]\,dy \\ \int_G k_2(x, y)[a_{21}(y)\varphi_1(y) + a_{22}(y)\varphi_2(y)]\,dy \end{pmatrix}. \tag{6.9}$$

Suppose that the spectral radius of K is not zero, i.e., $r(K) \neq 0$. Since $(C(G) \times C(G))^* = C^*(G) \times C^*(G)$, we know the linear conjugate operator K^* satisfies $K^*(C(G) \times C(G)) \subset C(G) \times C(G)$. Also $\forall \psi \in C(G) \times C(G)$,

$$K^*\psi = K^* \begin{pmatrix} \psi_1 \\ \psi_2 \end{pmatrix}(y) = \begin{pmatrix} \int_G [k_1(x, y)a_{11}(y)\psi_1(x) + k_2(x, y)a_{21}(y)\psi_2(x)]\,dx \\ \int_G [k_1(x, y)a_{12}(y)\psi_1(x) + k_2(x, y)a_{22}(y)\psi_2(x)]\,dx \end{pmatrix}.$$

$$\tag{6.10}$$

We know $r(K^*) = r(K) \neq 0$, in virtue of famous Krein–Rutman Theorem (Theorem 1.9.1), there exists $\psi^* \in P \times P,\ \psi^* \neq 0$ such that

$$\psi^* = \begin{pmatrix} \psi_1^* \\ \psi_2^* \end{pmatrix} = r^{-1}(K)K^*\psi^*. \tag{6.11}$$

Definition 6.1.1 We call linear integral operator K satisfies H-condition, if $\exists \psi^* \in P \times P,\ \psi^* \neq 0,\ \exists \beta > 0$ such that (6.11) is satisfied, and

$$\psi^*(y) \geq \beta \begin{pmatrix} k_1(\tau, y)a_{11}(y) + k_2(\tau, y)a_{21}(y) \\ k_1(\tau, y)a_{12}(y) + k_2(\tau, y)a_{22}(y) \end{pmatrix}, \qquad \forall \tau, y \in G. \tag{6.12}$$

Note that for single equation, there is a similar definition, see [111], p. 262.

Suppose $\psi^* \in P \times P \setminus \{0\}$ satisfies (6.11), for fixed $\delta > 0$ let

$$(P \times P)_{(\psi^*,\delta)} = \left\{ \varphi \in P \times P \ \Big| \ \int_G \psi^*(x)\varphi(x)\,dx \geq \delta \|\varphi\|_1 \right\} \qquad (6.13)$$

where

$$\int_G \psi^*(x)\varphi(x)\,dx = \int_G \psi_1^*(x)\varphi_1(x)\,dx + \int_G \psi_2^*(x)\varphi_2(x)\,dx.$$

It is easy to know that $(P \times P)_{(\psi^*,\delta)}$ is a cone of $C(G) \times C(G)$.

Lemma 6.1.1 (Zhitao Zhang [200]) *Linear integral operator K satisfies H-condition if and only if there exists $\psi^* \in P \times P \setminus \{0\}$, $\delta > 0$ such that (6.11) is satisfied and $K(P \times P) \subset (P \times P)_{(\psi^*,\delta)}$.*

Proof $\forall \varphi \in P \times P$, by (6.10) we have

$$\int_G \psi^*(x)K\varphi(x)\,dx$$

$$= \int_G \psi_1^*(x)\,dx \int_G k_1(x,y)\big[a_{11}(y)\varphi_1(y) + a_{12}(y)\varphi_2(y)\big]\,dy$$

$$+ \int_G \psi_2^*(x)\,dx \int_G k_2(x,y)\big[a_{21}(y)\varphi_1(y) + a_{22}(y)\varphi_2(y)\big]\,dy$$

$$= \int_G \varphi_1(y)\,dy \int_G \big[k_1(x,y)a_{11}(y)\psi_1^*(x) + k_2(x,y)a_{21}(y)\psi_2^*(x)\big]\,dx$$

$$+ \int_G \varphi_2(y)\,dy \int_G \big[k_1(x,y)a_{12}(y)\psi_1^*(x) + k_2(x,y)a_{22}(y)\psi_2^*(x)\big]\,dx$$

$$= r(K)\left[\int_G \psi_1^*(y)\varphi_1(y)\,dy + \int_G \psi_2^*(y)\varphi_2(y)\,dy\right] \qquad (6.14)$$

If (6.12) is satisfied, i.e.,

$$\psi_1^*(y) \geq \beta\big[k_1(\tau,y)a_{11}(y) + k_2(\tau,y)a_{21}(y)\big], \quad \forall \tau, y \in G;$$

$$\psi_2^*(y) \geq \beta\big[k_1(\tau,y)a_{12}(y) + k_2(\tau,y)a_{22}(y)\big], \quad \forall \tau, y \in G.$$

Thus,

$$\int_G \psi^*(x)K\varphi(x)\,dx$$

$$\geq \beta \cdot r(K)\left\{\int_G \big[k_1(\tau,y)a_{11}(y) + k_2(\tau,y)a_{21}(y)\big]\varphi_1(y)\,dy\right.$$

$$+ \int_G \left[k_1(\tau, y) a_{12}(y) + k_2(\tau, y) a_{22}(y) \right] \varphi_2(y) \, dy \bigg\}$$

$$= \beta \cdot r(K) \bigg\{ \int_G k_1(\tau, y) \left[a_{11}(y)\varphi_1(y) + a_{12}(y)\varphi_2(y) \right] dy$$

$$+ \int_G k_2(\tau, y) \left[a_{21}(y)\varphi_1(y) + a_{22}(y)\varphi_2(y) \right] dy \bigg\}$$

$$\geq \frac{\beta \cdot r(K)}{2} \| K\varphi \|_1. \tag{6.15}$$

Let $\delta = \frac{\beta \cdot r(K)}{2}$, then $K : P \times P \to (P \times P)_{(\psi^*, \delta)}$.

On the other hand, if $K : P \times P \subset (P \times P)_{(\psi^*, \delta)}$, by (6.14) we have $\forall \varphi(x) \in P \times P$

$$r(K) \left[\int_G \psi_1^*(y)\varphi_1(y) \, dy + \int_G \psi_2^*(y)\varphi_2(y) \, dy \right]$$

$$= \int_G \psi^*(x) K\varphi(x) \, dx$$

$$\geq \delta \bigg\{ \left\| \int_G k_1(\tau, y) \left[a_{11}(y)\varphi_1(y) + a_{12}(y)\varphi_2(y) \right] dy \right\|$$

$$+ \left\| \int_G k_2(\tau, y) \left[a_{21}(y)\varphi_1(y) + a_{22}(y)\varphi_2(y) \right] dy \right\| \bigg\}$$

$$\geq \delta \bigg\{ \int_G \left[k_1(\tau, y) a_{11}(y) + k_2(\tau, y) a_{21}(y) \right] \varphi_1(y) \, dy$$

$$+ \int_G \left[k_1(\tau, y) a_{12}(y) + k_2(\tau, y) a_{22}(y) \right] \varphi_2(y) \, dy \bigg\}. \tag{6.16}$$

Finally, since φ is arbitrary, we have

$$\psi_1^*(y) \geq \delta r^{-1}(K) \left[k_1(\tau, y) a_{11}(y) + k_2(\tau, y) a_{21}(y) \right], \tag{6.17}$$

$$\psi_2^*(y) \geq \delta r^{-1}(K) \left[k_1(\tau, y) a_{12}(y) + k_2(\tau, y) a_{22}(y) \right]. \tag{6.18}$$

$\square$

In the following lemma, we shall give some sufficient conditions for K satisfying H-condition.

Lemma 6.1.2 (Zhitao Zhang [200]) *Suppose one of the following conditions is satisfied*:

(i) $\exists v_i(x) \in P \setminus \{0\}$, $i = 1, 2$ *such that*

$$k_i(x, y) \geq v_i(x)k_i(\tau, y), \quad \forall x, y, \tau \in G;$$

and $\exists \psi^*(x) \in P \times P \setminus \{0\}$ *such that* $\psi^* = r^{-1}(K)K^*\psi^*$, $v_i(x) \cdot \psi_i^*(x) \not\equiv 0$.

(ii) $\exists v_i(x) \in P \setminus \{0\}$, $u_i(x) \in P \setminus \{0\}$ *and the functions* $T_i(x, y) \geq 0$, $i = 1, 2$, $\forall(x, y) \in G \times G$ *such that*

$$v_i(x)T_i(\tau, y) \leq k_i(x, y) \leq u_i(x)T_i(\tau, y), \quad \forall x, y, \tau \in G;$$

and $\exists \psi^*(x) \in P \times P \setminus \{0\}$ *such that* $\psi^* = r^{-1}(K)K^*\psi^*$, $v_i(x) \cdot \psi_i^*(x) \not\equiv 0$.

Then K satisfies H-condition.

Proof (i)

$$\psi^*(y) = r^{-1}(K) \cdot K^*\psi^*$$

$$= r^{-1}(K) \begin{pmatrix} \int_G [k_1(x, y)\psi_1^*(x)a_{11}(y) + k_2(x, y)\psi_2^*(x)a_{21}(y)]\,dx \\ \int_G [k_1(x, y)\psi_1^*(x)a_{12}(y) + k_2(x, y)\psi_2^*(x)a_{22}(y)]\,dx \end{pmatrix}$$

$$\geq r^{-1}(K)$$

$$\times \begin{pmatrix} \int_G [v_1(x)k_1(\tau, y)\psi_1^*(x)a_{11}(y) + v_2(x)k_2(\tau, y)\psi_2^*(x)a_{21}(y)]\,dx \\ \int_G [v_1(x)k_1(\tau, y)\psi_1^*(x)a_{12}(y) + v_2(x)k_2(\tau, y)\psi_2^*(x)a_{22}(y)]\,dx \end{pmatrix}$$

$$\geq r^{-1}(K) \min \left\{ \int_G v_i(x) \cdot \psi_i^*(x)\,dx, i = 1, 2 \right\}$$

$$\times \begin{pmatrix} k_1(\tau, y)a_{11}(y) + k_2(\tau, y)a_{21}(y) \\ k_1(\tau, y)a_{12}(y) + k_2(\tau, y)a_{22}(y) \end{pmatrix},$$

which implies (6.12) is satisfied, so K satisfies H-condition.

(ii) We know

$$\max\{\|u_i\|, i = 1, 2\} \cdot \psi^*(y) \geq \begin{pmatrix} u_1(x)\psi_1^*(y) \\ u_2(x)\psi_2^*(y) \end{pmatrix},$$

then similarly to the proof of (i), we can get the result. $\square$

Now we define

$$B\phi = \int_G [k_1(x, y)h_1(y) + k_2(x, y)h_2(y)]\phi(y)\,dy, \quad \forall \phi \in C(G).$$

Theorem 6.1.1 (Zhitao Zhang [200]) *Suppose* (H_1), (H_2), (H_3) *(or* (H_1), (H_2), (H_4) *and* (6.6)) *are satisfied,* $r(K) > 1 \geq r(B)$, *and K satisfies H-condition, then the system of equations* (6.1) *has a non-trivial continuous solution.*

Proof By $a_{ij}(x) > 0$, $i = 1, 2$, and (H_2), we know $\exists b_1 \geq 0$ such that

$$\frac{f_1(x, u, v)}{a_{ij}(x)} \geq -b_1, \quad \frac{f_2(x, u, v)}{a_{ij}(x)} \geq -b_1, \quad \forall x \in G, \ u, v \in \mathbb{R}. \tag{6.19}$$

When (H_1), (H_2), (H_3) are satisfied, $a_{11}(x) \equiv 0$, $a_{22}(x) \equiv 0$. Thus

$$K\varphi = K\begin{pmatrix} \varphi_1 \\ \varphi_2 \end{pmatrix}(x) = \begin{pmatrix} \int_G k_1(x, y)a_{12}(y)\varphi_2(y)\,dy \\ \int_G k_2(x, y)a_{21}(y)\varphi_1(y)\,dy \end{pmatrix} \tag{6.20}$$

It is easy to know that there exist functions $b_i'(x) \geq 0$ $(i = 1, 2)$ such that

$$f_1(x, u, v) \geq a_{12}(x)v - b_1'(x), \quad \forall x \in G, \ u, v \in \mathbb{R}, \tag{6.21}$$

$$f_2(x, u, v) \geq a_{21}(x)u - b_2'(x), \quad \forall x \in G, \ u, v \in \mathbb{R}. \tag{6.22}$$

Since K satisfies the H-condition, by Lemma 6.1.1 we know $\exists \psi^* \in P \times P \setminus \{0\}$, $\delta > 0$ such that

$$\psi^* = r^{-1}(K)K^*\psi^*;$$

moreover, $K : P \times P \to (P \times P)_{(\psi^*, \delta)}$. Let $\epsilon = r(K) - 1$,

$$d(x) = \begin{pmatrix} d_1(x) \\ d_2(x) \end{pmatrix} = \begin{pmatrix} \int_G k_1(x, y)a_{12}(y)b_1\,dy \\ \int_G k_2(x, y)a_{21}(y)b_1\,dy \end{pmatrix}, \tag{6.23}$$

$$R > \|d(x)\|_1 + \frac{1}{\epsilon\delta}\left[\epsilon \int_G \psi^*(x)d(x)\,dx \right.$$

$$\left. + \sum_{i=1}^{2} \iint_{G \times G} \psi_i^*(x)k_i(x, y)b_i'(y)\,dx\,dy\right]. \tag{6.24}$$

Suppose $\varphi^*(x)$ is the positive eigenfunction such that $\varphi^* = r^{-1}(K)K\varphi^*$ and $\|\varphi^*\|_1 = 1$. Now, for an arbitrary open bounded $\Omega \subset C(G) \times C(G)$ such that $B_R = \{\varphi \in C(G) \times C(G) | \|\varphi\|_1 \leq R\} \subset \Omega$, we will prove

$$\varphi - A\varphi \neq \lambda_0\varphi^*, \quad \forall \varphi \in \partial\Omega, \ \lambda_0 \geq 0. \tag{6.25}$$

If $\exists \bar{\varphi} \in \partial\Omega$, $\lambda_0 \geq 0$ such that $\bar{\varphi} - A\bar{\varphi} = \lambda_0\varphi^*$, we can suppose $\lambda_0 > 0$, otherwise, the conclusion is valid. Then

$$\bar{\varphi}(x) + d(x) = \lambda_0\varphi^*(x) + \begin{pmatrix} \int_G k_1(x, y)a_{12}(y)\left[\frac{f_1(y, \bar{\varphi}_1(y), \bar{\varphi}_2(y))}{a_{12}(y)} + b_1\right]dy \\ \int_G k_2(x, y)a_{21}(y)\left[\frac{f_2(y, \bar{\varphi}_1(y), \bar{\varphi}_2(y))}{a_{21}(y)} + b_1\right]dy \end{pmatrix}$$

$$= \lambda_0\varphi^*(x) + K\begin{pmatrix} \frac{f_1(x, \bar{\varphi}_1(x), \bar{\varphi}_2(x))}{a_{21}(x)} + b_1 \\ \frac{f_2(x, \bar{\varphi}_1(x), \bar{\varphi}_2(x))}{a_{12}(x)} + b_1 \end{pmatrix} \tag{6.26}$$

By $K : P \times P \subset (P \times P)_{(\psi^*, \delta)}$, we have $\bar{\varphi}(x) + d(x) \in (P \times P)_{(\psi^*, \delta)}$. Therefore, by (6.21)–(6.22) we have

$$
\int_G \psi^*(x) A\bar{\varphi}(x)\, dx - \int_G \psi^*(x)\bar{\varphi}(x)\, dx
$$

$$
\geq \int_G \psi^*(x) \begin{pmatrix} \int_G k_1(x,y)a_{12}(y)\bar{\varphi}_2(y)\,dy \\ \int_G k_2(x,y)a_{21}(y)\bar{\varphi}_1(y)\,dy \end{pmatrix} dx
$$

$$
- \int_G \psi^*(x) \begin{pmatrix} \int_G k_1(x,y)b_1'(y)\,dy \\ \int_G k_2(x,y)b_2'(y)\,dy \end{pmatrix} dx - \int_G \psi^*(x)\bar{\varphi}(x)\, dx
$$

$$
= \int_G \begin{pmatrix} \int_G k_2(x,y)a_{21}(y)\psi_2^*(x)\,dx \\ \int_G k_1(x,y)a_{12}(y)\psi_1^*(x)\,dx \end{pmatrix} \cdot \begin{pmatrix} \bar{\varphi}_1(y) \\ \bar{\varphi}_2(y) \end{pmatrix} dy
$$

$$
- \sum_{i=1}^{2} \iint_{G \times G} \psi_i^*(x) k_i(x,y) b_i'(y)\, dx\, dy - \int_G \psi^*(x)\bar{\varphi}(x)\, dx
$$

$$
= \int_G K^* \psi^*(y) \cdot \begin{pmatrix} \bar{\varphi}_1(y) \\ \bar{\varphi}_2(y) \end{pmatrix} dy - \sum_{i=1}^{2} \iint_{G \times G} \psi_i^*(x) k_i(x,y) b_i'(y)\, dx\, dy
$$

$$
- \int_G \psi^*(x)\bar{\varphi}(x)\, dx
$$

$$
= r(K) \int_G \begin{pmatrix} \psi_1^*(y) \\ \psi_2^*(y) \end{pmatrix} \cdot \begin{pmatrix} \bar{\varphi}_1(y) \\ \bar{\varphi}_2(y) \end{pmatrix} dy - \sum_{i=1}^{2} \iint_{G \times G} \psi_i^*(x) k_i(x,y) b_i'(y)\, dx\, dy
$$

$$
- \int_G \psi^*(x)\bar{\varphi}(x)\, dx
$$

$$
= \epsilon \int_G \psi^*(y)\bar{\varphi}(y)\, dy - \sum_{i=1}^{2} \iint_{G \times G} \psi^*(x) k_i(x,y) b_i'(y)\, dx\, dy
$$

$$
= \epsilon \int_G \psi^*(y)\big(\bar{\varphi}(y) + d(y)\big)\, dy \; \epsilon \int_G \psi^*(y) d(y)\, dy
$$

$$
- \sum_{i=1}^{2} \iint_{G \times G} \psi^*(x) k_i(x,y) b_i'(y)\, dx\, dy
$$

$$
\geq \epsilon\delta \|\bar{\varphi}\|_1 - \epsilon\delta \|d(x)\|_1 - \sum_{i=1}^{2} \iint_{G \times G} \psi^*(x) k_i(x,y) b_i'(y)\, dx\, dy
$$

$$
- \epsilon \int_G \psi^*(y) d(y)\, dy
$$

$$
> 0. \tag{6.27}
$$

But we also have

$$\int_G \psi^*(x)\bar{\varphi}(x)\,dx - \int_G \psi^*(x)A\bar{\varphi}(x)\,dx = \lambda_0 \int_G \psi^*(x)\bar{\varphi}(x)\,dx \geq 0 \qquad (6.28)$$

which contradicts (6.27). Thus $\deg(I - A, \Omega, 0) = 0$, by the definition of Ω we have

$$\mathrm{ind}(I - A, \infty) = 0. \qquad (6.29)$$

Let $B_{r_0} = \{\varphi \in C(G) \times C(G) | \|\varphi\|_1 < r_0\}$, without loss of generality, we suppose (6.1) has no solution on ∂B_{r_0}. Now we prove $\deg(I - A, B_{r_0}, 0) = 1$, we only need prove $\forall \varphi \in \partial B_{r_0}$, $\lambda_1 \geq 1$, $A\varphi \neq \lambda_1 \varphi$. If it is not the case, then $\exists \bar{\varphi} \in \partial B_{r_0}$, $\lambda_1 \geq 1$, $A\bar{\varphi} = \lambda_1 \bar{\varphi}$. Since A has no fixed point on ∂B_{r_0}, we know $\lambda_1 > 1$. Therefore,

$$\left|\lambda_1 \bar{\varphi}_1(x)\right| = \left|(A\bar{\varphi})_1\right| = \left|\int_G k_1(x, y) f_1\big(y, \bar{\varphi}_1(y), \bar{\varphi}_2(y)\big)\,dy\right|$$

$$\leq \int_G k_1(x, y) h_1(y)\big[\left|\bar{\varphi}_1(y)\right| + \left|\bar{\varphi}_2(y)\right|\big]\,dy, \qquad (6.30)$$

$$\left|\lambda_1 \bar{\varphi}_2(x)\right| \leq \int_G k_2(x, y) h_2(y)\big[\left|\bar{\varphi}_1(y)\right| + \left|\bar{\varphi}_2(y)\right|\big]\,dy, \qquad (6.31)$$

$$\lambda_1\big[\left|\bar{\varphi}_1(x)\right| + \left|\bar{\varphi}_2(x)\right|\big] \leq \int_G \big[k_1(x, y) h_1(y) + k_2(x, y) h_2(y)\big]$$

$$\times \big[\left|\bar{\varphi}_1(y)\right| + \left|\bar{\varphi}_2(y)\right|\big]\,dy. \qquad (6.32)$$

Let $p(x) = |\bar{\varphi}_1(x)| + |\bar{\varphi}_2(x)|$, then $\lambda_1 p(x) \leq (Bp)(x)$. So $r(B) \geq \lambda_1 > 1$, which contradicts $r(B) \leq 1$. Therefore,

$$\deg(I - A, B_{r_0}, 0) = 1. \qquad (6.33)$$

By (6.29) and (6.33), we know A has a fixed point $\varphi_0 \neq 0$, $\varphi_0 \in C(G) \times C(G)$, i.e., (6.1) has a non-trivial continuous solution.

Similarly, when (H_1), (H_2), (H_4) and (6.6) are satisfied, let

$$d(x) = \begin{pmatrix} d_1(x) \\ d_2(x) \end{pmatrix} = \begin{pmatrix} \int_G k_1(x, y) a_{11}(y) b_1\,dy \\ \int_G k_2(x, y) a_{22}(y) b_1\,dy \end{pmatrix} \qquad (6.34)$$

we have the same conclusion. $\qquad\qquad\qquad\qquad\qquad\qquad\qquad\qquad\qquad\qquad\qquad\square$

Theorem 6.1.2 (Zhitao Zhang [200]) *If $f_1(x, u, v) \equiv f_2(x, u, v) = f(x, u, v)$, and (H_1), (H_2) are satisfied, and*

$$f(x, u, v) \geq a_1(x) u + a_2(x) v - b(x), \quad \forall u \geq 0, \ v \geq 0, \ x \in G. \qquad (6.35)$$

where $a_1(x) > 0$, $a_2(x) > 0$, $b(x) \geq 0$. Moreover, K satisfies H-condition, $r(K) > 1 \geq r(B)$, Then (6.1) has a continuous non-trivial solution.

Proof Similarly to the proof of Theorem 6.1.1, there exist a constant $b_1 \geq 0$ and a function $b'(x) \geq 0$ such that

$$\frac{f(x,u,v)}{a_1(x)} \geq -b_1, \qquad \frac{f(x,u,v)}{a_2(x)} \geq -b_1, \qquad \forall x \in G, \ u, v \in \mathbb{R}. \tag{6.36}$$

$$f(x,u,v) \geq a_1(x)u + a_2(x)v - b'(x), \qquad \forall x \in G, \ u, v \in \mathbb{R}.$$

By (6.9) we have

$$K\begin{pmatrix} \varphi_1 \\ \varphi_2 \end{pmatrix}(x) = \begin{pmatrix} \int_G k_1(x,y)[a_1(y)\varphi_1(y) + a_2(y)\varphi_2(y)]\, dy \\ \int_G k_2(x,y)[a_1(y)\varphi_1(y) + a_2(y)\varphi_2(y)]\, dy \end{pmatrix}. \tag{6.37}$$

By (6.10) we get

$$K^*\psi = K^*\begin{pmatrix} \psi_1 \\ \psi_2 \end{pmatrix}(y)$$

$$= \begin{pmatrix} \int_G [k_1(x,y)a_1(y)\psi_1(x) + k_2(x,y)a_1(y)\psi_2(x)]\, dx \\ \int_G [k_1(x,y)a_2(y)\psi_1(x) + k_2(x,y)a_2(y)\psi_2(x)]\, dx \end{pmatrix}. \tag{6.38}$$

Let

$$d(x) = \begin{pmatrix} d_1(x) \\ d_2(x) \end{pmatrix} = \begin{pmatrix} \int_G k_1(x,y)[a_1(y) + a_2(y)] \cdot \frac{b_1}{2}\, dy \\ \int_G k_2(x,y)[a_1(y) + a_2(y)] \cdot \frac{b_1}{2}\, dy \end{pmatrix} \tag{6.39}$$

and let

$$R > \|d(x)\|_1 + \frac{1}{\epsilon\delta}\left[\epsilon \int_G \psi^*(x)d(x)\, dx \right.$$

$$\left. + \sum_{i=1}^{2} \iint_{G\times G} \psi_i^*(x)k_i(x,y)b'(y)\, dx\, dy \right]. \tag{6.40}$$

Similarly to (6.26) we obtain

$$\bar\varphi(x) + d(x)$$

$$= \lambda_0 \varphi^*(x)$$

$$+ \begin{pmatrix} \int_G k_1(x,y)\{a_1(y)[\frac{f(y,\bar\varphi_1(y),\bar\varphi_2(y))}{2a_1(y)} + \frac{b_1}{2}] + a_2(y)[\frac{f(y,\bar\varphi_1(y),\bar\varphi_2(y))}{2a_2(y)} + \frac{b_1}{2}]\}\, dy \\ \int_G k_2(x,y)\{a_1(y)[\frac{f(y,\bar\varphi_1(y),\bar\varphi_2(y))}{2a_1(y)} + \frac{b_1}{2}] + a_2(y)[\frac{f(y,\bar\varphi_1(y),\bar\varphi_2(y))}{2a_2(y)} + \frac{b_1}{2}]\}\, dy \end{pmatrix}$$

$$= \lambda_0 \varphi^*(x) + K\begin{pmatrix} \frac{f(x,\bar\varphi_1(x),\bar\varphi_2(x))}{2a_1(x)} + \frac{b_1}{2} \\ \frac{f(x,\bar\varphi_1(x),\bar\varphi_2(x))}{2a_2(x)} + \frac{b_1}{2} \end{pmatrix}. \tag{6.41}$$

By (6.37) and $K : P \times P \subset (P \times P)_{(\psi^*,\delta)}$, we have $\bar\varphi(x) + d(x) \in (P \times P)_{(\psi^*,\delta)}$. Other part of the proof is similar to that of Theorem 6.1.1. $\qquad\square$

Corollary 6.1.1 *Suppose* (H_1), (H_3) *are satisfied,* $\exists b_i^* > 0$ $(i = 1, 2)$ *such that*

$$f_1(x, u, v) \geq -\frac{b_1^*}{M_1}, \quad f_2(x, u, v) \geq -\frac{b_2^*}{M_2}, \quad \forall u \geq -b_1^*, \ v \geq -b_2^*, \ x \in G.$$

$$(6.42)$$

where $M_i = \max_{x \in G} \int_G k_i(x, y)\, dy$. *If* $r(K) > 1 \geq r(B)$, *and* K *satisfies H-condition, then* (6.1) *has a continuous non-trivial solution.*

Proof Define

$$\bar{f}_i = \begin{cases} f_i(x, u, v), & u \geq -b_1^*, \ v \geq -b_2^*, \ x \in G; \\ f_i(x, -2b_1^* - u, v), & u \leq -b_1^*, \ v \geq -b_2^*, \ x \in G; \\ f_i(x, u, -2b_2^* - v), & u \geq -b_1^*, \ v \leq -b_2^*, \ x \in G; \\ f_i(x, -2b_1^* - u, -2b_2^* - v), & u \leq -b_1^*, \ v \leq -b_2^*, \ x \in G. \end{cases}$$

Let

$$A_1\varphi(x) = \begin{pmatrix} \int_G k_1(x, y)\bar{f}_1(y, \varphi_1(y), \varphi_2(y))\, dy \\ \int_G k_2(x, y)\bar{f}_2(y, \varphi_1(y), \varphi_2(y))\, dy \end{pmatrix}, \tag{6.43}$$

then A_1 satisfies the conditions of Theorem 6.1.1, so A_1 has a non-zero fixed point $\varphi^*(x)$. Also

$$\varphi^*(x) = \begin{pmatrix} \varphi_1^*(x) \\ \varphi_2^*(x) \end{pmatrix} = \begin{pmatrix} \int_G k_1(x, y)\bar{f}_1(y, \varphi_1(y), \varphi_2(y))\, dy \\ \int_G k_2(x, y)\bar{f}_2(y, \varphi_1(y), \varphi_2(y))\, dy \end{pmatrix}$$

$$\geq \begin{pmatrix} -\frac{b_1^*}{M_1} \int_G k_1(x, y)\, dy \\ -\frac{b_2^*}{M_2} \int_G k_2(x, y)\, dy \end{pmatrix} \geq \begin{pmatrix} -b_1^* \\ -b_2^* \end{pmatrix} \tag{6.44}$$

By the definition of $\bar{f}(x, u, v)$, we know $\bar{f}_i(x, \varphi_1^*(x), \varphi_2^*(x)) = f_i(x, \varphi_1^*(x), \varphi_2^*(x))$, thus

$$\varphi^*(x) = \begin{pmatrix} \varphi_1^*(x) \\ \varphi_2^*(x) \end{pmatrix} = \begin{pmatrix} \int_G k_1(x, y) f_1(y, \varphi_1^*(y), \varphi_2^*(y))\, dy \\ \int_G k_2(x, y) f_2(y, \varphi_1^*(y), \varphi_2^*(y))\, dy \end{pmatrix}$$

i.e., $\varphi^*(x) = \begin{pmatrix} \varphi_1^*(x) \\ \varphi_2^*(x) \end{pmatrix}$ is a non-trivial solution of (6.1). $\square$

Corollary 6.1.2 *Suppose* $f_i(x, u, v) \geq 0$, $\forall u \geq 0$, $v \geq 0$, *and* (H_1), (6.4), (6.5) *[or* (H_1), (6.7), (6.8)*] are satisfied; moreover, there exist positive continuous functions* $h_i(x)$ *and a constant* $r > 0$ *such that* $f_i(x, u, v) \leq h_i(x)(u + v)$, $\forall x \in G$, $u \geq 0$, $v \geq 0$, $u + v \leq r$. *If* $r(K) > 1 \geq r(B)$, K *satisfies H-condition, then* (6.1) *has a positive continuous non-trivial solution.*

Proof Let $b_1^* = b_2^* = 0$, the proof is similar to that of Corollary 6.1.1. $\square$

Remark 6.1.1 When $a_{12}(x) \equiv 1$, $a_{21}(x) \equiv 1$, we can use the following conditions instead of the (H_3) of Theorem 6.1.1:

$$\liminf_{v \to +\infty} \frac{f_1(x, u, v)}{v} > r^{-1}(K), \quad \text{uniformly for } x \in G, \ u \in \mathbb{R},$$

$$\liminf_{u \to +\infty} \frac{f_2(x, u, v)}{u} > r^{-1}(K), \quad \text{uniformly for } x \in G, \ v \in \mathbb{R},$$

$$\lim_{|u|+|v| \to 0} \frac{|f_i(x, u, v)|}{|u| + |v|} = 0 \quad (i = 1, 2), \quad \text{uniformly for } x \in G,$$

then the conclusion is valid.

6.1.3 Application to Two-Point Boundary Value Problems

Consider the following ordinary differential system of equations:

$$\begin{cases} -\varphi_1''(x) = f_1\big(x, \varphi_1(x), \varphi_2(x)\big) & \text{in } (0, 1), \\ -\varphi_2''(x) = f_2\big(x, \varphi_1(x), \varphi_2(x)\big) & \text{in } (0, 1), \\ \varphi_1(0) = \varphi_1(1) = 0, \\ \varphi_2(0) = \varphi_2'(1) = 0. \end{cases} \tag{6.45}$$

It is well known that the solution to (6.45) in $C^2[0, 1] \times C^2[0, 1]$ is equivalent to the solution of the following integral system of (6.46) in $C[0, 1] \times C[0, 1]$:

$$\begin{cases} \varphi_1(x) = \displaystyle\int_0^1 k_1(x, y) f_1\big(y, \varphi_1(y), \varphi_2(y)\big) \, dy, \\ \varphi_2(x) = \displaystyle\int_0^1 k_2(x, y) f_2\big(y, \varphi_1(y), \varphi_2(y)\big) \, dy, \end{cases} \tag{6.46}$$

where

$$k_1(x, y) = \begin{cases} x(1 - y), & x \le y, \\ y(1 - x), & y < x; \end{cases} \quad \text{and} \quad k_2(x, y) = \begin{cases} x, & x \le y, \\ y, & y < x. \end{cases}$$

Theorem 6.1.3 (Zhitao Zhang [200]) *If $f_i : [0, 1] \times \mathbb{R} \times \mathbb{R} \to \mathbb{R}$ is continuous and $\exists b \in \mathbb{R}$ such that $f_i(x, u, v) \ge b, \forall x \in [0, 1], u, v \in \mathbb{R}$; and*

$$\liminf_{v \to +\infty} \frac{f_1(x, u, v)}{v} > \pi^2, \quad \textit{uniformly for } x \in [0, 1], \ u \in \mathbb{R};$$

$$\liminf_{u \to +\infty} \frac{f_2(x, u, v)}{u} > \pi^2, \quad \textit{uniformly for } x \in [0, 1], \ v \in \mathbb{R};$$

$$\lim_{|u|+|v| \to 0} \frac{|f_i(x, u, v)|}{|u| + |v|} = 0 \quad (i = 1, 2), \quad \textit{uniformly for } x \in [0, 1].$$

Then (6.45) has a non-trivial solution in $C^2[0, 1] \times C^2[0, 1]$.

Proof Let $G = [0, 1]$,

$$K \begin{pmatrix} \varphi_1 \\ \varphi_2 \end{pmatrix}(x) = \begin{pmatrix} \int_0^1 k_1(x, y)\varphi_2(y)\,dy \\ \int_0^1 k_2(x, y)\varphi_1(y)\,dy \end{pmatrix}, \tag{6.47}$$

$$K^* \begin{pmatrix} \psi_1 \\ \psi_2 \end{pmatrix}(y) = \begin{pmatrix} \int_0^1 k_2(x, y)\psi_2(x)\,dx \\ \int_0^1 k_1(x, y)\psi_1(x)\,dx \end{pmatrix}, \tag{6.48}$$

then $K : P \times P \to P \times P$ is completely continuous, and we know

$$\begin{pmatrix} \sin \pi x \\ \sin \pi x \end{pmatrix} \in P \times P \setminus \{0\}, \qquad \left\| \begin{pmatrix} \frac{1}{2}\sin \pi x \\ \frac{1}{2}\sin \pi x \end{pmatrix} \right\|_1 = 1.$$

Moreover,

$$K \begin{pmatrix} \frac{1}{2}\sin \pi x \\ \frac{1}{2}\sin \pi x \end{pmatrix} = \begin{pmatrix} \frac{1}{2}\int_0^1 k_1(x, y)\sin \pi y\,dy \\ \frac{1}{2}\int_0^1 k_2(x, y)\sin \pi y\,dy \end{pmatrix}$$

$$= \frac{1}{2} \begin{pmatrix} \frac{1}{\pi^2}\sin \pi x \\ \frac{x}{\pi} + \frac{1}{\pi^2}\sin \pi x \end{pmatrix} \geq \frac{1}{\pi^2} \begin{pmatrix} \frac{1}{2}\sin \pi x \\ \frac{1}{2}\sin \pi x \end{pmatrix}.$$

Then we can easily get $r(K) \geq \frac{1}{\pi^2}$, so $r(K) \neq 0$, and $r(K^*) \neq 0$, $\exists \psi^* \in P \times P \setminus \{0\}$, such that $\psi^* = r^{-1}(K)K^*\psi^*$. Since if $\psi_1^*(x) \equiv 0$ or $\psi_2^*(x) \equiv 0$, by (6.48) we get $\psi^*(x) \equiv 0$, so $\psi_i^*(x) \not\equiv 0$ $(i = 1, 2)$, Let

$$v_1(x) = \begin{cases} x, & 0 \leq x \leq \frac{1}{2}, \\ 1 - x, & \frac{1}{2} \leq x \leq 1; \end{cases} \quad \text{and} \quad v_2(x) = x,$$

then $k_i(x, y) \geq v_i(x)k_i(\tau, y)$ $(i = 1, 2)$, $\forall x, \tau, y \in [0, 1]$, and $v_i(x) \cdot \psi_i^*(x) \not\equiv 0$. Thus by Lemma 6.1.2 we know K satisfies H-condition. By Remark 6.1.1 we see that (6.45) has a non-trivial solution. $\qquad\square$

6.2 Existence of Positive Solutions for a Semilinear Elliptic System

6.2.1 Introduction

We consider the existence of (component-wise) positive solutions for the following elliptic system:

$$\begin{cases} -\Delta u = f_1(x, u, v) & \text{in } \Omega, \\ -\Delta v = f_2(x, u, v) & \text{in } \Omega, \\ u = v = 0 & \text{on } \partial\Omega, \end{cases} \tag{6.49}$$

where f_1, $f_2 \in C(\overline{\Omega} \times \mathbb{R}^+ \times \mathbb{R}^+, \mathbb{R}^+)$, $\mathbb{R}^+ = [0, +\infty)$, $\Omega \subset \mathbb{R}^n$ ($n \geq 3$) is a smooth bounded domain.

In [55] Clement et al. established the existence of positive solutions for the case that $f_1 = u^{\alpha} v^{\beta}$ and that $f_2 = u^{\gamma} v^{\delta}$; in [161] Serrin and Zou obtained positive solutions of the Lane–Emden system ($f_1 = v^p$, $f_2 = u^q$); in [167] M.A.S. Souto considered nonnegative non-trivial solutions for more general nonlinearities $f_1 = m_{11}(x)u + m_{12}(x)v + f(x, u, v)$, $f_2 = m_{21}(x)u + m_{22}(x)v + g(x, u, v)$ where f has asymptotic behavior at infinity as u^{σ} and g satisfies some subcritical growth, in particular obtained positive solutions as $f_1 = u^{\sigma} + v^q$, $f_2 = u^p$; in [212] H. Zou discussed nonnegative non-trivial solutions for the nonlinearities $f_1 = au^r + bv^q$, $f_2 = cu^p + dv^s$ ($a + b > 0$, $c + d > 0$ and $p, q > 1$) and then dealt with more general cases that f_1 and f_2 have asymptotic behavior at infinity as $a(x)u^r + b(x)v^q$ and $c(x)u^p + d(x)v^s$ (here the coefficients are nonnegative continuous functions), respectively, moreover positive solutions were obtained when $f_1(x, 0, v) \neq 0$ for $v > 0$ and $f_2(x, u, 0) \neq 0$ for $u > 0$.

In [5] Alves et al. have studied a large class of sublinear and superlinear non-variational elliptic systems and obtained the existence of nonnegative non-trivial solutions under the assumptions that there is an a priori bound on the nonnegative solutions of superlinear system.

Roughly speaking, they require the coupled nonlinearities in systems have some similar features, e.g., both nonlinearities are superlinear or sublinear. Usually ones change the problem into the fixed point problem of the corresponding compactly continuous mapping on a single cone K in product space $C(\overline{\Omega}) \times C(\overline{\Omega})$ and apply the classical fixed point index theory combining with some a priori estimates technique.

Consequently, if the coupled nonlinearities in systems have different features, how should we do?

Now we are mainly concerned with the existence of positive solutions for system (6.49) involving a new class of nonlinearities in which one is superlinear and the other is sublinear in the sense of the following definition.

Definition 6.2.1 If f_1, f_2 in system (6.49) satisfy the following assumptions:

$$(A_1) \quad \limsup_{u \to 0^+} \max_{x \in \overline{\Omega}} \frac{f_1(x, u, v)}{u} < \delta_1 < \liminf_{u \to +\infty} \min_{x \in \overline{\Omega}} \frac{f_1(x, u, v)}{u}$$

uniformly w.r.t. $v \in \mathbb{R}^+$;

$$(A_2) \quad \liminf_{v \to 0^+} \min_{x \in \overline{\Omega}} \frac{f_2(x, u, v)}{v} > \delta_1 > \limsup_{v \to +\infty} \max_{x \in \overline{\Omega}} \frac{f_2(x, u, v)}{v}$$

uniformly w.r.t. $u \in \mathbb{R}^+$,

where δ_1 is the first eigenvalue of the Laplacian subject to Dirichlet data, then we say that f_1 is superlinear with respect to u at the origin and infinity and that f_2 is sublinear with respect to v at the origin and infinity.

Lemma 6.2.1 [8, 122] *Let E be a Banach space and $K \subset E$ be a closed convex cone in E, denote $K_r = \{u \in K \mid \|u\| < r\}$ and $\partial K_r = \{u \in K \mid \|u\| = r\}$, where $r > 0$. Let $T : \overline{K}_r \to K$ be a compact mapping and $0 < \rho \le r$.*

(i) *If $Tx \ne tx$ for all $x \in \partial K_\rho$ and for all $t \ge 1$, then $i(T, K_\rho, K) = 1$.*
(ii) *If there exists a compact mapping $H : \overline{K}_\rho \times [0, \infty) \to K$ such that*
    (a) *$H(x, 0) = Tx$ for all $x \in \partial K_\rho$,*
    (b) *$H(x, t) \ne x$ for all $x \in \partial K_\rho$ and all $t \ge 0$,*
    (c) *there is a $t_0 > 0$, such that $H(x, t) = x$ has no solution $x \in \overline{K}_\rho$, for $t \ge t_0$,*
    then $i(T, K_\rho, K) = 0$.

Lemma 6.2.2 [52] *Let E be a Banach space and let $K_i \subset E$ $(i = 1, 2)$ be a closed convex cone in E. For $r_i > 0$ $(i = 1, 2)$, denote $K_{r_i} = \{u \in K_i \mid \|u\| < r_i\}$, $\partial K_{r_i} = \{u \in K_i \mid \|u\| = r_i\}$. Suppose $A_i : K_i \to K_i$ is completely continuous. If $u_i \ne A_i u_i$, $\forall u_i \in \partial K_{r_i}$, then*

$$i(A, K_{r_1} \times K_{r_2}, K_1 \times K_2) = i(A_1, K_{r_1}, K_1) \cdot i(A_2, K_{r_2}, K_2),$$

where $A(u, v) \stackrel{\text{def}}{=} (A_1 u, A_2 v)$, $\forall (u, v) \in K_1 \times K_2$.

Let

$$E = \{u \in C(\overline{\Omega}) \mid u = 0 \text{ on } \partial\Omega\}, \qquad K = \{u \in E \mid u(x) \ge 0, \ \forall x \in \overline{\Omega}\}.$$

Let us call $S : C(\overline{\Omega}) \to C(\overline{\Omega})$ the solution operator of the linear problem

$$\begin{cases} -\Delta u = \psi, & \text{in } \Omega, \\ u = 0, & \text{on } \partial\Omega, \end{cases} \tag{6.50}$$

where $\psi \in C(\overline{\Omega})$. It is well known that S takes $C(\overline{\Omega})$ into $C^{1,\alpha}(\overline{\Omega})$ $(0 < \alpha < 1)$ and then S is a linear compact mapping in the space $C(\overline{\Omega})$.

For $\lambda \in [0, 1]$ and $u, v \in K$, we define the mappings $T_{\lambda,1}(\cdot, \cdot), T_{\lambda,2}(\cdot, \cdot) :$ $K \times K \to K$ and $T_\lambda(\cdot, \cdot) : K \times K \to K \times K$ by

$$\begin{cases} T_{\lambda,1}(u, v) = S\big[\lambda f_1(x, u, v) + (1 - \lambda) f_1(x, u, 0)\big], \\ T_{\lambda,2}(u, v) = S\big[\lambda f_2(x, u, v) + (1 - \lambda) f_2(x, 0, v)\big], \\ T_\lambda(u, v) = \big(T_{\lambda,1}(u, v), T_{\lambda,2}(u, v)\big). \end{cases} \tag{6.51}$$

It is easy to see that mappings $T_{\lambda,1}(\cdot, v)$, $T_{\lambda,2}(u, \cdot)$ and $T_\lambda(\cdot, \cdot)$ are compact.

Now we recall global results of Liouville-type theorems for $\mathbb{R}^N$ and for $\overline{\mathbb{R}}_+^N$:

Theorem 6.2.1 (Gidas and Spruck [94]) *Assume $p \in (1, \frac{N+2}{N-2})$ for $N \ge 3$, and $p > 1$ for $N = 1, 2$. If $u \in C^2(\mathbb{R}^N)$ is a nonnegative solution of the equation:*

$$\Delta u + u^p = 0 \quad in \ \mathbb{R}^N, \tag{6.52}$$

then $u \equiv 0$.

Theorem 6.2.2 (Gidas and Spruck [94]) *Assume* $p \in (1, \frac{N+2}{N-2})$ *for* $N \geq 3$, *and* $p >$
1 for $N = 1, 2$. *If* $u \in C^2(\mathbb{R}^N) \cap C(\overline{\mathbb{R}}^N_+)$ *is a nonnegative solution of the equation:*

$$\begin{cases} \Delta u + u^p = 0 & in \ \mathbb{R}^N_+, \\ u = 0 & on \ x_N = 0, \end{cases} \tag{6.53}$$

then $u \equiv 0$ *(where* $\mathbb{R}^N_+ = \{x = (x_1, x_2, \ldots, x_N) \in \mathbb{R}^N : x_N > 0\}$).

6.2.2 Existence of Positive Solutions

Lemma 6.2.3 (Zhang and Cheng [203]) *Assume that* f_1 *satisfies* (A_1) *and* (H_1),
then there exist $R_0 > r_0 > 0$ *such that for all* $r \in (0, r_0]$ *and* $R \in [R_0, +\infty)$,

$$i\big(T_{0,1}(\cdot, v), K_R \backslash \overline{K}_r, K\big) = -1.$$

Proof From the definition of $T_{\lambda,1}$, we know that $T_{0,1}(u, v) = S[f_1(x, u, 0)]$.
   In view of assumption (A_1), there exist $\varepsilon \in (0, \delta_1)$ and $r_0 > 0$, such that

$$f_1(x, u, 0) \leq (\delta_1 - \varepsilon)u, \quad \forall(x, u) \in \overline{\Omega} \times [0, r], \text{ where } r \in (0, r_0]. \tag{6.54}$$

   We claim that $T_{0,1}(u, v) \neq tu$ for all $t \geq 1$ and all $u \in \partial K_r$. In fact, if there exist
$t_0 \geq 1$ and $u_0 \in \partial K_r$ such that $T_{0,1}(u_0, v) = t_0 u_0$, then u_0 satisfies the following
equation:

$$\begin{cases} -\Delta u_0 = t_0^{-1} f_1(x, u_0, 0), & \forall x \in \Omega, \\ u_0|_{\partial\Omega} = 0. \end{cases}$$

Multiplying both sides of the equation above by a positive eigenfunction φ_1 associated to the first eigenvalue δ_1 of $(-\Delta, H_0^1(\Omega))$ and integrating on Ω, we get

$$\int_\Omega (-\Delta u_0)\varphi_1 = \int_\Omega t_0^{-1} f_1(x, u_0, 0)\varphi_1.$$

Combining with (6.54), we have

$$\delta_1 \int_\Omega u_0\varphi_1 \leq (\delta_1 - \varepsilon) \int_\Omega u_0\varphi_1,$$

which is a contradiction! Hence, applying conclusion (i) of Lemma 6.2.1 we obtain

$$i\big(T_{0,1}(\cdot, v), K_r, K\big) = 1 \quad \text{for all } r \in (0, r_0]. \tag{6.55}$$

   By virtue of assumption (A_1) and continuity of f_1, there exist $\varepsilon > 0$ and $C > 0$
such that

$$f_1(x, u, 0) \geq (\delta_1 + \varepsilon)u - C, \quad \forall(x, u) \in \overline{\Omega} \times \mathbb{R}^+. \tag{6.56}$$

Next, we show that there exists $R_0 > r_0$ such that

$$i\big(T_{0,1}(\cdot,v), K_R, K\big) = 0 \quad \text{for all } R \in [R_0, \infty). \tag{6.57}$$

For this matter, we need to construct the homotopy $H : \overline{K}_R \times \mathbb{R}^+ \to K$ as follows:

$$H(u,t) = S\big[f_1(x, u+t, 0)\big].$$

Now we verify all the conditions of (ii) in Lemma 6.2.1 which yields (6.57).

First, it is obvious that condition (a) of Lemma 6.2.1 holds.

Second, we prove that there exists a $t_0 > 0$ such that equation $H(u,t) = u$ does not have solutions for $t \geq t_0$, which implies condition (c) of Lemma 6.2.1. Actually, let u be a solution for the following equation:

$$\begin{cases} -\Delta u = f_1(x, u+t, 0), & \forall x \in \Omega, \\ u|_{\partial\Omega} = 0. \end{cases}$$

In combination with (6.56), we have

$$-\Delta u \geq (\delta_1 + \varepsilon)(u+t) - C.$$

Multiplying both sides of the inequality above by φ_1 and integrating on Ω, we obtain

$$\int_{\Omega} (-\Delta u)\varphi_1 \geq (\delta_1 + \varepsilon) \int_{\Omega} (u+t)\varphi_1 - C \int_{\Omega} \varphi_1.$$

From the inequality above, it is easy to see that $t \leq C/(\delta_1 + \varepsilon)$. As a result, choosing $t_0 = C/(\delta_1 + \varepsilon) + 1$ we can conclude the desired conclusion.

Finally, we only need to verify condition (b) of Lemma 6.2.1. In fact, by the growth condition (H_1), we know that for all $t \in [0, t_0]$, the solutions for equation $H(u,t) = u$ have a uniform a priori bound R_0^* (based on the "blow up" a priori estimates in [94]). Hence, for all $R \geq R_0 \equiv \max\{r_0, R_0^*\} + 1$, we have $H(u,t) \neq u$ for all $u \in \partial K_R$.

Noticing (6.55) and (6.57), for all $r \in (0, r_0]$ and $R \in [R_0, +\infty)$ we have

$$i\big(T_{0,1}(\cdot,v), K_R \backslash \overline{K}_r, K\big) = -1. \tag{6.58}$$

$\square$

Lemma 6.2.4 (Zhang and Cheng [203]) *Assume that f_2 satisfies (A_2), then there exist $\overline{R}_0 > \overline{r}_0 > 0$ such that for all $\overline{r} \in (0, \overline{r}_0]$ and $\overline{R} \in [\overline{R}_0, +\infty)$,*

$$i\big(T_{0,2}(u,\cdot), K_{\overline{R}} \backslash \overline{K}_{\overline{r}}, K\big) = 1.$$

Proof By the definition of $T_{\lambda,2}$, we get $T_{0,2}(u,v) = S[f_2(x,0,v)]$.

From assumption (A_2), there exist $\varepsilon > 0$ and $\overline{r}_0 > 0$ such that

$$f_2(x,0,v) \geq (\delta_1 + \varepsilon)v, \quad \forall(x,v) \in \overline{\Omega} \times [0,\overline{r}], \text{ where } \overline{r} \in (0,\overline{r}_0]. \tag{6.59}$$

Now we show that

$$i\big(T_{0,2}(u,\cdot),K_{\bar r},K\big)=0 \quad\text{for all }\bar r\in(0,\bar r_0].\tag{6.60}$$

In fact, we only need to make the homotopy $H^*:\overline{K_{\bar r}}\times\mathbb{R}^+\to K$ as follows:

$$H^*(v,t)=S\big[f_2(x,0,v)\big]+\frac{t}{\delta_1}\varphi_1,$$

and then prove that H^* satisfies all the conditions of (ii) in Lemma 6.2.1.

First, it is clear that condition (a) of Lemma 6.2.1 is valid.

Second, we consider solutions for equation $H^*(v,t)=v$. Assume that v is a solution for it, then v satisfies the following equation:

$$\begin{cases}-\Delta v=f_2(x,0,v)+t\varphi_1, & \forall x\in\Omega,\\ v|_{\partial\Omega}=0.\end{cases}$$

Noticing (6.59), we have

$$-\Delta v\ge(\delta_1+\varepsilon)v+t\varphi_1.$$

Multiplying both sides of the above inequality by φ_1 and integrating on Ω, we know that

$$\int_\Omega(-\Delta v)\varphi_1\ge(\delta_1+\varepsilon)\int_\Omega v\varphi_1+t\int_\Omega\varphi_1^2,$$

which implies a contradiction $\delta_1\ge\delta_1+\varepsilon$! As a result, conditions (b) and (c) of Lemma 6.2.1 also hold.

By assumption (A_2) and continuity of f_2, there exist $\varepsilon\in(0,\delta_1)$ and $C>0$ such that

$$f_2(x,0,v)\le(\delta_1-\varepsilon)v+C,\quad\forall(x,v)\in\overline\Omega\times\mathbb{R}^+.\tag{6.61}$$

Next, we show that there exists $\vec R_0>\bar r_0$ such that

$$i\big(T_{0,2}(u,\cdot),K_{\overline R},K\big)=1\quad\text{for all }\overline R\in[\overline R_0,\infty).\tag{6.62}$$

With that purpose, suppose that there exist $t\ge1$ and $v\in\partial K_{\overline R}$ such that $T_{0,2}(u,v)=tv$, that is,

$$\begin{cases}-\Delta v=t^{-1}f_2(x,0,v), & \forall x\in\Omega,\\ v|_{\partial\Omega}=0.\end{cases}\tag{6.63}$$

In what follows, we prove that there exists a positive constant C (independent of t) such that $\|v\|_\infty\le C$ for all solutions v of (6.63). From (6.61) and (6.63), it follows that

$$\int_\Omega|\nabla v|^2\le(\delta_1-\varepsilon)\int_\Omega v^2+C\int_\Omega v,$$

combining with Poincaré's inequality and Hölder's inequality, which implies that

$$\|v\|_{L^2} \leq C, \quad \|v\|_{L^1} \leq C \quad \text{and} \quad \|v\|_{H_0^1} \leq C. \tag{6.64}$$

Furthermore, by (6.61) and Sobolev embedding theorem, we know that

$$\left\| f_2(x,0,v) \right\|_{L^{2*}} \leq C\|v\|_{H_0^1} + C. \tag{6.65}$$

By L^p-theory about elliptic equations, we get $v \in W^{2,2^*}(\Omega)$ and

$$\|v\|_{W^{2,2^*}} \leq C\left\| f_2(x,0,v) \right\|_{L^{2*}}. \tag{6.66}$$

In combination with (6.61), (6.64)–(6.66) and the boot-strap technique, it is not difficult to show that there is a positive constant C (independent of t) such that $\|v\|_{L^\infty} \leq C$, that is, all solutions of (6.63) have a uniform bound C. Choosing $\overline{R}_0 = \max\{\overline{r}_0, C\} + 1$, we have $T_{0,2}(u,v) \neq tv$ for all $t \geq 1$ and $v \in \partial B_{\overline{R}}$, $\forall \overline{R} \geq \overline{R}_0$. As a result, applying conclusion (i) of Lemma 6.2.1 we conclude that (6.62) is valid.

By (6.60) and (6.62), for all $\overline{r} \in (0, \overline{r}_0]$ and $\overline{R} \in [\overline{R}_0, +\infty)$ we have

$$i\left(T_{0,2}(u,\cdot), K_{\overline{R}}\backslash \overline{K}_{\overline{r}}, K\right) = 1. \tag{6.67}$$

$\square$

Lemma 6.2.5 (Zhang and Cheng [203]) *Suppose that f_1 satisfies (H_1) and that f_2 satisfies (A_2) and (H_2). Let $(u(x), v(x))$ be a positive solution of system (6.82), then there exists some uniform constant C (independent of λu and v) such that $\|u\|_{L^\infty} \leq C$ and $\|v\|_{L^\infty} \leq C$.*

Proof We will prove that there exist positive constants C_1 and C_2 (independent of λ, u and v) such that $\|v\|_{L^\infty} \leq C_1$ and $\|u\|_{L^\infty} \leq C_2$ according to the following two steps.

Step 1. We will show that there exists a positive constant C_1 (independent of λ, u and v) such that $\|v\|_{L^\infty} \leq C_1$, which is based on L^p-theory and boot-strap technique. Furthermore, there is a positive constant C^* (independent of λ, u and v) such that $\|v\|_{C^{1,\alpha}} \leq C^*$, here $\alpha \in (0,1)$.

Noticing that $(u(x), v(x))$ satisfies the following equation:

$$\begin{cases} -\Delta v(x) = \lambda f_2\big(x, u(x), v(x)\big) + (1-\lambda) f_2\big(x, 0, v(x)\big), & \forall x \in \Omega, \\ v|_{\partial\Omega} = 0. \end{cases} \tag{6.68}$$

By assumptions (A_2) and (H_2), there exist $\varepsilon \in (0, \delta_1)$ and $C > 0$ such that

$$\lambda f_2(x,u,v) + (1-\lambda) f_2(x,0,v) \leq (\delta_1 - \varepsilon)v + C,$$

$$\forall (x,u,v) \in \overline{\Omega} \times \mathbb{R}^+ \times \mathbb{R}^+. \tag{6.69}$$

By (6.68) and (6.69), it follows that

$$\int_{\Omega} |\nabla v(x)|^2 \le (\delta_1 - \varepsilon) \int_{\Omega} v^2(x) + C \int_{\Omega} v(x),$$

combining with Poincaré's inequality and Hölder's inequality, which implies that

$$\|v\|_{L^2} \le C, \quad \|v\|_{L^1} \le C \quad \text{and} \quad \|v\|_{H_0^1} \le C. \tag{6.70}$$

Furthermore, by (6.69) and Sobolev embedding theorem, we know that

$$\left\| \lambda f_2(x, u, v) + (1 - \lambda) f_2(x, 0, v) \right\|_{L^{2^*}} \le C \|v\|_{H_0^1} + C. \tag{6.71}$$

By L^p-theory about elliptic equations, we get $v \in W^{2,2^*}(\Omega)$ and

$$\|v\|_{W^{2,2^*}} \le C \left\| \lambda f_2(x, u, v) + (1 - \lambda) f_2(x, 0, v) \right\|_{L^{2^*}}. \tag{6.72}$$

In combination with (6.69)–(6.72) and boot-strap technique, it is not difficult to show that there is a positive constant C_1 (independent of λ, u and v) such that $\|v\|_{L^\infty} \le C_1$.

In addition, by L^p-theory and Sobolev embedding theorem, it is easy to prove that there is a positive constant C^* (independent of λ, u and v) such that $\|v\|_{C^{1,\alpha}} \le C^*$, here $\alpha \in (0, 1)$.

Step 2. We will prove that there exists a positive constant C_2 (independent of λ, u and v) such that $\|u\|_{L^\infty} \le C_2$, which is based on the "blow up" a priori estimates technique in [94].

Suppose, by contradiction, that there is no such a priori bound. That is, there exist a sequence of numbers $\{\lambda_k\}_{k=1}^{\infty} \subset [0, 1]$ and a sequence of positive solutions $\{(u_k, v_k)\}_{k=1}^{\infty}$ to a family of systems

$$\begin{cases} -\Delta u_k = \lambda_k f_1(x, u_k, v_k) + (1 - \lambda_k) f_1(x, u_k, 0), & \forall x \in \Omega, \\ -\Delta v_k = \lambda_k f_2(x, u_k, v_k) + (1 - \lambda_k) f_2(x, 0, v_k), & \forall x \in \Omega, \\ u_k|_{\partial\Omega} = v_k|_{\partial\Omega} = 0, \end{cases} \tag{6.73}$$

such that $\lim_{k \to \infty} \|u_k\|_{L^\infty} = \infty$.

By the maximum principle, there exists a sequence of points $\{P_k\}_{k=1}^{\infty} \subset \Omega$ such that

$$M_k \equiv \sup_{x \in \Omega} u_k(x) = u_k(P_k) \to \infty \quad \text{as } k \to \infty. \tag{6.74}$$

We may assume that $\lambda_k \to \lambda \in [0, 1]$ and $P_k \to P \in \overline{\Omega}$ as $k \to \infty$. The proof breaks down into two cases depending on whether $P \in \Omega$ or $P \in \partial\Omega$.

Case I: As $P \in \Omega$. Let $2d$ denote the distance of P to $\partial\Omega$, and $B_r(a)$ the ball of radius r and center $a \in \mathbb{R}^n$. Let μ_k be a sequence of positive numbers (to be defined below) and $y = \frac{x - P_k}{\mu_k}$. Define the scaled function

$$\bar{u}_k(y) = \mu_k^{\frac{2}{q-1}} u_k(x). \tag{6.75}$$

Choose μ_k such that

$$\mu_k^{\frac{2}{q-1}} M_k = 1. \tag{6.76}$$

Since $M_k \to \infty$, we have $\mu_k \to 0$ as $k \to \infty$. For large k, $\overline{u}_k(y)$ is well defined in $B_{d/\mu_k}(0)$, and

$$\sup_{y \in B_{d/\mu_k}(0)} \overline{u}_k(y) = \overline{u}_k(0) = 1. \tag{6.77}$$

Moreover, $\overline{u}_k(y)$ satisfies in $B_{d/\mu_k}(0)$

$$-\Delta \overline{u}_k(y) = \mu_k^{\frac{2q}{q-1}} \Big[\lambda_k f_1\big(\mu_k y + P_k, \overline{u}_k(y), v_k(\mu_k y + P_k)\big)$$
$$+ (1 - \lambda_k) f_1\big(\mu_k y + P_k, \overline{u}_k(y), 0\big) \Big]. \tag{6.78}$$

Note that v_k are uniformly bounded (see Step 1), and by assumption (H_1)

$$\begin{cases} \lim_{k \to \infty} \left| \mu_k^{\frac{2q}{q-1}} f_1\big(\mu_k y + P_k, \mu_k^{-\frac{2}{q-1}} \overline{u}_k(y), v_k(\mu_k y + P_k)\big) \right. \\ \qquad \left. - h_1\big(\mu_k y + P_k, v_k(\mu_k y + P_k)\big)\big(\overline{u}_k(y)\big)^q \right| = 0, \\ \lim_{k \to \infty} \left| \mu_k^{\frac{2q}{q-1}} f_1\big(\mu_k y + P_k, \mu_k^{-\frac{2}{q-1}} \overline{u}_k(y), 0\big) - h_1(\mu_k y + P_k, 0)\big(\overline{u}_k(y)\big)^q \right| = 0. \end{cases} \tag{6.79}$$

Therefore, given any radius R such that $B_R(0) \subset B_{d/\mu_k}(0)$, by L^p-theory we can find uniform bounds for $\|\overline{u}_k\|_{W^{2,p}(B_R(0))}$.

Choosing $p > n$ large, by Sobolev compact embedding theorem we find that $\{\overline{u}_k\}$ is precompact in $C^{1,\alpha}(B_R(0))$ $(0 < \alpha < 1)$. It follows that there exists a subsequence $\overline{u}_{k_j}$ converging to $\overline{u}$ in $W^{2,p}(B_R(0)) \cap C^{1,\alpha}(B_R(0))$. By Hölder continuity $\overline{u}(0) = 1$. From the result obtained in Step 1 and the Arzelà–Ascoli theorem, there exists a subsequence of $v_k(\mu_k y + P_k)$, relabeling $v_{k_j}(\mu_{k_j} y + P_{k_j})$, which converges to $v(P)$ in $C(B_R(0))$. Furthermore, since

$$\begin{cases} \lambda_{k_j} \to \lambda, \quad v_{k_j}(\mu_{k_j} y + P_{k_j}) \to v(P), \\ h_1\big(\mu_{k_j} y + P_{k_j}, v_{k_j}(\mu_{k_j} y + P_{k_j})\big) \to h_1\big(P, v(P)\big), \\ h_1(\mu_{k_j} y + P_{k_j}, 0) \to h_1(P, 0), \end{cases} \tag{6.80}$$

as $k_j \to \infty$, $\overline{u}(y)$ is a solution of

$$-\Delta \overline{u}(y) = \big[\lambda h_1\big(P, v(P)\big) + (1 - \lambda) h_1(P, 0) \big] \overline{u}^q(y). \tag{6.81}$$

We claim that $\overline{u}$ is well defined in all of $\mathbb{R}^n$ and $\overline{u}_{k_j} \to \overline{u}$ in $W^{2,p} \cap C^{1,\alpha}$ $(p > n)$ on any compact subset. To show this we consider $B_{R'}(0) \supset B_R(0)$. Repeating the above argument with $B_{R'}(0)$, the subsequence $\overline{u}_{k_j}$ has a convergent subsequence $\overline{u}_{k'_j} \to \overline{u}'$ on $B_{R'}(0)$. $\overline{u}'$ satisfies (6.76), and necessarily $\overline{u}'|_{B_R(0)} = \overline{u}$. By unique continuation the entire original subsequence $\overline{u}_{k_j}$ converges, so that $\overline{u}$ is well defined. By the global result of Liouville type (see Theorem 6.2.1) we have $\overline{u} = 0$, a contradiction, since $\overline{u}(0) = 1$.

Case II: As $P \in \partial\Omega$. By arguments similar to Case I, we can reduce the problem of a priori bounds to the global results of Liouville type (see Theorem 6.2.2) and deduce a contradiction. $\square$

Before proving Theorem 6.2.3, let us state our main idea of proof. First, we deal with the single equations $-\Delta u = f_1(x, u, 0)$ and $-\Delta v = f_2(x, 0, v)$ with Dirichlet boundary conditions, and then consider the following parameterized system:

$$\begin{cases} -\Delta u = \lambda f_1(x, u, v) + (1 - \lambda) f_1(x, u, 0), & \text{in } \Omega, \\ -\Delta v = \lambda f_2(x, u, v) + (1 - \lambda) f_2(x, 0, v), & \text{in } \Omega, \\ u = v = 0, & \text{on } \partial\Omega, \end{cases} \tag{6.82}$$

where parameter $\lambda \in [0, 1]$. Based on the preceding preliminaries, we only need to consider the fixed point index of compact mapping T_λ corresponding to system (6.82). Applying the homotopy invariance and product formula (see Lemma 6.2.2) of the fixed point index together with some fixed point index results (see Lemmas 6.2.3 and 6.2.4), we can compute the fixed point index of compact mapping T_1 corresponding to system (6.49), and establish the existence of positive solutions.

Theorem 6.2.3 (Zhang and Cheng [203]) *Assume that f_1 satisfies (A_1) and that f_2 satisfies (A_2). If there exist $q \in (1, \frac{n+2}{n-2})$, $h_1 \in C(\overline{\Omega} \times \mathbb{R}^+, \mathbb{R}^+\backslash\{0\})$ and $h_2 \in B_{\mathrm{loc}}(\mathbb{R}^+, \mathbb{R}^+)$ such that*

$$(H_1) \quad \lim_{u \to +\infty} \frac{f_1(x, u, v)}{u^q} = h_1(x, v)$$

uniformly with respect to $(x, v) \in \overline{\Omega} \times [0, M]$ $(\forall M > 0)$,

$$(H_2) \quad \limsup_{u \to +\infty} \max_{x \in \overline{\Omega}} f_2(x, u, v) = h_2(v)$$

uniformly with respect to $v \in [0, M]$ $(\forall M > 0)$,

then system (6.49) has at least one positive solution.

Proof We can seek the fixed points of T_1 in one certain open set $(K_{R_1}\backslash\overline{K}_{r_1}) \times (K_{R_2}\backslash\overline{K}_{r_2})$, where $r_1 \in (0, r_0]$, $R_1 \in [R_0, \infty)$, $r_2 \in (0, \overline{r}_0]$ and $R_2 \in [\overline{R}_0, \infty)$ will be determined later.

Combining Lemma 6.2.2 with (6.58) and (6.67), we know that

$$i\left(T_0, (K_{R_1}\backslash\overline{K}_{r_1}) \times (K_{R_2}\backslash\overline{K}_{r_2}), K \times K\right) = -1.$$

In order to seek the non-trivial fixed points of T_1, we want to prove that

$$i\big(T_1, (K_{R_1}\backslash\overline{K}_{r_1}) \times (K_{R_2}\backslash\overline{K}_{r_2}), K \times K\big)$$
$$= i\big(T_0, (K_{R_1}\backslash\overline{K}_{r_1}) \times (K_{R_2}\backslash\overline{K}_{r_2}), K \times K\big).$$

By the homotopy invariance of fixed point index, we only need to verify that

$$(u, v) \neq T_\lambda(u, v), \quad \forall \lambda \in [0, 1] \quad \text{and} \quad (u, v) \in \partial\big[(K_{R_1}\backslash\overline{K}_{r_1}) \times (K_{R_2}\backslash\overline{K}_{r_2})\big]. \tag{6.83}$$

First, by condition (A_1) there are $\varepsilon \in (0, \delta_1)$ and $r_1 \in (0, r_0]$ such that

$$\lambda f_1(x, u, v) + (1 - \lambda) f_1(x, u, 0) \leq (\delta_1 - \varepsilon)u,$$
$$\forall x \in \overline{\Omega}, \ u \in [0, r_1] \text{ and } v \in \mathbb{R}^+. \tag{6.84}$$

We claim that

$$(u, v) \neq T_\lambda(u, v), \quad \forall \lambda \in [0, 1] \quad \text{and} \quad (u, v) \in \partial K_{r_1} \times K. \tag{6.85}$$

In fact, if there exist $\lambda_0 \in [0, 1]$ and $(u_0, v_0) \in \partial K_{r_1} \times K$, such that $(u_0, v_0) = T_{\lambda_0}(u_0, v_0)$, then (u_0, v_0) satisfies the following equation:

$$\begin{cases} -\Delta u_0 = \lambda_0 f_1(x, u_0, v_0) + (1 - \lambda_0) f_1(x, u_0, 0), & \forall x \in \Omega, \\ u_0|_{\partial\Omega} = 0. \end{cases} \tag{6.86}$$

By (6.84) and (6.86), we have

$$-\Delta u_0 \leq (\delta_1 - \varepsilon)u_0.$$

Multiplying both sides of the inequality above by φ_1 and integrating on Ω, we get

$$-\int_\Omega \Delta u_0 \varphi_1 \leq (\delta_1 - \varepsilon) \int_\Omega u_0 \varphi_1,$$

which yields a contradiction $\delta_1 \leq \delta_1 - \varepsilon$!

Second, by assumption (A_2) we know that there exist $\varepsilon > 0$ and $r_2 \in (0, \overline{r}_0]$ such that

$$\lambda f_2(x, u, v) + (1 - \lambda) f_2(x, 0, v) \geq (\delta_1 + \varepsilon)v,$$
$$\forall x \in \overline{\Omega}, \ v \in [0, r_2] \text{ and } u \in \mathbb{R}^+. \tag{6.87}$$

By (6.87) and the proof similar to (6.85), we obtain

$$(u, v) \neq T_\lambda(u, v), \quad \forall \lambda \in [0, 1] \quad \text{and} \quad (u, v) \in K \times \partial K_{r_2}. \tag{6.88}$$

Finally, we consider the equation $T_\lambda(u, v) = (u, v)$, that is,

$$\begin{cases} -\Delta u = \lambda f_1(x, u, v) + (1 - \lambda) f_1(x, u, 0), & \forall x \in \Omega, \\ -\Delta v = \lambda f_2(x, u, v) + (1 - \lambda) f_2(x, 0, v), & \forall x \in \Omega, \\ u|_{\partial\Omega} = v|_{\partial\Omega} = 0. \end{cases} \tag{6.89}$$

From Lemma 6.2.5, all the solutions for (6.89) have a uniform a priori bound C independent of λ, u and v. Hence, choosing $R_1 \geq \max\{R_0, C + 1\}$ and $R_2 \geq \max\{\overline{R}_0, C + 1\}$ we have

$$\begin{cases} (u, v) \neq T_\lambda(u, v), & \forall \lambda \in [0, 1] \text{ and } (u, v) \in \partial K_{R_1} \times K, \\ (u, v) \neq T_\lambda(u, v), & \forall \lambda \in [0, 1] \text{ and } (u, v) \in K \times \partial K_{R_2}. \end{cases} \tag{6.90}$$

In combination with (6.85), (6.88) and (6.90), it is easy to see that (6.83) is valid. $\qquad\qquad\square$

Example 6.2.1 The system

$$\begin{cases} -\Delta u = \tan^{-1}(1 + v)\, u^2 & \text{in } \Omega, \\ -\Delta v = \delta_1 \cot^{-1}(-u)|\sin v| & \text{in } \Omega, \\ u = v = 0 & \text{on } \partial\Omega, \end{cases} \tag{6.91}$$

where $\Omega \subset \mathbb{R}^3$ is a smooth bounded domain, δ_1 is the first eigenvalue of the Laplacian subject to Dirichlet data. It is easy to verify that the nonlinearities in system (6.91) satisfy the conditions of Theorem 6.2.3. Hence, system (6.91) has at least one positive solution.

Remark 6.2.1 We point out that if f_1 (resp. f_2) is sublinear with respect to u (resp. v) at the origin and infinity in the sense of our definitions, in addition, there exist $g_1, g_2 \in B_{\text{loc}}(\mathbb{R}^+, \mathbb{R}^+)$ such that

(G_1) $\limsup\limits_{v \to +\infty} \max\limits_{x \in \overline{\Omega}} f_1(x, u, v) = g_1(u)$

       uniformly with respect to $u \in [0, M]$ $(\forall M > 0)$,

(G_2) $\limsup\limits_{u \to +\infty} \max\limits_{x \in \overline{\Omega}} f_2(x, u, v) = g_2(v)$

       uniformly with respect to $v \in [0, M]$ $(\forall M > 0)$,

then system (6.49) has at least one positive solution, which can be obtained by the proof similar to Theorem 6.2.3.

Remark 6.2.2 For the priori estimate and existence of positive solution $u \in C^2(\bar{\Omega})$ of

$$\begin{cases} -\Delta u = f(u) & \text{in } \Omega \subset \mathbb{R}^n, \\ u = 0 & \text{on } \partial\Omega, \\ u > 0 & \text{in } \Omega, \end{cases} \tag{6.92}$$

where Ω is some bounded regular domain, f is superlinear, please see [80].

Chapter 7
Dancer–Fučik Spectrum

7.1 The Spectrum of a Self-adjoint Operator

Let H be a Hilbert space and A a linear operator with a dense domain $D(A)$. The set of all Ax with $x \in D(A)$ is denoted by $R(A)$ or $\operatorname{im} A$ (see [91]). Let D^* be a linear manifold in H such that $v \in D^*$ if there is an element $w \in H$ such that

$$(w, u) = (v, Au) \quad \text{for all } u \in D(A).$$

Such w is unique, since $D(A)$ is dense in H, one defines $w = A^* v$ and $D(A^*) = D^*$. The operator A^* is called adjoint to A, its domain can be not dense in H, so $(A^*)^*$ is not always defined. Moreover, if it is defined, it can be different from A. An operator A is called self-adjoint if $A^* = A$.

If $A^* = A$ then $(Au, v) = (u, Av)$ for all $u, v \in D(A)$. However, this equality is not sufficient for self-adjointness, since $D(A^*)$ can be wider than $D(A)$. If $(Au, v) = (u, Av)$ for all $u, v \in D(A)$, the operator A is called symmetric.

Theorem 7.1.1 (See [91]) *If a linear operator A with a dense domain is invertible, then A^* is also invertible and $(A^*)^{-1} = (A^{-1})^*$.*

If an operator A is self-adjoint, bounded, and invertible, then A^{-1} is self-adjoint.

Definition 7.1.1 A resolvent set $\rho(A)$ of an operator A is a set of complex numbers λ for which the operator $(\lambda I - A)^{-1}$ is defined and bounded. The set $C \setminus \rho(A)$ is called the spectrum of the operator A and is denoted as $\sigma(A)$. If the map $\lambda I - A$ is not one-to-one, then λ belongs to the point spectrum $\sigma_p(A)$. Otherwise, if this map is one-to-one and $\operatorname{im}(\lambda I - A)$ is dense in H, but does not coincide with H, we say that λ belongs to the continuous spectrum $\sigma_c(A)$.

The notion essential spectrum $\sigma_{\mathrm{ess}}(A)$ consists of the points $\lambda \in \sigma(A)$ that are not isolated eigenvalue of a finite multiplicity. It is easy to say that $\lambda \in \sigma_{\mathrm{ess}}(A)$ if and only if there exists an infinite sequence $\{v_j\}$ of mutually orthogonal unit vectors such that $\|Av_j - \lambda v_j\| \to 0$ as $j \to \infty$.

Here are some useful theorems:

Z. Zhang, *Variational, Topological, and Partial Order Methods with Their Applications*,   199
Developments in Mathematics 29,
DOI 10.1007/978-3-642-30709-6_7, © Springer-Verlag Berlin Heidelberg 2013

Theorem 7.1.2 (See [91]) *A linear closed symmetric operator in H is self-adjoint if and only if its spectrum is lying on the real axis.*

7.2 Dancer–Fučik Spectrum on Bounded Domains

Let Ω be a bounded domain in $\mathbb{R}^N$, $N \geq 1$, and let A be a self-adjoint operator on $L^2(\Omega)$, we assume that $A \geq \lambda_1 > 0$ and that $C_0^\infty(\Omega) \subset D := D(A^{1/2})$. We assume further that A has compact resolvent and that λ_1 is a simple eigenvalue with corresponding eigenfunction φ_1 which does not vanish on Ω a.e. Thus the spectrum of A consists only of isolated eigenvalues λ_k of finite multiplicities (with eigenfunction φ_k) satisfying

$$0 < \lambda_1 < \lambda_2 < \cdots < \lambda_k < \cdots . \tag{7.1}$$

The Dancer–Fučik Spectrum Σ of A is the set of those points $(b, a) \in \mathbb{R}^2$ such that

$$Au = bu^+ - au^-, \quad u \in D \tag{7.2}$$

has a non-trivial solution, where $u^\pm = \max\{\pm u, 0\}$. This generalized notion of spectrum was introduced in the 1970s by Fučik [93] and Dancer [64] who considered the problem

$$-\Delta u = bu^+ - au^- \quad \text{in } \Omega, \qquad u|_{\partial\Omega} = 0.$$

It was recognized by Fučik [92], Dancer [64] and others that this set is an important factor in the study of semilinear elliptic boundary value problems with jumping nonlinearities.

$$Au = f(x, u), \quad u \in D, \tag{7.3}$$

when

$$\frac{f(x, t)}{t} \to a \quad \text{as } t \to -\infty; \qquad \frac{f(x, t)}{t} \to b \quad \text{as } t \to +\infty. \tag{7.4}$$

Several papers have been devoted since then to the Fučik spectrum of the Laplacian on a bounded domain Ω (e.g., [62, 78, 160]). In [160] Schechter obtained the existence of Fučik spectrum near the points (λ_k, λ_k), $k \in \mathbb{N}$, where λ_k are the eigenvalues of $-\Delta$. In [62], Cuesta et al. studied the Fučik spectrum of the p-Laplacian on a bounded domain in $\mathbb{R}^N$, and obtained the first non-trivial curve in the Fučik spectrum as well as some properties of the curve. In [16], Arias et al. studied the beginning of the Fučik spectrum with weights on the right side. In [30] it was proved that the eigenfunctions corresponding to eigenvalues (α, β) in the first non-trivial curve are foliated Schwarz symmetric, and in [24] it was proved that these eigenfunctions are not radially symmetric. The reader can find further references to the literature on the Fučik spectrum for the Laplace operator on bounded domains in the above mentioned papers.

In one dimension it was shown in [92] that Σ consists of a sequence of curves going through the points (λ_k, λ_k). Cac [37] showed in higher dimension that in the region $\lambda_{k-1} < a < \lambda_k$, $\lambda_k < b < \lambda_{k+1}$, Σ contains one or two curves emanating from (λ_k, λ_k) while the squares $\lambda_{k-1} < a, b < \lambda_k$, $\lambda_k < a, b < \lambda_{k+1}$ are free of the Fučik spectrum. M. Schechter extended the methods of Cac to show that for each $k \geq 1$, there are one or two continuous, non-increasing curves in Σ in the square $\lambda_{k-1} \leq a, b \leq \lambda_{k+1}$ passing through the point (λ_k, λ_k) and such that all point in the square except those on the curves and possibly those between the curves are not in Σ. This is true for elliptic boundary value problems of any order. E.g.

$$-\Delta u = bu^+ - au^-, \quad \text{in } \Omega, \qquad u|_{\partial\Omega} = 0. \tag{7.5}$$

$$-\Delta u = \lambda u, \quad \text{in } \Omega, \qquad u|_{\partial\Omega} = 0. \tag{7.6}$$

(7.6) has a sequence of eigenvalues $0 < \lambda_1 < \lambda_2 < \cdots < \lambda_k < \cdots$.

Σ of (7.5) clearly contains in particular (λ_1, λ_1), (λ_2, λ_2), $\ldots$, (λ_k, λ_k), $\ldots$ and the two lines $\{\lambda_1\} \times \mathbb{R}$ and $\mathbb{R} \times \{\lambda_1\}$. The weak solutions of (7.5) are critical points of the functional

$$I_{(b,a)}(u) = \frac{1}{2} \int_\Omega |\nabla u|^2 - b|u^+|^2 - a|u^-|^2, \quad \forall u \in H_0^1(\Omega). \tag{7.7}$$

If $(b, a) \notin \Sigma$, $u = 0$ is the only critical point of $I_{(b,a)}$, and we can define the critical group $C_q(I_{(b,a)}, 0)$. Dancer [66] shows that, for $\lambda_{l-1} < a < \lambda_l < b < \lambda_{l+1}$,

$$C_q(I_{(b,a)}, 0) = 0, \quad \text{if } q < d_{l-1} \text{ or } q > d_l,$$

where $d_l = \dim N_l$, N_l is the subspace spanned by the eigenfunctions corresponding to $\lambda_1, \lambda_2, \ldots, \lambda_l$. Moreover, we have

Theorem 7.2.1 (M. Schechter [160]) *There are sequences of non-increasing functions $\mu_k(a)$, $\nu_k(a)$, $a \in \mathbb{R}^+$, $k = 0, 1, 2, \ldots$, such that*

$$\mu_k(\lambda_k) = \lambda_k, \qquad \nu_k(\lambda_{k+1}) = \lambda_{k+1} \tag{7.8}$$

and for $k > 0$,

(a) *if a, $\nu_k(a) \geq \lambda_k$, then $(a, \nu_k(a)) \in \Sigma$;*
(b) *if a, $\mu_k(a) \leq \lambda_k$, then $(a, \mu_k(a)) \in \Sigma$;*
(c) *if $a, b \geq \lambda_k$, $a + b > 2\lambda_k$ and $b < \nu_k(a)$, then $(a, b) \notin \Sigma$;*
(d) *if $a, b \leq \lambda_{k+1}$, $a + b < 2\lambda_{k+1}$ and $b > \mu_k(a)$, then $(a, b) \notin \Sigma$;*
(e) *if $\mu_k(a) < b < \nu_k(a)$ and either $a, b \geq \lambda_k$ or $a, b \leq \lambda_{k+1}$, then $(a, b) \notin \Sigma$;*
(f) *for a, $\nu_k(a) \geq \lambda_k$, $\nu_k(a)$ is continuous and strictly decreasing;*
(g) *for a, $\mu_k(a) \leq \lambda_{k+1}$, $\mu_k(a)$ is continuous and strictly decreasing;*
(h) *$\mu_k(a) \leq \nu_k(a)$, $a \in \mathbb{R}^+$.*

Let $Q_l = (\lambda_{l-1}, \lambda_{l+1}) \times (\lambda_{l-1}, \lambda_{l+1})$. By Theorem 1 of [67], we find that as $(b, a) \in Q_l \setminus \Sigma$, and (b, a) in the type (II) region (between the two curves $\nu_{l-1}(a), \mu_l(a)$ through (λ_l, λ_l))

$$C_q(I_{(b,a)}, 0) = 0, \quad \text{if } q \leq d_{l-1} \text{ or } q \geq d_l.$$

For the smoothness of the curve on the Dancer–Fučik spectrum, we have

Theorem 7.2.2 (Li, Li, Liu and Pan [130]) *Assume λ_l with $l \geq 2$ is a simple eigenvalue and $\varphi_l \in S := \{u \mid \|u\| = 1, \ u \in H_0^1(\Omega)\}$ is a corresponding eigenfunction such that $\int_\Omega (\varphi_l^+)^2 \neq \int_\Omega (\varphi_l^-)^2$. Then there exist two C^1 curves (u_i, b_i) : $I_i \to S \times [\lambda_{l-1}, \lambda_{l+1}]$, $I_i \subset [\lambda_{l-1}, \lambda_{l+1}]$ being closed intervals $(i = 1, 2)$, such that $\Sigma \cap Q_l = C_1 \cup C_2$ (where $C_i = \{(b_i(a), a) : a \in I_i\})$, $f(u_i(a), b_i(a), a) = 0$,*

$$u_1(\lambda_l) = \varphi_l, \qquad u_2(\lambda_l) = -\varphi_l, \qquad b_1(\lambda_l) = b_2(\lambda_l) = \lambda_l,$$

$$b_i'(a) = -\frac{\int_\Omega (u_i^-(a))^2}{\int_\Omega (u_i^+(a))^2}.$$

where $f(u, b, a) = u - (-\Delta)^{-1}(bu^+ - au^-)$ for $u \in L_p(\Omega)$, $p > 1$.

By the proof in [130] we find that C_1, C_2 cross at (λ_l, λ_l). That is, if we assume $\int_\Omega (\varphi_l^+)^2 > \int_\Omega (\varphi_l^-)^2$, then in a neighborhood of (λ_l, λ_l), C_2 is above C_1 on the left side of (λ_l, λ_l), C_2 is below C_1 on the right side of (λ_l, λ_l).

Next we study eigenvalues of $-\Delta_p := -\operatorname{div}(|\nabla u|^{p-2}\nabla u)$ in $W_0^{1,p}(\Omega)$ with Dirichlet boundary condition,

$$-\Delta_p u = \lambda |u|^{p-2} u \quad \text{in } \Omega, \qquad u|_{\partial\Omega} = 0. \tag{7.9}$$

Then we have

$$\lambda_1 = \min\left\{\int_\Omega |\nabla u|^p : u \in W_0^{1,p}(\Omega) \text{ and } \int_\Omega |u|^p = 1\right\}. \tag{7.10}$$

It is well known that $\lambda_1 > 0$ is the first, simple, and that admits an eigenfunction $\phi_1(x) > 0$ in Ω and $\phi_1(x) \in C_0^1(\bar{\Omega})$ (by Proposition 2.1 of [61], we know $\phi_1 \in L^\infty(\Omega)$ and by Lemma 1.1 of [18], $(-\Delta_p)^{-1} : L^\infty(\Omega) \to C_0^1(\bar{\Omega})$ is continuous, compact, order-preserving). A value $\lambda \in \mathbb{R}$ is an eigenvalue of (7.9) if and only if there exists $u \in W_0^{1,p}(\Omega) \setminus \{0\}$ such that

$$\int_\Omega |\nabla u|^{p-2} \nabla u \nabla \varphi \, dx = \lambda \int_\Omega |u|^{p-2} u \varphi \, dx \tag{7.11}$$

for all $\varphi \in C_0^\infty(\Omega)$. u is called an eigenfunction corresponding to λ. The spectrum is a closed set in $\mathbb{R}$. Define

$$\Phi(u) = \int_\Omega |\nabla u|^p \, dx \quad \text{and} \quad J(u) = \int_\Omega |u|^p \, dx, \quad \forall u \in W_0^{1,p}(\Omega).$$

So a real value λ is an eigenvalue of problem (7.9) if and only if there exists $u \in W_0^{1,p}(\Omega) \setminus \{0\}$ such that $\Phi'(u) = \lambda J'(u)$. Let $\mathcal{M} = \{u \in W_0^{1,p}(\Omega) : J(u) = 1\}$, then $\mathcal{M}$ is a manifold in $W_0^{1,p}(\Omega)$ of class C^1. For any $u \in \mathcal{M}$, the tangent space of $\mathcal{M}$ at u is denoted by $T_u\mathcal{M}$, that is, $T_u\mathcal{M} = \{w \in W_0^{1,p}(\Omega) : \langle J'(u), w \rangle = 0\}$. Let $\Phi_{\mathcal{M}}$ be the restriction of Φ to $\mathcal{M}$. We recall that a value c is a critical value of $\Phi_{\mathcal{M}}$ if $\Phi'(u)|_{T_u\mathcal{M}} \equiv 0$, $\Phi_{\mathcal{M}} = c$ for some $u \in \mathcal{M}$.

It follows from standard arguments that positive eigenvalues of (7.9) correspond to positive critical values of $\Phi_{\mathcal{M}}$. A first sequence of positive critical values of $\Phi_{\mathcal{M}}$ comes from the Ljusternik–Schnirelman critical point theory on C^1 manifolds proved by Szulkin [174]. That is, if $\gamma(A)$ denotes the Krasnosel'skii genus on $W_0^{1,p}(\Omega)$ and for any $k \in \mathbb{N}_* := \{1, 2, \ldots\}$, we set $\Gamma_k := \{A \subset \mathcal{M} : A \text{ is compact, symmetric and } \gamma(A) \geq k\}$. Then the value

$$\lambda_k := \inf_{A \in \Gamma_k} \max_{u \in A} \Phi(u) \tag{7.12}$$

is an eigenvalue of (7.9), and λ_1 is isolated, $0 < \lambda_1 < \lambda_2 \leq \lambda_3 \leq \cdots$, $\lim_{k \to +\infty} \lambda_k = +\infty$ (also see [61, 133]). But the question of the structure of the spectrum of (7.9) is still open.

Remark 7.2.1 By (1.18), (1.19) and (7.10), and if Ω is a ball, then the first eigenfunction ϕ_1 according to λ_1 is radially symmetric.

Proposition 7.2.1 (See [61, 133]) *Let Ω_1 be a proper open subset of $\Omega_2 \subset \mathbb{R}^N$, then $\lambda_1(\Omega_2) < \lambda_1(\Omega_1)$.*

λ_2 can also be defined as

$$\lambda_2 = \inf\{\lambda > \lambda_1; \lambda \text{ is an eigenvalue of } -\Delta_p \text{ in } W_0^{1,p}(\Omega)\}.$$

The Fučik spectrum of the p-Laplacian on $W_0^{1,p}(\Omega)$ is defined as the set Σ_p of those $(a, d) \in \mathbb{R}^2$ such that

$$-\Delta_p u = a\left(u^+\right)^{p-1} - d\left(u^-\right)^{p-1} \quad \text{in } \Omega, \qquad u|_{\partial\Omega} = 0, \tag{7.13}$$

has a non-trivial solution, where $u^+ = \max\{u, 0\}$, $u^- = \max\{-u, 0\}$. Here $1 < p < \infty$. By [62], the Fučik spectrum of p-Laplacian on $W_0^{1,p}(\Omega)$ is defined as the set of Σ_p of those $(a, d) \in \mathbb{R}^2$ such that (7.13) has a non-trivial solution u. Σ_p clearly contains in particular (λ_k, λ_k), $k = 1, 2, \ldots$ and the two lines $\{\lambda_1\} \times \mathbb{R}$ and $\mathbb{R} \times \{\lambda_1\}$.

For the first non-trivial curve Γ in Σ_p (see [62]), there exists a continuous function $\eta(t)$ defined on $(\lambda_1, \lambda_2]$ in ad-plane, such that

(a) (7.13) has a non-trivial solution for $(a, d) \in \Gamma = \{(a, \eta(a)); \lambda_1 < a \leq \lambda_2\} \cup \{(\eta(d), d); \lambda_1 < d \leq \lambda_2\}$;
(b) (7.13) has no non-trivial solution for $\lambda_1 < d < \eta(a)$, $a \in (\lambda_1, \lambda_2]$ or $\lambda_1 < a < \eta(d)$, $d \in (\lambda_1, \lambda_2]$.

Remark 7.2.2 Certainly it is true for $p = 2$. Moreover, $(\lambda_k, \lambda_k) \in \Sigma$, $\forall k = 1, 2, \ldots$.

7.3 Dancer–Fučik Point Spectrum on $\mathbb{R}^N$

7.3.1 Introduction

We are concerned with the Fučik point spectrum for Schrödinger operators $-\Delta + V$ in $L^2(\mathbb{R}^N)$ for certain types of potential $V : \mathbb{R}^N \to \mathbb{R}$. It is well known that

Theorem 7.3.1 (See [91]) *If $V(x) \to 0$ as $|x| \to \infty$, then the essential spectrum of the Schrödinger operator $L = -\Delta + V(x)$ coincides with $[0, \infty)$.*

Theorem 7.3.2 (See [91]) *If a measure locally bounded function $V(x)$ is such that*

$$\liminf_{|x| \to \infty} V(x) \geq a,$$

then the operator $L = -\Delta + V(x)$ is semi-bounded from below (that is, L is symmetric such that $(Lu, u) \geq m\|u\|^2$ for all $u \in D(L)$, where $m \in \mathbb{R}$) and has a discrete spectrum on $(-\infty, a)$, so that for any $\varepsilon > 0$ the spectrum of L on $(-\infty, a - \varepsilon)$ consists of a finite number of eigenvalues of finite multiplicities.

The Fučik point spectrum is defined as the set Σ of all $(\alpha, \beta) \in \mathbb{R}^2$ such that

$$-\Delta u + V(x)u = \alpha u^+ + \beta u^-, \quad x \in \mathbb{R}^N, \tag{7.14}$$

has a non-trivial solution u in the form domain of $-\Delta + V$; here $u^+ = \max\{u, 0\}$, $u^- = \min\{u, 0\}$.

These results on bounded domains can be extended to (7.14) if the potential satisfies

(V_0) $V \in C(\mathbb{R}^N, \mathbb{R})$ is coercive: $\lim_{|x| \to \infty} V(x) = +\infty$.

It is well known that then $-\Delta + V$ has compact resolvent, the spectrum consists only of a sequence of eigenvalues $\lambda_k \to \infty$ of finite multiplicity, and the first eigenfunction $e_1 > 0$ is strictly positive. The solutions of (7.14) lie in the form domain $\{u \in H^1(\mathbb{R}^N) : \int_{\mathbb{R}^N} Vu^2 < \infty\}$ of the Schrödinger operator $-\Delta + V$. We state one result when (V_0) holds.

Theorem 7.3.3 (Bartsch, Wang and Zhang [31]) *Suppose (V_0) holds. Then*

(a) $\{\lambda_1\} \times \mathbb{R} \cup \mathbb{R} \times \{\lambda_1\} \subset \Sigma$ *is an isolated subset of the Fučik spectrum of* (7.14).

(b) *There exists a continuous, strictly decreasing function $\vartheta : (\lambda_1, \infty) \to (\lambda_1, \infty)$ such that $\Theta := \{(\alpha, \vartheta(\alpha)) : \alpha > \lambda_1\} \subset \Sigma$. Moreover, $\vartheta(\alpha) \to \infty$ as $\alpha \to \lambda_1$, and $\vartheta(\alpha) \to \lambda_1$ as $\alpha \to \infty$. The Fučik eigenfunctions corresponding to $(\alpha, \beta) \in \Theta$ change sign.*

(c) *If V is radially symmetric, then the non-trivial solutions $u : \mathbb{R}^N \to \mathbb{R}$ of (7.14) with $(\alpha, \beta) \in \Theta$ are foliated Schwarz symmetric but not radially symmetric.*

The proof of Theorem 7.3.3 is an adaptation of the proofs of the corresponding results on bounded domains in [16, 24, 30, 62]. In fact, (V_0) may be weakened as in [29]. We leave this to the interested reader and consider instead a more complicated setting.

We deal with potentials which have a potential well whose steepness is controlled by a real parameter λ. More precisely we require:

(V_1) $V = V_\lambda = a + \lambda g$ with $a, g \in C(\mathbb{R}^N, \mathbb{R})$ satisfying $\inf a > -\infty$, $g \geq 0$ and $\Omega := \operatorname{int}(g^{-1}(0)) \neq \emptyset$.

(V_2) There exist $M_0 > 0$ such that $A := \{x \in \mathbb{R}^N : g(x) < M_0\}$ is bounded.

(V_3) $\overline{\Omega} = g^{-1}(0)$ and $\partial\Omega$ is locally Lipschitz.

We write Σ_λ for the set of those $(\alpha, \beta) \in \mathbb{R}^2$ such that

$$-\Delta u + a(x)u + \lambda g(x)u = \alpha u^+ + \beta u^-, \quad x \in \mathbb{R}^N, \tag{7.15}$$

has a non-trivial solution u in the form domain of $-\Delta + a + \lambda g$. We obtain results in this setting for $\lambda > 0$ large, that is, when the potential well is steep.

We construct the first non-trivial curve Θ_λ in the Fučik point spectrum Σ_λ by minimax methods using the approach from [62]. We also study qualitative properties of the curve and of the corresponding eigenfunctions. Finally we apply these results and establish some existence results for multiple solutions of nonlinear Schrödinger equations with jumping nonlinearity.

Throughout the section we assume $(V_{1,2,3})$. Replacing a by $a + 1 - \inf a$ we may also assume that $\inf V_\lambda \geq \inf a = 1$ for $\lambda \geq 0$. This translation corresponds to a translation of the Fučik point spectrum without changing the eigenfunctions.

7.3.2 The Trivial Part of the Fučik Point Spectrum

We first investigate that part of the Fučik point spectrum which corresponds to positive or negative eigenfunctions. The form domain of $-\Delta + V_\lambda$ is

$$H = H_\lambda := \left\{ u \in H^1(\mathbb{R}^N) : \int_{\mathbb{R}^N} (a(x) + \lambda g(x))u^2\, dx < \infty \right\}.$$

This is a Hilbert space with the inner product

$$(u, v)_\lambda := \int_{\mathbb{R}^N} \left(\nabla u \cdot \nabla v + (a(x) + \lambda g(x))uv \right) dx,$$

and corresponding norm

$$\|u\|_\lambda = \left(\int_{\mathbb{R}^N} \left(|\nabla u|^2 + (a(x) + \lambda g(x))u^2 \right) dx \right)^{1/2}.$$

We begin with some basic facts on the spectrum of $-\Delta + V_\lambda$.

Remark 7.3.1 For any given elements $\varphi_1, \ldots, \varphi_k$ of H_λ, we set

$$U_\lambda(\varphi_1, \ldots, \varphi_k) := \inf\{\|u\|_\lambda^2 : \|u\|_{L^2(\mathbb{R}^N)} = 1, \ (u, \varphi_i)_\lambda = 0 \text{ for } i = 1, 2, \ldots, k\}.$$

For $k \in \mathbb{N}$, the kth Rayleigh quotient is defined as

$$\mu_k^\lambda = \sup_{\varphi_1, \ldots, \varphi_{k-1} \in H_\lambda} U_\lambda(\varphi_1, \ldots, \varphi_{k-1}).$$

Of course, if $k = 1$ this has to be understood as $\mu_1^\lambda = \inf\{\|u\|_\lambda^2 : \|u\|_{L^2(\mathbb{R}^N)} = 1\}$. Clearly $\mu_k^\lambda < \mu_k$ where $0 < \mu_1 < \mu_2 \leq \cdots$ are the Dirichlet eigenvalues of $-\Delta + a$ in Ω, counted with multiplicities. μ_k^λ is increasing in λ, and the results from [29, 142] imply $\mu_k^\lambda \to \mu_k$. Since $\inf \sigma_{\mathrm{ess}}(-\Delta + V_\lambda) \geq 1 + \lambda M_0$ as a consequence of (V_2) it follows that $\mu_1^\lambda, \ldots, \mu_k^\lambda < \inf \sigma_{\mathrm{ess}}(-\Delta + V_\lambda)$ are achieved for $\lambda \geq (\mu_k - 1)/M_0$. μ_1^λ is a simple eigenvalue with normalized eigenfunction $e_1^\lambda > 0$, and $e_1^\lambda \to e_1$ in $L^2(\mathbb{R}^N)$ as $\lambda \to \infty$; $e_1 > 0$ the normalized Dirichlet eigenfunction of $-\Delta + a$ in Ω.

An obvious consequence of Remark 7.3.1 is the following.

Proposition 7.3.1 (Bartsch, Wang and Zhang [31])

(a) $\Sigma_\lambda \subset [\mu_1^\lambda, \infty) \times [\mu_1^\lambda, \infty)$.
(b) *If μ_1^λ is achieved then $\{\mu_1^\lambda\} \times \mathbb{R} \cup \mathbb{R} \times \{\mu_1^\lambda\} \subset \Sigma_\lambda$.*

We now show that any subset of $\{\mu_1^\lambda\} \times \mathbb{R} \cup \mathbb{R} \times \{\mu_1^\lambda\}$ with both components bounded from above is isolated in Σ_λ for λ large.

Proposition 7.3.2 (Bartsch, Wang and Zhang [31]) *Given $(\alpha, \beta) \in \{\mu_1^\lambda\} \times \mathbb{R} \cup \mathbb{R} \times \{\mu_1^\lambda\}$, then if $\lambda > \frac{\max\{\alpha, \beta\} - 1}{M_0}$ there does not exist a sequence $(\alpha_n, \beta_n) \in \Sigma_\lambda \setminus (\{\mu_1^\lambda\} \times \mathbb{R} \cup \mathbb{R} \times \{\mu_1^\lambda\})$ such that $(\alpha_n, \beta_n) \to (\alpha, \beta)$.*

Proof Arguing by contradiction we fix $\lambda > \frac{\max\{\alpha, \beta\} - 1}{M_0}$ and assume the existence of a sequence $(\alpha_n, \beta_n) \in \Sigma_\lambda$ such that $\alpha_n \neq \mu_1^\lambda \neq \beta_n$ and $(\alpha_n, \beta_n) \to (\mu_1^\lambda, \beta)$, for instance. By Proposition 7.3.1 this implies $\alpha_n, \beta_n > \mu_1^\lambda$, hence $\beta \geq \mu_1^\lambda$. Let $u_n \in H_\lambda$ be a solution of

$$-\Delta u_n + V_\lambda u_n = \alpha_n u_n^+ + \beta_n u_n^- \quad \text{in } \mathbb{R}^N \tag{7.16}$$

with $\|u_n\|_{L^2(\mathbb{R}^N)} = 1$. It follows that u_n must change sign and that (u_n) remains bounded in H_λ. Consequently, we have along a subsequence,

$$u_n \rightharpoonup u \quad \text{in } H_\lambda, \qquad u_n \to u \quad \text{in } L^2_{\mathrm{loc}}(\mathbb{R}^N), \tag{7.17}$$

hence

$$-\Delta u + V_\lambda u = \mu_1^\lambda u^+ + \beta u^- \quad \text{in } \mathbb{R}^N.$$

Multiplying by u^+ and integrating, we obtain

$$\int_{\mathbb{R}^N} \left(|\nabla u^+|^2 + V_\lambda (u^+)^2\right) = \mu_1^\lambda \int_{\mathbb{R}^N} (u^+)^2,$$

which implies $u^+ = \gamma e_1^\lambda$ for some $\gamma \geq 0$.

We distinguish the cases $\gamma > 0$ and $\gamma = 0$.

Case 1: $\gamma > 0$. In this case we have $u = u^+ = \gamma e_1^+ > 0$ and thus

$$\mathrm{mes}\big(A \cap \{u_n < 0\}\big) \to 0 \quad \text{as } n \to \infty; \tag{7.18}$$

here mes denotes the Lebesgue measure. Since

$$\int_{\mathbb{R}^N \setminus A} V_\lambda (u_n^-)^2 \geq (1 + \lambda M_0) \int_{\mathbb{R}^N \setminus A} (u_n^-)^2,$$

we have

$$\text{either} \quad u_n^-|_{\mathbb{R}^N \setminus A} = 0 \quad \text{or} \quad \int_{\mathbb{R}^N \setminus A} V_\lambda (u_n^-)^2 > \beta_n \int_{\mathbb{R}^N \setminus A} (u_n^-)^2 \quad \text{for } n \text{ large}$$

because $1 + \lambda M_0 > \beta_n$ for n large.

Subcase 1a: $u_n^-|_{\mathbb{R}^N \setminus A} = 0$ along a subsequence. Here it follows from (7.17) that $u_n^- \to u^- = 0$ in $L^2(\mathbb{R}^N)$. Using (7.16) we see that $u_n^- \to 0$ in H_λ. This implies $\|u_n^+\|_{L^2(\mathbb{R}^N)} \to 1$ and, again by (7.16)

$$\|u_n^+\|_\lambda^2 = \|u_n\|_\lambda^2 + o(1) = \alpha_n \|u_n^+\|_{L^2(\mathbb{R}^N)}^2 + o(1) \to \alpha.$$

This contradicts

$$\|u_n^+\|_\lambda^2 \geq \int_{\mathbb{R}^N \setminus A} V_\lambda (u_n^+)^2 \geq (1 + \lambda M_0) \int_{\mathbb{R}^N \setminus A} (u_n^+)^2$$

because $1 + \lambda M_0 > \alpha_n$ for n large.

Subcase 1b: $\int_{\mathbb{R}^N \setminus A} V_\lambda (u_n^-)^2 > \beta_n \int_{\mathbb{R}^N \setminus A} (u_n^-)^2$ along a subsequence. Multiplying (7.16) by u_n^- and integrating yields

$$\int_{\mathbb{R}^N} \left(|\nabla u_n^-|^2 + V_\lambda (u_n^-)^2\right) = \beta_n \int_{\mathbb{R}^N} (u_n^-)^2.$$

This together with the Hölder and Sobolev inequalities for some $2 < q < 2N/(N-2)$ implies in the present subcase

$$\|\nabla u_n^-\|_{L^2(A)}^2 \leq \int_A \left(|\nabla u_n^-|^2 + V_\lambda (u_n^-)^2\right) < \beta_n \int_A (u_n^-)^2$$

$$\leq \beta_n \operatorname{mes}\big(A \cap \{u_n < 0\}\big)^{1-\frac{2}{q}} \big\|u_n^-\big\|_{L^q(A)}^2$$

$$\leq c\beta_n \operatorname{mes}\big(A \cap \{u_n < 0\}\big)^{1-\frac{2}{q}} \big\|\nabla u_n^-\big\|_{L^2(A)}^2.$$

Now (7.18) implies $u_n^-|_A = 0$ for n large. However, $u_n^- \neq 0$ because u_n changes sign, so that we obtain for n large the contradiction

$$\beta_n \big\|u_n^-\big\|_{L^2(\mathbb{R}^N)}^2 = \big\|u_n^-\big\|_\lambda^2 \geq (1 + \lambda M_0)\big\|u_n^-\big\|_{L^2(\mathbb{R}^N)}^2 > \beta_n \big\|u_n^-\big\|_{L^2(\mathbb{R}^N)}^2.$$

Case 2: $\gamma = 0$. In that case we have $u = u^- \leq 0$, hence $\|u\|_\lambda^2 = \beta \|u\|_{L^2(\mathbb{R}^N)}^2$. This is only possible if $u = \gamma e_1^\lambda$ with $\gamma \leq 0$. If $\gamma < 0$ then $\beta = \mu_1^\lambda$, and we can argue analogously to case 1. The case $\gamma = 0$, i.e. $u = 0$, cannot occur. In fact, we must have $u|_A \neq 0$ because otherwise $u_n|_A \to 0$ in $L^2(A)$, hence

$$\int_{\mathbb{R}^N} V_\lambda u_n^2 \geq (1 + \lambda M_0) \int_{\mathbb{R}^N \setminus A} u_n^2 \to 1 + \lambda M_0 \quad \text{as } n \to \infty.$$

On the other hand, (7.16) yields

$$\limsup_{n \to \infty} \int_{\mathbb{R}^N} V_\lambda u_n^2 \leq \limsup_{n \to \infty} \|u_n\|_\lambda^2 = \limsup_{n \to \infty} \int_{\mathbb{R}^N} \big(\alpha_n \big(u_n^+\big)^2 + \beta_n \big(u_n^-\big)^2\big)$$

$$\leq \max\{\alpha, \beta\},$$

so we obtain a contradiction because $1 + \lambda M_0 > \max\{\alpha, \beta\}$. $\qquad\square$

7.3.3 Non-trivial Fučik Eigenvalues by Minimax Methods

This section is devoted to the construction of the first non-trivial curve in the Fučik point spectrum of $-\Delta + V_\lambda$. In order to find non-trivial elements in Σ_λ we follow the approach of [62] and consider for $s \in \mathbb{R}$ the functional

$$I_{s,\lambda}(u) = \int_{\mathbb{R}^N} \big(|\nabla u|^2 + V_\lambda(x)u^2\big) - s \int_{\mathbb{R}^N} \big(u^+\big)^2.$$

$I_{s,\lambda}$ is a C^1-functional on H. We write $J_{s,\lambda}$ for the restriction of $I_{s,\lambda}$ to

$$S := \left\{ u \in H : I(u) = \int_{\mathbb{R}^N} u^2 = 1 \right\}.$$

By the Lagrange multiplier rule, $u \in S$ is a critical point of $J_{s,\lambda}$ if and only if there exists $t \in \mathbb{R}$ such that $I'_{s,\lambda}(u) = tI'(u)$, i.e.,

$$\int_{\mathbb{R}^N} \big(\nabla u \cdot \nabla v + V_\lambda(x)uv\big) - s \int_{\mathbb{R}^N} u^+ v = t \int_{\mathbb{R}^N} uv, \quad \text{for all } v \in H. \qquad (7.19)$$

This implies that

$$-\Delta u + V_\lambda(x)u = (s+t)u^+ + tu^- \quad \text{in } \mathbb{R}^N$$

holds in the weak sense. i.e., $(s+t,t) \in \Sigma_\lambda$. Taking $v = u$ in (7.19), one also sees that the Lagrange multiplier t is equal to the corresponding critical value $J_{s,\lambda}(u)$. Therefore we have the following.

Lemma 7.3.1 (Bartsch, Wang and Zhang [31]) *For every $s \in \mathbb{R}$ we have*

$$\left\{ (s+\alpha, \alpha) \in \Sigma_\lambda : \alpha \in \mathbb{R} \right\} = \left\{ \left(s + J_{s,\lambda}(u), J_{s,\lambda}(u) \right) : u \in S, \, J'_{s,\lambda}(u) = 0 \right\}.$$

In the sequel, we may clearly assume $s \geq 0$, because Σ_λ is symmetric with respect to the diagonal.

Proposition 7.3.3 (Bartsch, Wang and Zhang [31])

(a) *If μ_1^λ is achieved then e_1^λ is a strict global minimum of $J_{s,\lambda}$ with $J_{s,\lambda}(e_1^\lambda) = \mu_1^\lambda - s$. The corresponding point in Σ_λ is $(\mu_1^\lambda, \mu_1^\lambda - s) \in \{\mu_1^\lambda\} \times \mathbb{R}$.*
(b) *If $\lambda \geq (\mu_1 + s - 1)/M_0$ with M_0 from (V_2) then $-e_1^\lambda$ is a strict local minimum of $J_{s,\lambda}$, and $J_{s,\lambda}(-e_1^\lambda) = \mu_1^\lambda$. The corresponding point in Σ_λ is $(s + \mu_1^\lambda, \mu_1^\lambda) \in \mathbb{R} \times \{\mu_1^\lambda\}$.*

Recall from Remark 7.3.1 that μ_1^λ is achieved if $\lambda \geq (\mu_1 - 1)/M_0$. We do not know whether 7.3.3(b) holds if one just assumes that μ_1^λ is achieved.

Proof (a) μ_1^λ is achieved by $\pm e_1^\lambda$. For $u \in S \setminus \{\pm e_1^\lambda\}$ we have

$$J_{s,\lambda}(u) > \mu_1^\lambda \int_{\mathbb{R}^N} u^2 - s \int_{\mathbb{R}^N} \left(u^+\right)^2 \geq \mu_1^\lambda - s.$$

The result follows from $J_{s,\lambda}(e_1^\lambda) = \mu_1^\lambda - s$ and $J_{s,\lambda}(-e_1^\lambda) = \mu_1^\lambda$.

(b) Suppose there exists a sequence $u_n \in S$ with $u_n \neq -e_1^\lambda$, $u_n \to -e_1^\lambda$ in H_λ and $J_{s,\lambda}(u_n) \leq \mu_1^\lambda$. Observe that $u_n^+ \neq 0$ because $u_n \leq 0$ a.e. in $\mathbb{R}^N$, $u_n \neq -e_1^\lambda$, yields the contradiction

$$\mu_1^\lambda \geq J_{s,\lambda}(u_n) = \int_{\mathbb{R}^N} \left(|\nabla u_n|^2 + V_\lambda(x)u_n^2 \right) > \mu_1^\lambda.$$

Setting

$$\gamma_n^\lambda := \frac{\int_{\mathbb{R}^N} \left(|\nabla u_n^+|^2 + V_\lambda(x)|u_n^+|^2 \right)}{\int_{\mathbb{R}^N} |u_n^+|^2}$$

we obtain

$$J_{s,\lambda}(u_n) = \int_{\mathbb{R}^N} \left(|\nabla u_n^+|^2 + V_\lambda(x)|u_n^+|^2\right) + \int_{\mathbb{R}^N} \left(|\nabla u_n^-|^2 + V_\lambda(x)|u_n^-|^2\right)$$

$$- s \int_{\mathbb{R}^N} |u_n^+|^2$$

$$> (\gamma_n^\lambda - s) \int_{\mathbb{R}^N} |u_n^+|^2 + \mu_1^\lambda \int_{\mathbb{R}^N} |u_n^-|^2.$$

Now

$$J_{s,\lambda}(u_n) \leq \mu_1^\lambda = \mu_1^\lambda \int_{\mathbb{R}^N} |u_n^+|^2 + \mu_1^\lambda \int_{\mathbb{R}^N} |u_n^-|^2$$

implies

$$\gamma_n^\lambda - s < \mu_1^\lambda. \tag{7.20}$$

Recall $M_0 > 0$ and the bounded set $A = \{x \in \mathbb{R}^N : g(x) < M_0\}$ from (V_2), and recall that $V_\lambda \geq 1 + \lambda g$. Clearly we have

$$\int_{\mathbb{R}^N \setminus A} \left(|\nabla u_n^+|^2 + V_\lambda(x)|u_n^+|^2\right) \geq (1 + \lambda M_0) \int_{\mathbb{R}^N \setminus A} |u_n^+|^2. \tag{7.21}$$

We claim that

$$\int_A \left(|\nabla u_n^+|^2 + V_\lambda(x)|u_n^+|^2\right) \geq (1 + \lambda M_0) \int_A |u_n^+|^2 \quad \text{for } n \text{ large.} \tag{7.22}$$

If not, then $\int_A |u_n^+|^2 > 0$ and, setting $w_n = u_n/(\int_A |u_n^+|^2)^{1/2}$, we see that $\int_A (|\nabla w_n^+|^2 + V_\lambda(x)(w_n^+)^2)$ is bounded. This implies $w_n^+|_A \rightharpoonup w$ in H^1 and $w_n^+|_A \to w$ in $L^2(A)$ along a subsequence because A is bounded. Clearly $w \geq 0$ a.e. and $\int_A w^2 \geq 1$. Thus for some $\varepsilon > 0$, $\rho = \mathrm{mes}(\{x \in A : w(x) > \varepsilon\}) > 0$. We deduce that

$$\mathrm{mes}\big(\{x \in A : u_n(x) > 0\}\big) = \mathrm{mes}\big(\{x \in A : w_n(x) > 0\}\big) > \rho/2$$

for n sufficiently large. This, however, contradicts $u_n \to -e_1^\lambda$.

Now (7.21) and (7.22) imply $\gamma_n^\lambda \geq 1 + \lambda M_0$ for n sufficiently large. Together with (7.20) this yields $\lambda < (\mu_1^\lambda + s - 1)/M_0 < (\mu_1 + s - 1)/M_0$. $\qquad \square$

For $\lambda \geq (\mu_1 - 1)/M_0$ we define a mountain pass value of $J_{s,\lambda}$ by setting

$$\Gamma := \big\{\gamma \in C\big([0,1], S\big) : \gamma(0) = -e_1^\lambda, \gamma(1) = e_1^\lambda\big\}$$

and

$$c_\lambda(s) := \inf_{\gamma \in \Gamma} \max_{t \in [0,1]} J_{s,\lambda}\big(\gamma(t)\big). \tag{7.23}$$

Lemma 7.3.2 (Bartsch, Wang and Zhang [31]) $c_\lambda(0) = \mu_2^\lambda < \mu_2$ and $c_\lambda(s) \le \mu_2^\lambda$ for all $s \ge 0$, $\lambda > (\mu_1 - 1)/M_0$.

Proof We choose a minimizing sequence (v_n) for μ_2^λ, i.e. $v_n \in S$, $\langle e_1^\lambda, v_n \rangle_\lambda = 0$, $\|v_n\|_\lambda^2 \to \mu_2^\lambda$, and consider the paths

$$\gamma_n : [0, 1] \to S, \quad \gamma_n(t) = -e_1^\lambda \cos \pi t + v_n \sin \pi t.$$

The lemma follows from $J_{s,\lambda}(\gamma_\lambda(t)) \le \|v_n\|_\lambda^2 \to \mu_2^\lambda$. $\qquad\square$

Now we check the Palais–Smale condition in order to obtain a third critical point of $J_{s,\lambda}$.

Lemma 7.3.3 (Bartsch, Wang and Zhang [31]) *Given $s > 0$ and $c \in \mathbb{R}$, $J_{s,\lambda}$ satisfies the (PS)$_c$-condition on S provided $\lambda > \frac{s+c-1}{M_0}$. In particular, this holds for $c = c_\lambda(s)$ and $\lambda \ge \frac{\mu_2+s-1}{M_0}$.*

Proof We fix $s > 0$, $c \in \mathbb{R}$, $\lambda > \frac{s+c-1}{M_0}$, and consider a (PS)$_c$-sequence (u_n) in S. Thus

$$J_{s,\lambda}(u_n) = \|u_n\|_\lambda^2 - s\|u_n^+\|_{L^2}^2 = c + o(1) \quad \text{as } n \to \infty \tag{7.24}$$

and

$$\langle u_n, v \rangle_\lambda - s\langle u_n^+, v \rangle_{L^2} = t_n \langle u_n, v \rangle_{L^2} + o(1)\|v\|_\lambda$$

$$\text{as } n \to \infty \text{ uniformly in } v \in H_\lambda, \tag{7.25}$$

where $t_n \in \mathbb{R}$ is the Lagrange multiplier. (7.24) implies that $\|u_n\|_\lambda^2 \le c + s + o(1)$, hence along a subsequence

$$u_n \rightharpoonup u_0 \quad \text{in } H_\lambda \quad \text{and} \quad u_n \to u_0 \quad \text{in } L_{\text{loc}}^2(\mathbb{R}^N).$$

Setting $v = u_n$ in (7.25) and using (7.24) once more, we see that $|t_n| \le c + o(1)$, so $t_n \to t$ along a subsequence and $|t| \le c$. Setting $v = u_n - u_0$ in (7.25), we obtain for $n \to \infty$:

$$\|u_n - u_0\|_\lambda^2 \le s \int_{\mathbb{R}^N} (u_n^+ - u_0^+)(u_n - u_0) + t_n \int_{\mathbb{R}^N} u_n(u_n - u_0) + o(1)$$

$$\le s \int_{\mathbb{R}^N} (u_n - u_0)^2 + t \int_{\mathbb{R}^N} (u_n - u_0)^2 + o(1)$$

$$= (s + t) \int_{\mathbb{R}^N \setminus A} (u_n - u_0)^2 + o(1)$$

$$\le \frac{s + c}{1 + \lambda M_0} \|u_n - u_0\|_\lambda^2 + o(1).$$

It follows that $u_n \to u_0$ in H_λ because $1 + \lambda M_0 > s + c$ by our choice of λ. $\qquad\square$

A first consequence of Lemma 7.3.3 is a lower bound for $c_\lambda(s)$.

Lemma 7.3.4 (Bartsch, Wang and Zhang [31]) *If* $\lambda \geq (\mu_1 + s - 1)/M_0$ *then*

$$c_\lambda(s) > \max\left\{J_{s,\lambda}\left(-e_1^\lambda\right), J_{s,\lambda}\left(e_1^\lambda\right)\right\} = \mu_1^\lambda > 0.$$

Proof This follows from Proposition 7.3.3 and [16, Lemma 6]. The proof of [16, Lemma 6] requires the Palais–Smale condition at the level $J_{s,\lambda}(-e_1^\lambda) = \mu_1^\lambda < \mu_1$. $\square$

Now we obtain the following.

Theorem 7.3.4 (Bartsch, Wang and Zhang [31]) $(s + c_\lambda(s), c_\lambda(s)) \in \Sigma_\lambda$ *for* $s \geq 0$ *and* $\lambda \geq \frac{\mu_2 + s - 1}{M_0}$.

Proof We can apply the mountain pass theorem on C^1-manifolds (see [16, Proposition 4]) because $J_{s,\lambda}$ satisfies the $(PS)_c$-condition by Lemma 7.3.3 for $c = c_\lambda(s) < \mu_2$ and $\lambda \geq \frac{\mu_2 + s - 1}{M_0}$. $\square$

For $\lambda \geq \frac{\mu_2 - 1}{M_0}$ we set $s_\lambda := 1 + \lambda M_0 - \mu_2$ and define the non-trivial curve $\Theta_\lambda := \Theta_\lambda^+ \cup \Theta_\lambda^- \subset \Sigma_\lambda$ by setting

$$\Theta_\lambda^+ := \left\{\left(s + c_\lambda(s), c_\lambda(s)\right) : 0 \leq s \leq s_\lambda\right\}$$

and

$$\Theta_\lambda^- := \left\{\left(c_\lambda(s), s + c_\lambda(s)\right) : 0 \leq s \leq s_\lambda\right\}.$$

In the next section we investigate Θ_λ and the corresponding eigenfunctions.

7.3.4 Some Properties of the First Curve and the Corresponding Eigenfunctions

Θ_λ is the first non-trivial curve in Σ_λ in the following sense:

Proposition 7.3.4 (Bartsch, Wang and Zhang [31]) *For* $0 \leq s \leq s_\lambda$

$$c_\lambda(s) = \min\left\{\beta > \mu_1^\lambda : (s + \beta, \beta) \in \Sigma_\lambda\right\}.$$

Proof The theorem can be proved as [62, Theorem 3.1] using that $J_{s,\lambda}$ satisfies $(PS)_c$ condition for $c \leq c_\lambda(s)$ and $\lambda \geq \frac{\mu_2 + s - 1}{M_0}$. $\square$

Lemma 7.3.5 (Bartsch, Wang and Zhang [31])

(a) *The map*

$$\left[(\mu_2 - 1)/M_0, \infty\right) \times [0, s_\lambda] \to \mathbb{R}, \quad (\lambda, s) \mapsto c_\lambda(s),$$

is continuous.
(b) $\lambda_1 < \lambda_2$ *implies* $c_{\lambda_1}(s) < c_{\lambda_2}(s)$.
(c) $s_1 < s_2$ *implies* $c_\lambda(s_1) > c_\lambda(s_2)$ *and* $s + c_\lambda(s_1) < s + c_\lambda(s_2)$.

Proof (a) and (b) are straightforward to prove; (c) can be proved as in [62, Proposition 4.1]. $\qquad\square$

Proposition 7.3.5 (Bartsch, Wang and Zhang [31]) *There exists a continuous, strictly decreasing function*

$$\vartheta_\lambda : \left[c_\lambda(s_\lambda), s_\lambda + c_\lambda(s_\lambda)\right] \to \left[c_\lambda(s_\lambda), s_\lambda + c_\lambda(s_\lambda)\right]$$

such that $\Theta_\lambda = \{(\alpha, \vartheta_\lambda(\alpha)) : \alpha \in [c_\lambda(s_\lambda), s_\lambda + c_\lambda(s_\lambda)]\}$. *In particular,* $\vartheta_\lambda(\mu_2^\lambda) = \mu_2^\lambda$.
Moreover, ϑ_λ *is onto, hence* $\vartheta_\lambda(c_\lambda(s_\lambda)) = s_\lambda + c_\lambda(s_\lambda)$ *and* $\vartheta_\lambda(s_\lambda + c_\lambda(s_\lambda)) = c_\lambda(s_\lambda)$.

Proof It follows from Lemma 7.3.5 that for $\alpha \in [\mu_2^\lambda, s_\lambda + c_\lambda(s_\lambda)]$ there exists a unique $s \in [0, s_\lambda]$ with $\alpha = s + c_\lambda(s)$. Therefore we can define $\vartheta(\alpha) := c_\lambda(s)$. For $\alpha \in [c_\lambda(s_\lambda), \mu_2^\lambda]$ there exists a unique $s \in [0, s_\lambda]$ with $\alpha = c_\lambda(s)$, and we define $\vartheta(\alpha) := s + c_\lambda(s)$. The properties of ϑ_λ follow easily from Lemmas 7.3.5 and 7.3.2. $\qquad\square$

Remark 7.3.2 From the above results it is clear that whenever the spectrum curve Θ_λ exists it is below the first non-trivial spectrum curve Θ for $-\Delta + a$ in Ω.

Proposition 7.3.6 (Bartsch, Wang and Zhang [31]) $c_\lambda(s_\lambda) \to \mu_1$ *as* $\lambda \to +\infty$.

Proof We first observe that $\liminf_{\lambda \to \infty} c_\lambda(s_\lambda) \geq \mu_1$ because $c_\lambda(s_\lambda) > \mu_1^\lambda \to \mu_1$ as $\lambda \to +\infty$.

Now assume by contradiction that there exist $\lambda_n \to \infty$ and $\delta > 0$ such that

$$c_{\lambda_n}(s_{\lambda_n}) \geq \mu_1 + \delta \quad \text{for all } n. \tag{7.26}$$

We set $s_n := s_{\lambda_n} = 1 + \lambda_n M_0 - \mu_2$ and $J_n := J_{s_n, \lambda_n}$. According to [62, Lemma 4.3] there exists $\phi \in S \cap H_0^1(\Omega)$ with $\operatorname{ess\,sup}_\Omega(\phi/e_1) = \infty$; here $e_1 > 0$ is the first Dirichlet eigenfunction of $-\Delta + a$ in Ω, as in Remark 7.3.1. Consider the paths

$$\gamma_n : [0, 1] \to S, \quad \gamma_n(t) := -e_1^{\lambda_n} \cos \pi t + \phi \sin \pi t.$$

Let $J_n \circ \gamma_n$ achieve its maximum at $t_n \in [0, 1]$ and set $v_n := \gamma_n(t_n)$. We may assume that $t_n \to \tau \in [0, 1]$, which implies $v_n \to -e_1 \cos \pi \tau + \phi \sin \pi \tau$ in $L^2(\mathbb{R}^N)$

by Remark 7.3.1. In addition, $\|v_n\|_{\lambda_n}$ remains bounded because $\|e_1^\lambda\|_\lambda = \mu_1^\lambda \to \mu_1$ as $\lambda \to \infty$. As a consequence of Proposition 7.26 we have

$$J_n(v_n) = \|v_n\|_{\lambda_n}^2 - s_n \|v_n^+\|_{L^2}^2 \geq c_{\lambda_n}(s_{\lambda_n}) \geq \mu_1 + \delta.$$

It follows that $\|v_n^+\|_{L^2} \to 0$, hence $(-e_1 \cos \pi \tau + \phi \sin \pi \tau)^+ = 0$. By our choice of ϕ this is only possible for $\tau = 0$. This in turn implies the contradiction

$$\mu_1 + \delta \leq \|v_n\|_{\lambda_n}^2 \leq \mu_1^{\lambda_n} |\cos \pi t_n| + \|\phi\|_0 \sin \pi t_n \to \mu_1. \qquad \square$$

Remark 7.3.3 We mention another asymptotic property of Fučik spectra as λ tends to infinity. Let Σ and Θ be the Fučik spectra set and the first non-trivial curve for $-\Delta + a$ on Ω with zero Dirichlet condition. Then if $(\alpha, \beta) \notin \Sigma$, then for λ large $(\alpha, \beta) \notin \Sigma_\lambda$. Indeed, let $\lambda_n \to \infty$ and u_n be such that $\|u_n\|_{L^2} = 1$ and

$$-\Delta u_n + a(x)u_n + \lambda_n g(x)u_n = \alpha u_n^+ + \beta u_n^-, \quad x \in \mathbb{R}^N.$$

Then up to a subsequence for some $u \in H^1(\mathbb{R}^N)$, $u_n \rightharpoonup u$ in $H^1(\mathbb{R}^N)$ and $u_n \to u$ strongly in $L^2(\mathbb{R}^N)$. Since $\lambda_n \to \infty$, $u(x) = 0$ for $x \in \mathbb{R}^N \setminus \Omega$ and $u \in H_0^1(\Omega)$. Then testing the above equation with functions in $C_0^\infty(\Omega)$ we see u is a nonzero solution of

$$-\Delta u + a(x)u = \alpha u^+ + \beta u^-, \quad x \in \Omega.$$

We conclude this section with a symmetry result in case a and g, hence V_λ are radially symmetric. We first recall the notion of foliated Schwarz symmetric functions. Fix $P \in S^{N-1} = \{x \in \mathbb{R}^N : |x| = 1\}$, and for $r > 0$ let μ_r denote the standard measure on ∂B_r. The symmetrization A^P of a set $A \subset \partial B_r$ with respect to P is defined as the closed geodesic ball in ∂B_r centered at rP which satisfies $\mu_r(A^P) = \mu_r(A)$. For a continuous function $u : \mathbb{R}^N \to \mathbb{R}$, the foliated Schwarz symmetrization $u_P : \mathbb{R}^N \to \mathbb{R}$ of u with respect to P is defined by the condition

$$\{u_P \geq t\} \cap \partial B_r = \left[\{u \geq t\} \cap \partial B_r\right]^P \quad \text{for every } r > 0, t \in \mathbb{R}.$$

If $u = u_P$ for some $P \in S^{N-1}$, then we say that u is foliated Schwarz symmetric (with respect to P).

Theorem 7.3.5 (Bartsch, Wang and Zhang [31]) *If a and g are radially symmetric then every eigenfunction u of (7.15) with $(\alpha, \beta) \in \Theta_\lambda$ is foliated Schwarz symmetric but not radially symmetric.*

Proof That u is foliated Schwarz symmetric follows as in [30]. And that u is not radially symmetric can be shown as in [24]. $\qquad \square$

7.4 Dancer–Fučik Spectrum and Asymptotically Linear Elliptic Problems

7.4.1 Introduction

We consider the existence of one-signed solutions for the following Dirichlet problem:

$$\begin{cases} -\Delta u = f(x, u), & x \in \Omega, \\ u = 0, & x \in \partial\Omega, \end{cases} \tag{7.27}$$

where Ω is a bounded domain in $\mathbb{R}^N$ ($N \geq 1$) with smooth boundary $\partial\Omega$. The conditions imposed on $f(x, t)$ are as follows:

(f_1) $\quad f \in C(\Omega \times \mathbb{R}, \mathbb{R});$ $\quad f(x, 0) = 0, \quad \forall x \in \Omega.$

(f_2) $\quad \lim_{|t| \to 0} \dfrac{f(x, t)}{t} = \mu, \quad \lim_{|t| \to \infty} \dfrac{f(x, t)}{t} = \ell \quad$ uniformly in $x \in \Omega.$

Since we assume (f_2), problem (7.27) is called *asymptotically linear* at both zero and infinity. This kind of problems have captured great interest since the pioneer work of Amann [9]. For more information, see [25, 48, 49, 62, 71, 127, 128, 205, 210] etc. and the references therein.

Obviously, the constant function $u = 0$ is a trivial solution of problem (7.27). Therefore we are interested in finding non-trivial solutions. Let $F(x, u) = \int_0^u f(x, s) \, ds$. It follows from (f_1), (f_2) that the functional

$$J(u) = \frac{1}{2} \int_\Omega |\nabla u|^2 \, dx - \int_\Omega F(x, u) \, dx \tag{7.28}$$

is of class C^1 on the Sobolev space $H_0^1 := H_0^1(\Omega)$ with norm

$$\|u\| := \left(\int_\Omega |\nabla u|^2 \right)^{1/2},$$

and the critical points of J are solutions of (7.27). Thus we will try to find critical points of J. In doing so, one has to prove that the functional J satisfies the (PS) condition.

We denote by $0 < \lambda_1 < \lambda_2 \leq \lambda_3 \leq \cdots \leq \lambda_i \leq \cdots$ the eigenvalues of $(-\Delta, H_0^1)$ with eigenfunctions ϕ_i. If ℓ is an eigenvalue of $(-\Delta, H_0^1(\Omega))$, then the problem is *resonant* at infinity. This case is more delicate. To ensure that J satisfies the (PS) condition usually one needs to assume additional conditions, such as the well-known Landesman–Lazer condition, see e.g. [48, 49]; the angle condition at infinity, see [25].

Recently, in the case $0 \leq \mu < \lambda_1 < \ell$, Zhou [210] obtained a positive solution of problem (7.27) under (f_2) and the following conditions:

(H_1) $f \in C(\Omega \times \mathbb{R}, \mathbb{R})$; $f(x,t) \geq 0$, $\forall t \geq 0$, $x \in \Omega$ and

$$f(x,t) \equiv f(x,0) \equiv 0, \quad \forall t \leq 0, x \in \Omega.$$

(H_2) $\dfrac{f(x,t)}{t}$ is nondecreasing with respect to $t \geq 0$, a.e. on $x \in \Omega$.

Note that our assumption (f_1) is weaker than (H_1). And the condition (H_2) is a strong assumption.

We will prove that (f_1) and (f_2) are sufficient to obtain a positive solution and a negative solution of problem (7.27). Our main result is the following.

Theorem 7.4.1 (Zhang, Li, Liu and Feng [208]) *Assume that f satisfies (f_1), (f_2). If $\mu < \lambda_1 < \ell$, then problem (7.27) has at least two non-trivial solutions, one is positive, the other is negative.*

Note that in Theorem 7.4.1, even in the resonant case, we do not need to assume any additional conditions to ensure that J satisfies the (PS) condition. Thus Theorem 7.4.1 improve previous results, such as Zhou's [210]. This fact is interesting.

We can also consider the asymptotically linear Dirichlet problem for the p-Laplacian:

$$\begin{cases} -\Delta_p u = f(x,u), & x \in \Omega, \\ u = 0, & x \in \partial\Omega, \end{cases} \tag{7.29}$$

where $1 < p < +\infty$. Let $0 < \lambda_1^p < \lambda_2^p \leq \lambda_3^p \leq \cdots$ be the sequence of variational eigenvalues of the eigenvalue problem

$$\begin{cases} -\Delta_p u = \lambda |u|^{p-2} u, & x \in \Omega, \\ u = 0, & x \in \partial\Omega. \end{cases} \tag{7.30}$$

It is known that $-\Delta_p$ has a smallest eigenvalue, (see [62]) i.e. the principle eigenvalue, λ_1^p, which is simple and has an associated eigenfunction $\varphi_1 \in W_0^{1,p}(\Omega) \cap C^1(\Omega)$ that is strictly positive in Ω and $\int_\Omega \varphi_1^p = 1$. λ_1^p is defined as

$$\lambda_1^p = \min\left\{ \int_\Omega |\nabla u|^p : u \in W_0^{1,p}(\Omega) \text{ and } \int_\Omega |u|^p = 1 \right\}.$$

Assuming (f_1) and the following condition:

(f_2') $\displaystyle\lim_{|t|\to 0} \frac{f(x,t)}{|t|^{p-2}t} = \mu, \quad \lim_{|t|\to\infty} \frac{f(x,t)}{|t|^{p-2}t} = \ell$ uniformly in $x \in \Omega$,

we obtain the following.

Theorem 7.4.2 (Zhang, Li, Liu and Feng [208]) *Assume that f satisfies (f_1), (f_2'). If $\mu < \lambda_1^p < \ell$, then problem (7.29) has at least two non-trivial solutions, one is positive, the other is negative.*

Remark 7.4.1

(1) The existence of a positive solution of problem (7.29) was obtained by Li and Zhou [128, Theorem 1.1], under (H_1), (f_2') with $\mu = 0$ and

$$\left(H_2'\right) \quad \frac{f(x,t)}{t^{p-1}} \quad \text{is nondecreasing in } t > 0, \text{ for } x \in \Omega.$$

Condition (H_2) is a strong assumption. Moreover, if ℓ is an eigenvalue of (7.30), they need another condition

$$(fF) \quad \lim_{t \to \infty} \left\{ f(x,t)t - pF(x,t) \right\} = +\infty \quad \text{uniformly a.e. } x \in \Omega.$$

to product a positive solution. Thus our Theorem 7.4.2 extends [128, Theorem 1.1] greatly.
(2) Obviously, Theorem 7.4.1 is a special case of Theorem 7.4.2. But we would rather state the proof of Theorem 7.4.1 separately, because the proof is very simple and clear.

7.4.2 Proofs of Main Theorems

In this section, we will always assume that (f_1), (f_2) hold and give the proof of Theorem 7.4.1.

Consider the following problem:

$$\begin{cases} -\Delta u = f_+(x,u), & x \in \Omega, \\ u = 0, & x \in \partial\Omega, \end{cases} \tag{7.31}$$

where

$$f_+(x,t) = \begin{cases} f(x,t,) & t \geq 0, \\ 0, & t \leq 0. \end{cases} \tag{7.32}$$

Define a functional $J_+ : H_0^1 \to \mathbf{R}$,

$$J_+(u) = \frac{1}{2} \int_\Omega |\nabla u|^2 \, dx - \int_\Omega F_+(x,u) \, dx,$$

where $F_+(x,t) = \int_0^t f_+(x,s) \, ds$. We know $J_+ \in C^1(H_0^1, \mathbb{R})$.

Lemma 7.4.1 J_+ *satisfies the* (PS) *condition.*

Proof Let $\{u_n\} \subset H_0^1$ be a sequence such that

$$\left| J_+(u_n) \right| \leq c, \qquad J_+'(u_n) \to 0. \tag{7.33}$$

It is easy to see that

$$\left| f_+(x, u)u \right| \leq C\left(1 + |u|^2\right).$$

Now (7.33) implies that $\forall \phi \in H_0^1$

$$\int_\Omega \left(\nabla u_n \nabla \phi - f_+(x, u_n)\phi \right) dx \to 0. \tag{7.34}$$

Set $\phi = u_n$ we have

$$\begin{aligned}
\|u_n\|^2 &= \int_\Omega f_+(x, u_n)u_n \, dx + \left\langle J'_+(u_n), u_n \right\rangle \\
&\leq \int_\Omega f_+(x, u_n)u_n \, dx + o(1)\|u_n\| \\
&\leq C + C\|u_n\|_2^2 + o(1)\|u_n\|,
\end{aligned} \tag{7.35}$$

where $\| \cdot \|_2$ is the standard norm in $L^2 := L^2(\Omega)$. We claim that $\|u_n\|_2$ is bounded. For otherwise, we may assume that $\|u_n\|_2 \to +\infty$. Let $v_n = \frac{u_n}{\|u_n\|_2}$, then $\|v_n\|_2 = 1$. Moreover, from (7.35) we have

$$\begin{aligned}
\|v_n\|^2 &\leq o(1) + C + \frac{o(1)}{\|u_n\|_2} \cdot \frac{\|u_n\|}{\|u_n\|_2} \\
&= o(1) + C + o(1)\|v_n\|.
\end{aligned}$$

That is, $\|v_n\|$ is bounded. So, up to a subsequence, we have

$$v_n \rightharpoonup v \quad \text{in } H_0^1, \quad v_n \to v \quad \text{in } L^2, \quad \text{for some } v \text{ with } \|v\|_2 = 1.$$

From (7.34) it follows that

$$\int_\Omega \left(\nabla v \nabla \phi - \ell v^+ \phi \right) dx = 0, \quad \forall \phi \in H_0^1,$$

where $v^+ = \max\{0, v\}$. From this and the regularity theory we have

$$\begin{cases} -\Delta v = \ell v^+, & x \in \Omega, \\ v = 0, & x \in \partial\Omega. \end{cases} \tag{7.36}$$

The maximum principle implies that $v = v^+ \geq 0$. But $\ell > \lambda_1$ and hence $v \equiv 0$ which contradicts with $\|v\|_2 = 1$.

Since $\|u_n\|_2$ is bounded, from (7.35) we get the boundedness of $\|u_n\|$. A standard argument shows that $\{u_n\}$ has a convergent subsequence. Therefore, J_+ satisfies the (PS) condition. $\qquad \square$

Lemma 7.4.2 *Let $\phi_1 > 0$ be a λ_1-eigenfunction of $(-\Delta, H_0^1)$ with $\|\phi_1\| = 1$, if $\mu < \lambda_1 < \ell$, then we have*

(a) *there exist $\rho, \beta > 0$ such that $J_+(u) \geq \beta$ for all $u \in H_0^1$ with $\|u\| = \rho$.*
(b) $J_+(t\phi_1) \to -\infty$ *as $t \to +\infty$.*

Proof See the proof of [210, Lemma 2.5]. $\square$

Now we are in a position to state the proof of Theorem 7.4.1.

Proof of Theorem 7.4.1 By Lemma 7.4.1, Lemma 7.4.2, and the Mountain Pass Theorem [159, Theorem 2.2], the functional J_+ has a critical point u_+ with $J_+(u_+) \geq \beta$. But $J_+(0) = 0$, that is, $u_+ \neq 0$. Then u_+ is a non-trivial solution of (7.31). From the strong maximum principle, $u_+ > 0$. Hence u_+ is also a positive solution of (7.27).

Similarly, we obtain a negative solution u_- of (7.27).

The proof is completed. $\square$

Remark 7.4.2 If we assume further that $f \in C^1(\Omega \times \mathbb{R}, \mathbb{R})$ and ℓ is not an eigenvalue of $(-\Delta, H_0^1)$, that is, $\ell \in (\lambda_i, \lambda_{i+1})$ for some $i \geq 2$. Then the functional J defined in (7.28) satisfies the (PS) condition. Using Morse theory, one can prove that problem (7.27) has one more non-trivial solution u with $C_i(J, u) \neq 0$, where $C_i(J, u)$ is the ith *critical group* of J at u.

Remark 7.4.3 If we assume $\mu = \mu(x)$, $\ell = \ell(x)$, and $\mu(x) < \lambda_1$, $\ell(x) \in L^\infty(\Omega)$, $\ell(x) \geq \lambda_1$, $\mathrm{mes}\{x \in \Omega : \ell(x) > \lambda_1\} > 0$, then the conclusion of Theorem 7.4.1 is valid too. Since under this assumption, by (7.36) we get $\lambda_1 \int_\Omega v\phi_1 = \int_\Omega \nabla v \nabla \phi_1 = \int_\Omega \ell(x) v\phi_1$, thus $v \equiv 0$.

Next we sketch the proof of Theorem 7.4.2 and give some remarks. First, we recall the concept Fučik spectrum and a related result.

The Fučik spectrum of p-Laplacian with Dirichlet boundary condition is defined as the set Σ_p of those $(a, d) \in \mathbb{R}^2$ such that

$$\begin{cases} -\Delta_p u = a(u_+)^{p-1} - d(u_-)^{p-1}, & x \in \Omega, \\ u = 0, & x \in \partial\Omega \end{cases}$$

has a non-trivial solution, where $u^+ = \max\{u, 0\}$, $u^- = \max\{-u, 0\}$. By [62] we know that if $(a, d) \in \Sigma_p$ and $(a, d) \notin \mathbb{R} \times \lambda_1^p$, $(a, d) \notin \lambda_1^p \times \mathbb{R}$, then $a > \lambda_1^p$, $d > \lambda_1^p$. We will also need the following lemma, which is due to Zhang and Li [205, Lemma 3].

Lemma 7.4.3 *Assume that $h \in C(\Omega \times \mathbb{R}, \mathbb{R})$, $\lim_{t \to +\infty} \frac{h(t)}{|t|^{p-2}t} = a$, $\lim_{t \to -\infty} \frac{h(t)}{|t|^{p-2}t}$
$= d$. If $(a, d) \notin \Sigma_p$, then the functional $\varphi : W_0^{1,p}(\Omega) \to \mathbb{R}$,*

$$\varphi(u) = \frac{1}{p} \int_\Omega |\nabla u|^p \, dx - \int_\Omega H(u) \, dx$$

satisfies the (PS) condition, where $H(u) = \int_0^u h(t) \, dt$.

Sketch of the proof of Theorem 7.4.2 As in last section, consider the truncated problem

$$\begin{cases} -\Delta_p u = f_+(x,u), & x \in \Omega, \\ u = 0, & x \in \partial\Omega, \end{cases} \tag{7.37}$$

where f_+ is defined as in (7.32). Due to the maximum principle (see [184]), solutions of (7.37) are positive, thus are solutions of (7.29). We have

$$\lim_{t\to-\infty} \frac{f_+(x,t)}{|t|^{p-2}t} = 0, \qquad \lim_{t\to+\infty} \frac{f_+(x,t)}{|t|^{p-2}t} = \ell.$$

Since $\ell > \lambda_1^p$, one deduces directly from the definition of Fučik spectrum that $(\ell, 0) \notin \Sigma_p$, thus by Lemma 7.4.3, we deduce that the C^1-functional

$$J_+(u) = \frac{1}{p} \int_\Omega |\nabla u|^p \, dx - \int_\Omega F_+(x,u) \, dx$$

satisfies the (PS) condition on the Sobolev space $W_0^{1,p}(\Omega)$ with norm

$$\|u\|_{1,p} = \left(\int_\Omega |\nabla u|^p \, dx \right)^{1/p},$$

where $F_+(x,t) = \int_0^t f_+(x,s) \, ds$.

As [128, Lemma 2.3], the functional J_+ admits the "Mountain Pass Geometry". Thus J_+ has a nonzero critical point, which is a non-trivial solution of (7.37). From the strong maximum principle(see [184]), it is also a positive solution of (7.29).

Similarly, we obtain a negative solution of (7.29). $\square$

Remark 7.4.4 The problems (7.27) and (7.29) can be resonant at infinity, this is the main difficulty in verifying the (PS) condition. But after truncating, the problems are not resonant with respect to the Fučik spectrum. Thus, from the Fučik spectrum point of view, the corresponding functionals of the truncated problems satisfies the (PS) condition naturally. And our limit conditions at zero allow us to use the truncation technique and apply the Mountain Pass Theorem.

These are the main ingredient of this work.

Remark 7.4.5 In fact, let $P := \{u \in H_0^1 : u(x) \geq 0, \text{a.e.}\}$, the functional J does not satisfies the (PS) condition on the whole space H_0^1 whenever $\ell = \lambda_i$, $i > 1$, but from our proof J satisfies the (PS) condition on P, i.e., the unbounded (PS) sequences do not belong to P. This idea may be used to weaken the compact conditions for other problems.

Chapter 8
Sign-Changing Solutions

8.1 Sign-Changing Solutions for Superlinear Dirichlet Problems

8.1.1 Nehari Manifold and Sign-Changing Solutions

Assume that $\varphi \in C^1(H, \mathbb{R})$ is such that $\varphi'(0) = 0$. A necessary condition for $u \in H$ to be a critical point of φ is that $(\varphi'(u), u) = 0$. This condition defines the Nehari manifold

$$S := \left\{ u \in H : (\varphi'(u), u) = 0, u \neq 0 \right\}.$$

The Sobolev space $H := H_0^1(\Omega)$ with the inner product

$$(u, v) = \int_\Omega \nabla u \cdot \nabla v$$

and the norm

$$\|u\| = \left(\int_\Omega |\nabla u|^2 \right)^{\frac{1}{2}},$$

i.e., the space of functions whose (weak) derivative belongs to the space $L^2(\Omega)$. We first introduce one result of Castro et al. [44]. They studied

$$\begin{cases} -\Delta u = f(u) & \text{in } \Omega, \\ u = 0 & \text{on } \partial\Omega, \end{cases} \tag{8.1}$$

where Ω be a smooth bounded domain in $\mathbb{R}^N$, $N \geq 2$.

Assume that

(A_1) $f \in C^1(\mathbb{R}, \mathbb{R})$ such that $f(0) = 0$. There exist constants $A > 0$ and $p \in (1, \frac{N+2}{N-2})$ such that $|f'(u)| \leq A(|u|^{p-1} + 1)$ for all $u \in \mathbb{R}$, resp. $p \in (2, \infty)$ in case $N = 2$.

Z. Zhang, *Variational, Topological, and Partial Order Methods with Their Applications*,
Developments in Mathematics 29,
DOI 10.1007/978-3-642-30709-6_8, © Springer-Verlag Berlin Heidelberg 2013

(A_2) There exists $m \in (0, 1)$ such that for all $u \in \mathbb{R}$

$$f(u)u - 2F(u) \geq mf(u)u,$$

where $F(u) = \int_0^u f(s)\,ds$.

(A_3) $f'(u) > \frac{f(u)}{u}$, $\forall u \neq 0$ and $\lim_{|u|\to\infty} \frac{f(u)}{u} = \infty$.

We know (A_1) implies that f is subcritical, i.e., there exists $B > 0$ such that $|f(u)| \leq B(|u|^p + 1)$; (A_3) implies that f is superlinear.

Let $0 < \lambda_1 < \lambda_2 \leq \lambda_3 \leq \cdots$ be the eigenvalues of $-\Delta$ with zero Dirichlet boundary condition in Ω, i.e. they are eigenvalues of the following problems:

$$-\Delta u = \lambda u \quad \text{in } \Omega, \qquad u|_{\partial\Omega} = 0.$$

Define the functional $J : H \to \mathbb{R}$ by

$$J(u) = \frac{1}{2} \int_\Omega |\nabla u|^2\,dx - \int_\Omega F(u)\,dx. \tag{8.2}$$

By regularity theory for elliptic boundary value problems the weak solutions of (8.2) (i.e., the critical points of J) are classical solutions under the condition $f \in C^1$ (see [95]). They get the following.

Theorem 8.1.1 (Castro, Cossio and Neuberger [44]) *Under assumptions (A_1)–(A_3), if $f'(0) < \lambda_1$, then (8.1) has at least three nontrivial solutions: $\omega_1 > 0$ in Ω, $\omega_2 < 0$ in Ω, and ω_3. The function ω_3 changes sign exactly once in Ω, i.e., $(\omega_3)^{-1}(\mathbb{R} - \{0\})$ has exactly two connected components. If nondegenerate, the one-sign solutions are Morse index 1 critical points of J, and the sign-changing solution has Morse index 2. Furthermore,*

$$J(\omega_3) \geq J(\omega_1) + J(\omega_2).$$

Remark 8.1.1 One can relax (A_2) to hold only for $|u|$ sufficiently large. This is necessary in order to consider nonlinearities f such that $f'(0) > 0$.

Remark 8.1.2 If $f'(0) > \lambda_1$, then by multiplying (8.1) by an eigenfunction corresponding to λ_1 and integrating by parts, it is easily seen that (8.1) does not have one-signed solutions.

Our assumptions on f imply that $J \in C^2(H, \mathbb{R})$, and that

$$J'(u)(v) = \big(J'(u), v\big) = \int_\Omega \nabla u \cdot \nabla v\,dx - \int_\Omega f(u)v\,dx, \quad \forall v \in H. \tag{8.3}$$

Define $\gamma : H \to \mathbb{R}$ by $\gamma(u) = (\nabla J(u), u) = \int_\Omega [|\nabla u|^2 - f(u)u]\,dx$ and compute

$$\gamma'(u)(v) = \big(\nabla\gamma(u), v\big) = 2\int_\Omega \nabla u \cdot \nabla v\,dx - \int_\Omega f(u)v\,dx - \int_\Omega f'(u)uv\,dx,$$

$$\forall v \in H. \tag{8.4}$$

Definition 8.1.1 For $u \in L^1(\Omega)$, we define $u^+(x) = \max\{u(x), 0\} \in L^1(\Omega)$, and $u^-(x) = \min\{u(x), 0\} \in L^1(\Omega)$. If $u \in H$ then $u^+, u^- \in H$. We say that $u \in L^1(\Omega)$ changes sign if $u^+ \neq 0$ and $u^- \neq 0$. For $u \neq 0$ we say that u is positive ($u > 0$) if $u^- = 0$, and similarly, u is negative ($u < 0$) if $u^+ = 0$.

We use S to denote the Nehari manifold, $S \subset H$ here and give various subsets of S:

$$S = \{u \in H \setminus \{0\} : \gamma(u) = 0\}, \qquad \hat{S} = \{u \in S : u^+ \neq 0, u^- \neq 0\},$$

$$S_1 = \{u \in \hat{S} : \gamma(u^+) = 0\},$$

$$G^+ = \{u \in S : u > 0\}, \qquad \hat{S}^+ = \{u \in S : \gamma(u^+) < 0\}, \qquad W^+ = G^+ \cup \hat{S}^+,$$

$$G^- = \{u \in S : u < 0\}, \qquad \hat{S}^- = \{u \in S : \gamma(u^+) > 0\}, \qquad W^- = G^- \cup \hat{S}^-.$$

We have the disjoint unions $S = G^+ \cup \hat{S} \cup G^-$ and $\hat{S} = \hat{S}^+ \cup S_1 \cup \hat{S}^-$. We note that nontrivial solutions to (8.1) are in S, one-sign solution are in $G^+ \cup G^-$, and sign-changing solutions are in S_1. We define $S^\infty = \{u \in H : \|u\| = 1\}$ (notice that 0 is sometimes denoted as the zero element of H).

Lemma 8.1.1 (Castro, Cossio and Neuberger [44]) *Under the above assumptions we have*:

(a) *0 is a local minimum of J. If $u \in H \setminus \{0\}$, then there exists a unique $\bar{\lambda} = \bar{\lambda}(u) \in (0, \infty)$ such that $\bar{\lambda}u \in S$. Moreover, $J(\bar{\lambda}u) = \max_{\lambda > 0} J(\lambda u) > 0$. If $\gamma(u) < 0$ then $\bar{\lambda} < 1$; and if $\gamma(u) > 0$ then $\bar{\lambda} > 1$ and $J(u) > 0$.*
(b) *The function $\bar{\lambda} \in C^1(S^\infty, (0, \infty))$. The set S is closed, unbounded, and a connected C^1-submanifold of H diffeomorphic to S^∞.*
(c) *$u \in S$ is a critical point of J on $H \Leftrightarrow u$ is a critical point of $J|_S$.*
(d) *$J|_S$ is coercive, i.e., $J(u) \to \infty$ as $\|u\| \to \infty$ in S. Also, $0 \notin S$ and $\inf_S J > 0$.*

Lemma 8.1.2 (Castro, Cossio and Neuberger [44]) *The function $h : H \to H$ defined by $h(u) = u^+$ is continuous. Also, h defines a continuous function from $L^{p+1}(\Omega)$ into itself.*

Lemma 8.1.3 (Castro, Cossio and Neuberger [44]) *Given $w \in \hat{S}$, there exists a path $r_\omega \equiv r \in C^1([0, 1], S)$ such that*

(a) *$r(0) = aw^+ \in G^+$ for some $a > 0$, $r(1) = bw^- \in G^-$ for some $b > 0$, $r(\frac{a}{a+b}) = w$.*
(b) *$a > 1 \Leftrightarrow b < 1 \Leftrightarrow w \in \hat{S}^+$, $\gamma(r(t)^+) < 0 \Leftrightarrow t \in (0, \frac{1}{2}) \Leftrightarrow r(t) \in \hat{S}^+$.*
(c) *$r(\frac{1}{2}) = aw^+ + bw^- \in S_1$, $r([0, 1]) \cap S_1 = \{r(\frac{1}{2})\}$.*
(d) *$J(r(0)) < J(r(t)) < J(r(\frac{1}{2}))$, for $t \in (0, \frac{1}{2})$.*

Lemma 8.1.4 (Castro, Cossio and Neuberger [44]) *The sets G^+, $\hat{S}^+$, W^+, S_1, W^-, $\hat{S}^-$ and G^- satisfy:*

(a) *G^+, S_1 and G^- are closed, and G^+, G^- are connected.*
(b) *$\hat{S}$ is open, and the subsets $\hat{S}^+$ and $\hat{S}^-$ are open and separated by S_1.*
(c) *W^+ and W^- are the only two components of $S \setminus S_1$. In particular, G^+ and G^- are separated by S_1.*
(d) *If $w \in G^+$ and $J(w) = \min_{G^+} J$, then $J(w) = \min_{W^+} J$ and w is a critical point of J.*

Lemma 8.1.5 (Castro, Cossio and Neuberger [44]) *If $w \in S_1$ and $J(w) = \min_{S_1} J$, then w is a critical point of J.*

Now the proof of the main Theorem 8.1.1 is as follows.

Proof We first establish the existence of one-sign solution $w_1 > 0$ (one finds $w_2 < 0$ similarly). Furthermore, we show that $J|_S$ has local minima at w_1 and w_2, and hence these two critical points are of Morse index 1 (if nondegenerate).

We define $c_1 = \inf_{G^+} J$ and take $\{u_n\} \subset G^+$ with $\lim_{n \to \infty} J(u_n) = c_1$. The coercivity of J and the subcritical condition on f allow us to apply the Sobolev Imbedding Theorem, whence we find $\bar{u} \in H \subset L^{p+1}(\Omega)$ such that (without loss of generality)

$$u_n \rightharpoonup \bar{u} \quad \text{in } H, \qquad u_n \to \bar{u} \quad \text{in } L^{p+1}(\Omega),$$

$$\int_\Omega u_n f(u_n)\,dx \to \int_\Omega \bar{u} f(\bar{u})\,dx, \qquad \int_\Omega F(u_n)\,dx \to \int_\Omega F(\bar{u})\,dx.$$

That $\bar{u} \neq 0$ is evident, as $\int_\Omega \bar{u} f(\bar{u})\,dx = \lim_{n \to \infty} \int_\Omega u_n f(u_n)\,dx = \lim_{n \to \infty} \|u_n\|^2 > 0$ (see Lemma 8.1.1(d)). By the continuity of $h : L^{p+1}(\Omega) \to L^{p+1}(\Omega)$, we see that $\bar{u} > 0$. We wish to show that $u_n \to \bar{u}$ in H. If we assume to the contrary that $u_n \not\to \bar{u}$ in H, then without loss of generality we may assume that $\|\bar{u}\| < \liminf_{n \to \infty} \|u_n\|^2$. It follows that $\gamma(\bar{u}) < \liminf_{n \to \infty} \gamma(u_n) = 0$, so by Lemma 8.1.1(a) there exists $0 < \alpha < 1$ such that $\alpha\bar{u} \in G^+$. Consequentially, we get the following contradiction:

$$J(\alpha\bar{u}) < \liminf_{n \to \infty} J(\alpha u_n) \le \liminf_{n \to \infty} J(u_n) = \inf_{G^+} J = c_1.$$

We conclude that $u_n \to \bar{u}$ in H, $\alpha = 1$, $\bar{u} \in G^+$, and $J(\bar{u}) = \min_{G^+} J = c_1$. By Lemma 8.1.4(d) we see that $w_1 = \bar{u} \in G^+$ is a critical point of J, and hence a positive solution to (8.1).

We can obtain the negative solution $w_2 \in G^- \subset W^-$ in the same fashion, whence we can define

$$c_2 = \inf_{G^-} J = \inf_{W^-} J = J(w_2).$$

Now we show that there exists a solution w_3 to (8.1) which changes sign exactly once. If the solution is a non-degenerate critical point, then it has Morse index 2.

Let $c_3 = \inf_{S_1} J$ and $\{u_n\} \subset S_1$ be such that $J(u_n) \to c_3$. Using $\gamma((u_n)^+) = 0$, we see that $\{(u_n)^+\} \subset G^+$ and $\{(u_n)^-\} \subset G^-$. Since $J|_S$ is coercive, $\{u_n\}$ is bounded. By the Sobolev Imbedding Theorem, without loss of generality, we can assume that there exist $u, v, w \in H$ such that

$$u_n \rightharpoonup u, \quad (u_n)^+ \rightharpoonup v, \quad (u_n)^- \rightharpoonup w \quad \text{in } H,$$

$$u_n \to u, \quad (u_n)^+ \to v, \quad (u_n)^- \to w \quad \text{in } L^{p+1}(\Omega).$$

By Lemma 8.1.2 we know that $h : L^{p+1}(\Omega) \to L^{p+1}(\Omega)(u \to u^+)$ is a continuous transformation, so we see that $u^+ = v \geq 0$ and $u^- = w \leq 0$. Let us see that $u \in S_1$. Since $(u_n)^+ \to u^+$ in $L^{p+1}(\Omega)$ and f is subcritical,

$$\int_\Omega F((u_n)^+)\, dx \to \int_\Omega F(u^+)\, dx \quad \text{and}$$

$$\int_\Omega f((u_n)^+)(u_n)^+\, dx \to \int_\Omega f(u^+)u^+\, dx$$

By $(u_n)^+ \in S$ and Lemma 8.1.1(d), we see that

$$\int_\Omega u^+ f(u^+)\, dx = \lim_{n\to\infty} \int_\Omega f((u_n)^+)(u_n)^+\, dx = \lim_{n\to\infty} \|(u_n)^+\|^2 > 0,$$

consequentially $u^+, u^- \neq 0$ and $u = u^+ + u^-$ is sign-changing.

Let us see that $(u_n)^+ \to u^+$ in H. If we suppose not, then without loss of generality we may assume that $\|u^+\|^2 < \liminf_{n\to\infty} \|(u_n)^+\|^2$, whence

$$\gamma(u^+) = \|u^+\|^2 - \int_\Omega u^+ f(u^+)\, dx$$

$$< \liminf_{n\to\infty} \|(u_n)^+\|^2 - \lim_{n\to\infty} \int_\Omega (u_n)^+ f((u_n)^+)\, dx = 0.$$

From Lemma 8.1.1(a) we see that there exists $0 < \alpha < 1$ such that $\alpha u^+ \in G^+$, and similarly that there exists $0 < \beta \leq 1$ such that $\beta u^- \in G^-$. We conclude that $\alpha u^+ + \beta u^- \in S_1$. This provides a contradiction, since

$$J(\alpha u^+ + \beta u^-) < \liminf_{n\to\infty} J(\alpha(u_n)^+ + \beta(u_n)^-)$$

$$= \liminf_{n\to\infty} \{J(\alpha(u_n)^+) + J(\beta(u_n)^-)\}$$

$$\leq \liminf_{n\to\infty} \{J((u_n)^+) + J((u_n)^-)\}$$

$$= \liminf_{n\to\infty} J(u_n) = \inf_{S_1} J = c_3. \tag{8.5}$$

Hence $(u_n)^+ \to u^+$ in H and $\alpha = 1$. Similarly, we conclude that $(u_n)^- \to u^-$ in H and $\beta = 1$, which proves that $u \in S_1$, $u_n \to u$ in H, and $J(u) = c_3$. Letting $w_3 = u$, we see that $J|_{S_1}$ attains its minimum at w_3. We find that w_3 is a critical point of J by Lemma 8.1.5, hence a solution to (8.1).

Let us see that w_3 changes sign exactly once. Since w_3 is of class C^2, hence continuous, $E = \{x \in \Omega : u(x) \neq 0\}$ is open. Suppose E has more than two components. Since w_3 changes sign, without loss of generality we can assume that there exist connected components A, B and C of E such that $u > 0$ in A and $u < 0$ in B. Let u_A, u_B and u_C be the zero extensions of $w_3|_A$, $w_3|_B$, and $w_3|_C$ to all of Ω. Since $\Delta w_3 + f(w_3) = 0$ on Ω it follows that $\gamma(u_A) = \gamma(u_B) = \gamma(u_C) = 0$. Hence $J > 0$ on S implies $J(u_A + u_B) < J(u_A + u_B + u_C) \leq J(w_3) = c_3$, a contradiction since $u_A + u_B \in S_1$. We conclude that E has exactly two components.

If w_3 is non-degenerate critical point, we see that w_3 has Morse index 2 in H by observing that $J''(w_3)(v, v) < 0$ for $v = (w_3)^+$ and $v = (w_3)^-$, $((w_3)^+, (w_3)^-) = 0$, and $J|_{S_1}$ has a minimum at w_3.

It is clear that

$$c_3 = J(w_3) = J\left(w_3^+\right) + J\left(w_3^-\right) \geq J(w_1) + J(w_2) = c_1 + c_2.$$

This completes the proof. $\qquad\qquad\qquad\qquad\qquad\qquad\qquad\qquad\qquad\qquad\qquad\square$

8.1.2 Additional Properties of Sign-Changing Solutions to Superlinear Elliptic Equations

Now we give Bartsch–Weth's results on additional properties of sign-changing solutions to superlinear elliptic equations (see [26])

$$\begin{cases} -\Delta u = f(x, u) & \text{for } x \in \Omega, \\ u|_{\partial\Omega} = 0 \end{cases} \tag{8.6}$$

on a smooth, bounded domain $\Omega \subset \mathbb{R}^N$, $N \geq 2$.

We make the follow assumptions similar to the above:

(f_1) $f \in C^1(\Omega \times \mathbb{R}, \mathbb{R})$, $f(x, 0) = 0$ for all $x \in \Omega$.

(f_2) There exist constants $p \in (2, \frac{2N}{N-2})$, resp. $p \in (2, \infty)$ in case $N = 2$, such that
$|f'(x, t)| \leq C(|u|^{p-2} + 1)$ for all $x \in \Omega$, $t \in \mathbb{R}$, where $f' := \partial f/\partial t$.

(f_3) $f'(x, t) > \frac{f(x,t)}{t}$, $\forall x \in \Omega$, $t \neq 0$.

(f_4) There exist $R > 0$ and $\theta > 0$ such that for all $x \in \Omega$, $|t| \geq R$

$$0 < \theta F(x, t) \leq tf(x, t),$$

where $F(x, t) = \int_0^t f(x, s)\, ds$.

Clearly these assumptions hold for

$$f(x,t) = \sum_{i=1}^{d} a_i(x)|t|^{p_i-2}t, \qquad (8.7)$$

where $2 \le p_1 < \cdots < p_d < 2N/(N-2)$, $a_1,\ldots,a_d$ are bounded nonnegative C^1-functions and a_d is bounded from below by a positive constant. In fact, $a_i \in L^\infty$ is sufficient, since the differentiability of f with respect to x is not necessary in (f_1). They gain further information on sign-changing solutions, in particular on the nodal structure, extremality properties and the Morse index with respect to the energy functional

$$\Phi(u) = \frac{1}{2}\int_\Omega |\nabla u(x)|^2\, dx - \int_\Omega F(x,u(x))\, dx. \qquad (8.8)$$

It is a well-known consequence of (f_1), (f_2) that $\Phi \in C^2(H)$ and that critical points of Φ are weak solutions of (8.6).

Consider the set

$$M := \left\{ u \in H : u^+ \ne 0,\, u^- \ne 0,\, \Phi'(u)u^+ = \Phi'(u)u^- = 0 \right\}.$$

The set M is not a manifold and we do not expect it to be a complete metric space if $\mu_1 < 0$ (where $\mu_1 < \mu_2 \le \cdots$ are the Dirichlet eigenvalues of the operator $-\Delta - f'(x,0)$). The condition $\mu_1 > 0$ required in [44] implies that M is a closed subset of H, hence a complete metric space. Obviously, all sign-changing solutions of (8.6) are contained in M. Now put

$$\beta := \inf_{u \in M} \Phi(u). \qquad (8.9)$$

We will show that $\Phi(\bar{u}) = \beta$, i.e., $\bar{u}$ is a least energy sign-changing solution. Thus we also obtain the infimum of Φ on M is achieved by a critical point of Φ. We then conclude by the following result.

Theorem 8.1.2 (Bartsch and Weth [26]) *Suppose (f_1)–(f_4) hold. Then every weak solution $u \in M$ with $\Phi(u) = \beta$ has Morse index 2 and has precisely two nodal domains.*

(f_1)–(f_4) can be weaken by the following hypotheses:

(A_1) $f : \Omega \times \mathbb{R} \to \mathbb{R}$ is a Caratheodory and $f(x,0) = 0$ for all $x \in \Omega$.
(A_2) There exist $p \in (2, 2N/(N-2)]$, resp. $p \in (2, \infty)$ in case $N = 2$, and $C > 0$ such that $f(x,t) \le C(|t| + |t|^{p-1})$ for all $t \in \mathbb{R}$ and a.e. $x \in \Omega$.
(A_3) The function $t \to f(x,t)/|t|$ is nondecreasing on $\mathbb{R} \setminus \{0\}$ for a.e. $x \in \Omega$.

Stronger variant of (A_3):

$(\tilde{A}_3)$ The function $t \to f(x,t)/|t|$ is strictly increasing on $\mathbb{R} \setminus \{0\}$ for a.e. $x \in \Omega$.

Note that for nonlinearities of the form (8.7) condition (A_3) is a consequence of the sign condition

$$a_i(x) \geq 0 \quad \text{for } x \in \Omega, \quad i = 1, 2, \dots, d.$$

In view of (A_2) the nonlinearity f may have critical growth at infinity. Nevertheless (A_2) ensures that every weak solution u of (8.6) is at least continuous in Ω. Indeed, combining (A_2) with Sobolev embeddings, we deduce that

$$\frac{f(\cdot, u(\cdot))}{u} \in L^{N/2}(\Omega).$$

Hence the Brezis–Kato Theorem yields $u \in L^q_{\mathrm{loc}}(\Omega)$ for every $2 \leq q < \infty$. In particular there is a number $s > N/2$ such that $f(\cdot, u(\cdot)) \in L^s_{\mathrm{loc}}(\Omega)$, thus u is continuous by elliptic regularity.

Lemma 8.1.6 (Bartsch and Weth [26]) *Suppose (A_1)–(A_3) hold, and consider $u \in H \setminus \{0\}$ with $\Phi'(u)u = 0$. Then*

$$0 \leq \Phi(u) = \sup_{t \geq 0} \Phi(tu). \tag{8.10}$$

If in addition $(\tilde{A}_3)$ is satisfied, then $\Phi(u) > 0$.

Proof For $t \geq 0$ we define $h(t) := \Phi(tu)$. Then $h(0) = 0$, and

$$h'(t) = \Phi'(tu)u = t \int_\Omega \left(|\nabla u|^2 - \frac{f(x, tu)}{tu} u^2 \right)$$

for $t > 0$. Hence (A_3) implies that $t \to \frac{h'(t)}{t}$ is non-increasing on $(0, \infty)$, and thus the set $S := \{t > 0 : h'(t) = 0\}$ is a subinterval of $(0, \infty)$ which contains $t = 1$ by assumption. Let $b \leq \infty$ be the right endpoint of S. Then h is strictly decreasing on (b, ∞), whereas

$$0 \leq \max_{t \in [0,b]} h(t) \leq \max_{t \in S} h(t) = h(1).$$

This yields (8.10). If in addition $(\tilde{A}_3)$ holds, then $t \to h'(t)/t$ is strictly decreasing on $(0, \infty)$. Hence $S = \{1\}$, and $h'(t) > 0$ for $0 < t < 1$. We conclude that $\Phi(u) = h(1) > h(0) = 0$, as claimed. $\qquad\square$

Theorem 8.1.3 (Bartsch and Weth [26]) *Suppose (A_1), (A_2) and $(\tilde{A}_3)$ hold. Then every weak solution $u \in M$ of (8.6) with $0 < \Phi(u) \leq \beta$ has precisely 2 nodal domains.*

Proof Suppose in contradiction that u has at least three nodal domains. We choose nodal domains Ω_1, Ω_2 such that $u(x) \geq 0$, $x \in \Omega_1$; $u(x) \leq 0$, $x \in \Omega_2$. The associated functions $v_1, v_2 \in H$ are defined as

$$v_i(x) := \begin{cases} u(x), & \forall x \in \Omega_i, \\ 0, & \forall x \in \Omega \setminus \Omega_i. \end{cases} \tag{8.11}$$

Clearly $v_1 + v_2 \in M$ and the function $v := u - v_1 - v_2$ satisfies $\Phi'(v)v = 0$. This implies $\Phi(v) > 0$ by Lemma 8.1.6, hence $\beta \leq \Phi(v_1 + v_2) < \Phi(u)$, which contradicts the assumption. $\qquad\square$

For the Morse index of sign-changing solutions with least energy, now we use the assumptions (f_1)–(f_4). Consider the Hilbert space $H_1 := H \cap H^2(\Omega)$, endowed with the scalar product from $H^2(\Omega)$. Moreover, denote by $\|\cdot\|_1$ the induced norm. We need the following technical lemma concerning the functionals:

$$Q_\pm : H \to \mathbb{R}, \quad Q_\pm(u) = \int_\Omega |\nabla u^\pm| \, dx = \int_\Omega \nabla u \cdot \nabla u^\pm \, dx,$$

$$\Psi_\pm : H \to \mathbb{R}, \quad \Psi_\pm(u) = \int_\Omega f(x, u)u^\pm \, dx.$$

Lemma 8.1.7 (Bartsch and Weth [26])

(a) $Q_\pm$ *is differentiable at* $u \in H_1$ *with derivative* $Q_\pm'(u) \in H^{-1}$ *given by*

$$Q_\pm'(u)v = \int_{\pm u > 0} \left((-\Delta u)v + \nabla u \nabla v \right) dx.$$

(b) $Q_\pm'|_{H_1} \in C^1(H_1)$.
(c) $\Psi_\pm \in C^1(H)$ *with derivative given by*

$$\Psi_\pm'(u)v = \int_\Omega f'(x, u^\pm)u^\pm v \, dx + \int_\Omega f(x, u^\pm)v \, dx.$$

Lemma 8.1.8 (Bartsch and Weth [26]) *The set* $M \cap H_1$ *is a* C^1*-manifold of codimension two in* H.

Proof Define

$$g_\pm : H \to \mathbb{R}, \quad g_\pm(u) = \Phi'(u)u^\pm,$$

so that $M \cap H_1 = \{u \in H_1 : u^+ \neq 0, u^- \neq 0, g_+(u) = 0 = g_-(u)\}$.
    Lemma 8.1.7 implies that $g_\pm|_{H_1} \in C^1(H_1)$. For $u \in M \cap H_1$ we obtain

$$g_+'(u)u^+ = \int_\Omega \left(|\nabla u^+|^2 - f'(x, u)(u^+)^2 \right) dx, \qquad g_+'(u)u^- = 0,$$

$$g'_-(u)u^- = \int_\Omega \left(|\nabla u^-|^2 - f'(x, u)(u^-)^2\right) dx, \qquad g'_-(u)u^+ = 0.$$

Hence (f_3) yields $g'_+(u)u^+ < 0$ and $g'_-(u)u^- < 0$ for $u \in M \cap H_1$. Approximating u^+ and u^- by functions in H_1, we conclude that $(g'_+(u), g'_-(u)) \in \mathbb{L}(H_1, \mathbb{R}^2)$ is onto for every $u \in M \cap H_1$. From this the assertion follows. $\qquad\square$

In the following, if $u \in H$ is a critical point of Φ, we denote by $m(u)$ the Morse index of u.

Proposition 8.1.1 (Bartsch and Weth [26]) *Let $u \in M$ be a critical point of Φ with $\Phi(u) = \beta$. Then $m(u) = 2$.*

Proof By (f_3) there holds

$$\Phi''(u)\left(u^\pm, u^\pm\right) = \int_\Omega \left(|\nabla u^\pm|^2 - f'(x, u)(u^\pm)^2\right) < 0,$$

hence $m(u) \geq 2$. To show $m(u) \leq 2$, note first that $u \in H_1$ by elliptic regularity. Denote by $T \subset H_1$ the tangent space of the manifold $M \cap H_1$ at u. We show that

$$\Phi''(u)(v, v) \geq 0 \quad \text{for all } v \in T. \tag{8.12}$$

Indeed, by Lemma 8.1.8 there exists for every $v \in T$ a C^1-curve $\gamma : [-1, 1] \to M \cap H$ such that $\gamma(0) = u$ and $\dot\gamma(0) = v$. Since $\Phi'(u)v = 0$, we calculate that $\Phi \circ \gamma : [-1, 1] \to \mathbb{R}$ is even twice differentiable at $t = 0$ with derivative

$$\left.\frac{\partial^2}{\partial t^2} \Phi \circ \gamma\right|_{t=0} = \Phi''(u)(v, v).$$

Recalling that $\Phi(u) = \min_{v \in M \cap H_1} \Phi(v)$, we infer that $\partial^2(\Phi \circ \gamma)/\partial t^2|_{t=0} \geq 0$, and hence (8.12) follows. Since $T \subset H_1$ has codimension two and H_1 is dense in H, we conclude $m(u) \leq 2$, as required. $\qquad\square$

The proof of Theorem 8.1.2 follows from Theorem 8.1.3 and Proposition 8.1.1.

Remark 8.1.1 In [28], T. Bartsch, K.C. Chang, and Z.Q. Wang also have some results on the Morse indices of sign-changing solutions of nonlinear elliptic problems. See also [213].

8.2 Sign-Changing Solutions for Jumping Nonlinear Problems

8.2.1 On Limit Equation of Lotka–Volterra Competing System with Two Species

The two species Lotka–Volterra competing equation system is

$$\begin{cases} -\Delta u = au - u^2 - cuv & \text{in } \Omega, \\ -\Delta v = dv - v^2 - euv & \text{in } \Omega, \\ u|_{\partial\Omega} = v|_{\partial\Omega} = 0. \end{cases} \tag{8.13}$$

Here $a, d > \lambda_1$, $c, e > 0$, Ω is a smooth,bounded domain in $\mathbb{R}^N$, $N \geq 1$.

Let λ_j be the jth eigenvalue of $-\Delta$ with zero Dirichlet boundary data, $0 < \lambda_1 < \lambda_2 < \cdots < \lambda_k < \cdots$, denote by n_k the dimension of $\ker(-\Delta - \lambda_k I)$ and let $\{\varphi_{k,1}, \varphi_{k,2}, \ldots, \varphi_{k,n_k}\}$ be an orthogonal basis of $\ker(-\Delta - \lambda_k I)$ with the property that $\int_\Omega |\nabla \varphi_{k,j}|^2 = 1$ for $j = 1, 2, \ldots, n_k$ and $k = 1, 2, \ldots$. It is well known that $\lambda_1 > 0$, λ_1 is simple, i.e., $n_{1,1} = 1$, and $\varphi_{1,1} > 0$. We denote $\varphi_{1,1}$ by φ_1, i.e.,

$$-\Delta\varphi_1 = \lambda_1\varphi_1, \quad \varphi_1|_{\partial\Omega} = 0.$$

As the interaction parameters c, e go to infinity, we get the following limiting equation:

$$-\Delta u = h(u), \quad u|_{\partial\Omega} = 0, \tag{8.14}$$

where $h : \mathbb{R} \to \mathbb{R}$ is defined by

$$h(u) = \begin{cases} au - \alpha u^2, & u \geq 0, \\ du + \beta u^2, & u \leq 0, \end{cases}$$

and $\alpha \geq 0$, $\beta > 0$. In [68], Dancer and Du have shown that if (8.14) has a sign-changing solution u_0 which is isolated and has nonzero fixed point index, then for c, e both large with c/e close to α/β, (8.13) has a positive solution (u, v) which is close to $(u_0^+/\alpha, -u_0^-/\beta)$. They also proved if $a, d > \lambda_2$, then (8.14) has at least one sign-changing solution, moreover if its nontrivial solution is isolated then (8.14) has a solution which changes sign and has fixed point index -1. In [69] they find the exact set of points (a, d) in $\mathbb{R}^2$ where the statement holds. It is completely determined by the homogeneous jumping nonlinear problem

$$-\Delta u = au^+ + du^-, \quad u|_{\partial\Omega} = 0. \tag{8.15}$$

Nontrivial solutions of (8.15) correspond to nontrivial critical points of the following functional on H:

$$J_{(a,d)}(u) = \frac{1}{2}\int_\Omega |\nabla u|^2 \, dx - \frac{a}{2}\int_\Omega (u^+)^2 \, dx - \frac{d}{2}\int_\Omega (u^-)^2 \, dx. \tag{8.16}$$

We define the Fučik spectrum or Fučik–Dancer spectrum

$$\Sigma = \big\{(a, d) \in \mathbb{R}^2 : (8.15) \text{ has a nontrivial solution}\big\}$$

and let

$$S_k = \big\{(a, d) \in \mathbb{R}^2, \lambda_k \le a, d \le \lambda_{k+1}, (a, d) \ne (\lambda_k, \lambda_k), (a, d) \ne (\lambda_{k+1}, \lambda_{k+1})\big\}.$$

By [160] we know $S_k \not\subset \Sigma$.

Define

$$D_k = \big\{(a, d) \in \mathbb{R}^2 : \text{ there exists a path } \gamma(t) = \big(\gamma_1(t), \gamma_2(t)\big), t \in [0, 1], \text{ such that}$$
$$\big(\gamma_1(t), \gamma_2(t)\big) \notin \Sigma \text{ and } \gamma(0) = (a, d), \gamma(1) \in S_k\big\}.$$

From the results cited in [69], there exists a continuous function $\eta(t)$ defined on $(\lambda_1, \lambda_2]$ with the properties that

(a) η is strictly decreasing, $\eta(\lambda_2) = \lambda_2$, $\lim_{\lambda \to \lambda_1 + 0} \eta(\lambda) = +\infty$;
(b) equation (8.15) has a nontrivial solution for $(a, d) = (a, \eta(a))$, $a \in (\lambda_1, \lambda_2]$ and $(a, d) = (\eta(d), d)$, $d \in (\lambda_1, \lambda_2]$;
(c) equation (8.15) has no nontrivial solution for $\lambda_1 < d < \eta(a)$, $a \in (\lambda_1, \lambda_2]$ or $\lambda_1 < a < \eta(d)$, $d \in (\lambda_1, \lambda_2]$.

We denote by Γ the curve $\{(a, \eta(a)) : \lambda_1 < a \le \lambda_2\} \cup \{(\eta(d), d) : \lambda_1 < d \le \lambda_2\}$.

For convenience, we denote by $\tilde{S}$ the set of points (a, d) which are above Γ in the a–d plane.

We define

$$\Delta(a, d) = \big\{u \in H : J_{(a,d)}(u) < 0\big\}.$$

Let

$$S := \big\{(a, d) : a, d > \lambda_1, \Delta(a, d) \text{ is path-connected}\big\}.$$

It is proved in [69] that

$$\tilde{S} = S.$$

Lemma 8.2.1 (Dancer and Du [69]) *$\Delta(a, d)$ with $a, d > \lambda_1$ is path-connected if and only if there exists a path in $\Delta(a, d)$ connecting φ_1 and $-\varphi_1$.*

Theorem 8.2.1 (Dancer and Du [69]) *If $(a, d) \in S$, then (8.14) has at least one sign-changing solution. Moreover, if each nontrivial solution of (8.14) is isolated, then there exists a sign-changing solution with fixed point index -1. Here the fixed point refers to that of the mapping $u \to (-\Delta)^{-1} h(u)$ from $W_0^{1,2}(\Omega)$ to itself.*

Proof We consider the cases $\alpha > 0$ and $\alpha = 0$ separately. Suppose first that $\alpha > 0$. Since $h(u)$ is concave on $(0, \infty)$, by a well-known result, one knows that (8.14) has a unique positive solution u^+. Similarly, (8.14) has a unique positive solution u^-.

An easy upper and lower solution argument shows that $0 \le u^+ \le \alpha^{-1}a$, $-\beta d \le u^- \le 0$, and any other solution of (8.14) satisfies $-\beta d \le \alpha^{-1}a$. Moreover, except for the trivial solution $u \equiv 0$ and u^+ and u^-, any other solution of (8.14) changes sign. Therefore, to prove the theorem, we need only find a nontrivial solution other than u^+ and u^-.

Now we modify $h(u)$ outside $[-\beta^{-1}, \alpha^{-1}a]$ in such a way that the modified problem

$$-\Delta u = \tilde{h}(u), \quad u|_{\partial\Omega} = 0 \tag{8.17}$$

has the same solution set as (8.14), that $\tilde{h}$ is smooth except at $u = 0$, and that $\tilde{h}$ is bounded on $\mathbb{R}$. It is easily seen that this is possible.

Define

$$H(u) = \int_0^u \tilde{h}(s)\,ds.$$

Then for $u \in [-\beta^{-1}d, \alpha^{-1}a]$,

$$H(u) = \begin{cases} \frac{1}{2}au^2 - \frac{1}{3}\alpha u^3, & 0 \le u \le \alpha^{-1}a, \\ \frac{1}{3}du^2 + \frac{1}{3}\beta u^3, & 0 \ge u \ge -\beta^{-1}d. \end{cases}$$

Define on H a functional I by

$$I(u) = \frac{1}{2}\int_\Omega |\nabla u|^2\,dx - \int_\Omega H(u)\,dx.$$

It is well known that I is C^1 and satisfies the PS condition, and critical points of I are solutions of (8.14). By the proof of Theorem 4.2 of [68], u^+, u^- are two strict local minima of I. Hence, define

$$c = \inf_{L \in \Delta} \max_{u \in L} I(u),$$

where Δ denotes the set of all continuous paths in H joining u^+ and u^-, it follows from the Mountain Pass Theorem that c is a critical value of I, and $c > \max\{I(u^+), I(u^-)\}$. Therefore, if we can show that $c \ne 0$, then c corresponds to a nontrivial critical point of I as $I(0) = 0$. This must be a sign-changing solution of (8.14) as it differs from 0, u^+ and u^-.

Now we construct a path in H joining u^+ and u^-, and on which $I(u) < 0$. If this is done, then by the definition, $c < 0$ and (8.14) will have a sign-changing solution.

A simple calculation shows that

$$I(tu^+) = \frac{t^2}{2}\alpha\left(\frac{2}{3}t - 1\right)\int_\Omega (u^+)^3\,dx < 0 \quad \text{for } t \in (0, 1] \tag{8.18}$$

and

$$I(tu^-) = \frac{t^2}{2}\beta\left(1 - \frac{2}{3}t\right)\int_\Omega (u^-)^3\,dx < 0 \quad \text{for } t \in (0, 1]. \tag{8.19}$$

In particular,

$$J_{(a,d)}\big(tu^+\big) < I\big(tu^+\big) < 0, \quad J_{(a,d)}\big(tu^-\big) < I\big(tu^-\big) < 0, \quad \forall t \in (0, 1].$$

Hence, u^+, u^- belong to $\Delta(a, d)$, and by the path-connectedness of $\Delta(a, d)$, we can find a path $L_0 \subset \Delta(a, d)$ connecting u^+ and u^-. Moreover, since $(a, d) \in S$, L_0 can be chosen L^∞ bounded. We show that for $\varepsilon > 0$ small, $I(u) < 0$ for $u \in \varepsilon L_0 \equiv \{\varepsilon u : u \in L_0\}$. In fact, since L_0 is L^∞ bounded, there exists $\varepsilon_0 > 0$ such that

$$-\beta^{-1}d \le \varepsilon u \le \alpha^{-1}a \quad \text{for } u \in L_0 \text{ and } 0 < \varepsilon \le \varepsilon_0.$$

Since L_0 is compact in H, there exists $\delta > 0$ such that

$$J_{a,d}(u) \le -\delta \quad \text{for } u \in L_0,$$

and hence

$$J_{a,d}(u)(\varepsilon u) = \varepsilon^2 J(u) \le -\varepsilon^2\delta \quad \text{for } u \in L_0 \text{ and } \varepsilon > 0.$$

Therefore,

$$I(\varepsilon u) = J_{a,d}(u)(\varepsilon u) + \frac{1}{3}\alpha\varepsilon^3 \int_\Omega \big(u^+\big)^3 dx - \frac{1}{3}\beta\varepsilon^3 \int_\Omega \big(u^-\big)^3 dx$$

$$\le -\varepsilon^2\delta + \frac{1}{3}(\alpha + \beta)\varepsilon^3 \int_\Omega |u|^3 dx < 0, \tag{8.20}$$

for $u \in L_0$ and

$$0 < \varepsilon < \min\left\{ \frac{3}{2(\alpha + \beta)} \cdot \delta \cdot \left(\max_{u \in L_0} \int_\Omega |u|^3 dx \right)^{-1}, \varepsilon_0 \right\}.$$

Thus, by (8.18)–(8.20), for $\varepsilon > 0$ small, the path

$$\big\{tu^+ : 1 \ge t \ge \varepsilon\big\} \cup \varepsilon L_0 \cup \big\{tu^- : \varepsilon \le t \le 1\big\}$$

meets our requirement. This proves the first part of the theorem for the case $\alpha > 0$.

The second part of the theorem for case $\alpha > 0$ follows from the proof Theorem 4.2 of [68]. (Note that, since $c < 0$, any solution with $I(u) = c$ is nonzero.)

The proof for the case $\alpha = 0$ is only a modification of the proof above. Therefore we only point out the differences. In this case, (8.14) has no positive solution, it has the trivial solution $u \equiv 0$ and a unique negative solution u^-, all other solutions change sign and satisfy

$$u \ge -\beta^{-1}d.$$

We now modify h outside $(-\beta^{-1}d, \infty)$ such that $\tilde{h}$ is bounded on $(-\infty, 0)$, is smooth except at $u = 0$ and that the modified problem (8.17) has the same solution set as (8.14). Then u^- is a strict local minimum of I. Since $I(t\varphi_1) \to -\infty$

as $t \to \infty$, we can find that $t_0 > 0$ such that $I(t_0\varphi_1) < I(u^-)$. Therefore, if I satisfies the (PS) condition, then we are again in the mountain pass setting and a sign-changing solution is found once we can construct a path in H connecting $t_0\varphi_1$ and u^-. Then the path

$$\{t\varphi_1 : t_0 \geq t \geq \varepsilon\} \cup \varepsilon L_1 \cup \{tu^- : \varepsilon \leq t \leq 1\}$$

with $\varepsilon > 0$ small meets our requirements.

It remains to show that I satisfies the (PS) condition. This is a consequence of the fact that $\tilde{h}$ has the form $\tilde{h}(u) = au^+ + h^-(u)$, where h^- is bounded on $(-\infty, +\infty)$. We omit the details. This completes the proof of the first part of the theorem for $\alpha = 0$. The second part in this case is proved in the same way as for the case $\alpha > 0$. The proof of Theorem 8.2.1 is now complete. $\qquad\square$

Remark 8.2.1 In [72], dynamics of two species Lotka–Volterra competing equations system with diffusion and large interaction is studied by Dancer and Zhang, the solutions approach the stationary state as t tends to infinity. The sign-changing solution of the limit equation is used.

8.2.2 On General Jumping Nonlinear Problems

We consider general jumping nonlinear elliptic boundary value problem with resonance at infinity:

$$\begin{cases} -\Delta u = f(x, u) & \text{in } \Omega, \\ u|_{\partial\Omega} = 0. \end{cases} \tag{8.21}$$

We assume that $\Omega \subset \mathbb{R}^N$ ($N \geq 1$) is a smooth, bounded domain, $f(x, 0) = 0$, $f(x, u)u \geq 0$, and $f \in C(\bar{\Omega} \times \mathbb{R})$ is locally Lipschitz continuous in u uniformly in x here. Furthermore, we assume $f(x, u)$ has jumping nonlinearities at zero or infinity,

$$(H_1) \quad \lim_{u \to 0^+} \frac{f(x, u)}{u} = a_1, \quad \lim_{u \to 0^-} \frac{f(x, u)}{u} = d_1, \quad \text{uniformly for } x \in \Omega,$$

$$(H_2) \quad \lim_{u \to +\infty} \frac{f(x, u)}{u} = a_2, \quad \lim_{u \to -\infty} \frac{f(x, u)}{u} = d_2, \quad \text{uniformly for } x \in \Omega.$$

Let

$$p(x, t) := f(x, t) - a_2 t^+ - d_2 t^-, \qquad P(x, t) = \int_0^t p(x, s)\, ds.$$

We assume that either

$$H(x, t) := 2P(x, t) - tp(x, t) \leq W(x) \in L^1(\Omega), \tag{8.22}$$

and

$$H(x,t) \to -\infty \quad \text{a.e. as } |t| \to \infty; \tag{8.23}$$

or

$$H(x,t) := 2P(x,t) - tp(x,t) \geq W(x) \in L^1(\Omega), \tag{8.24}$$

and

$$H(x,t) \to +\infty \quad \text{a.e. as } |t| \to \infty. \tag{8.25}$$

We sometimes assume:

(H_3) Problem (8.21) has a negative strict subsolution $\varphi(x)$ and a positive strict super solution $\psi(x)$, that is,

$$-\Delta\varphi < f(x,\varphi), \quad \varphi < 0 \quad \text{in } \Omega, \qquad \varphi|_{\partial\Omega} = 0.$$

$$-\Delta\psi > f(x,\psi), \quad \psi > 0 \quad \text{in } \Omega, \qquad \psi|_{\partial\Omega} = 0.$$

Nontrivial solutions of (8.21) correspond to nontrivial critical points of the following functional on H:

$$J(u) = \frac{1}{2}\int_\Omega |\nabla u|^2 \, dx - \int_\Omega F(x,u) \, dx, \tag{8.26}$$

where $F(x,u) = \int_0^u f(x,s) \, ds$.

We will find that the existence of solutions of (8.21) depends on the homogeneous jumping nonlinear problem (8.15), its nontrivial solutions correspond to nontrivial critical points of the following functional $J_{(a,d)}(u)$ (see (8.16)) on H.

We use the following "compactness condition": Let $\psi(t)$ be a positive non-increasing function on $(0, \infty)$ satisfying

$$\int_1^\infty \psi(t) \, dt = \infty. \tag{8.27}$$

We say that J satisfies $(\psi)_c$ if any sequence $\{u_k\} \subset H$ satisfying

$$J(u_k) \to c, \qquad \frac{J'(u_k)}{\psi(\|u_k\|)} \to 0, \tag{8.28}$$

has a convergent subsequence. If this holds for every $c \in \mathbb{R}$, we say that J satisfies (ψ). This reduces to the usual Palais–Smale condition for $\psi(t) \equiv 1$ and to a condition introduced by Cerami [47] for $\psi(t) = 1/(1+t)$.

We use the following lemma.

Lemma 8.2.2 (Perera and Schechter [152]) *If* (8.22), (8.23) *hold, then for any* $c \in \mathbb{R}$,

$$J(u_k) \to c, \qquad (1 + \|u_k\|) J'(u_k) \to 0$$

implies that $\{u_k\}$ has a convergent subsequence, i.e., J satisfies the compactness condition (ψ) with $\psi(t) = 1/(1+t)$.

Remark 8.2.2   As shown in [153], if (8.24), (8.25) hold, then for any $c \in \mathbb{R}$,

$$J(u_k) \to c, \qquad \left(1 + \|u_k\|\right) J'(u_k) \to 0$$

implies that $\{u_k\}$ has a convergent subsequence, i.e., J satisfies the compactness condition (ψ) with $\psi(t) = 1/(1+t)$ too. In fact we can prove it easily as in the proof of Lemma 5.1 of [152]. (See Remark 8.2.7 in the following.)

Now we give some definitions: Since $f(x,u)$ is locally Lipschitz continuous for u, so J is a C^{2-0} functional defined on $C_0^1(\bar{\Omega})$. Let $M = \{u \in H \mid J'(u) = 0\}$ and $X = C_0^1(\bar{\Omega})$ with the usual norm $\|u\|_X = \max_{0 \le |\alpha| \le 1} \sup_{x \in \bar{\Omega}} |D^\alpha u(x)|$. It is well known that $X \subset H$ is densely embedded into H, the critical points set $M \subset X$. Let $u(t, u_0), 0 \le t < \eta(u_0)$ ($\eta(u_0)$ is the maximum existence interval) be the forward strong solution of the initial value problem in X:

$$u' = -J'(u), \quad u(0) = u_0 \in X. \tag{8.29}$$

Definition 8.2.1   We call $N \subset X$ is an invariant set of descent flow of J if $\{u(t, u_0) \mid 0 \le t < \eta(u_0), u_0 \in N\} \subset N$. (It is similar to a definition given in Sun and Liu [172].)

Let $K = (-\Delta)^{-1}$, $\bar{F}(u) = f(x, u(x))$, for $x \in \bar{\Omega}$, $u \in H$. Then $J'(u) = u - K\bar{F}(u)$. The forward strong solution of the initial value problem (8.29) satisfies that

$$u(t, u_0) = e^{-t}\left[u_0 + \int_0^t e^s K\bar{F}\left(u(s, u_0)\right) ds \right], \quad \text{for } t \ge 0. \tag{8.30}$$

Since $f(x, t)$ is local Lipschitz continuous in $t \in \mathbb{R}$ uniformly for x, $K : L^\infty(\Omega) \to C_0^1(\bar{\Omega})$ is a bounded linear operator, and $\bar{F} : C_0^1(\bar{\Omega}) \hookrightarrow L^\infty(\Omega)$ is local Lipschitz continuous, for every $u_0 \in X$, there exists $B_\delta(u_0)$, $l_1 > 0$ such that

$$\left\| K\bar{F}(u) - K\bar{F}(u_0) \right\|_X \le l_1 \|u - u_0\|_X, \quad \forall u \in B_\delta(u_0), \tag{8.31}$$

where $B_\delta(u_0) = \{u \in X : \|u - u_0\|_X < \delta, \delta > 0\}$. Thus the forward strong solution $u(t, u_0)$ of (8.29) exists and is unique in X.

Let $P_H := \{u \in H : u \ge 0 \text{ almost everywhere}\}$, $P := P_H \cap X = \{u \in C_0^1(\bar{\Omega}) \mid u \ge 0\}$, P has nonempty interior $\overset{\circ}{P}$. Since $K(P) \subset P$, $K(-P) \subset (-P)$ and $P, -P$ are convex closed sets in X, by the theory of ordinary differential equations in Banach spaces, we know $P, -P$ are both invariant sets of descent flow of J under the condition $f(x, u)u \ge 0$. Noticing $K\bar{F} : P \to \overset{\circ}{P}$ and (8.30), we know that $u(t, u_0) \in \overset{\circ}{P}$, $\forall t > 0$, for $u_0 \in P$, and $\overset{\circ}{P}, -\overset{\circ}{P}$ are both invariant sets of descent flow of J.

Definition 8.2.2 Suppose that $\Phi \in C^1(X, \mathbb{R})$, $a \in \mathbb{R}$. We say that Φ has the retracting property for a on X, if $\forall b > a$, $\Phi^{-1}[a, b] \cap X \cap M = \emptyset$, then $\Phi^a \cap X$ is a retract of $\Phi^b \cap X$, i.e. there exists $\eta : \Phi^b \cap X \to \Phi^a \cap X$ continuous in the topology of X, such that $\eta(\Phi^b \cap X) \subset \Phi^a \cap X$, $\eta|_{\Phi^a \cap X} = id|_{\Phi^a \cap X}$.

Lemma 8.2.3 (Dancer and Zhang [71]) *Suppose that* (8.22), (8.23) *or* (8.24), (8.25) *are satisfied, then for any* $a \in \mathbb{R}$, *Then* J *has the retracting property for* a *on* X.

Lemma 8.2.4 (Dancer and Zhang [71]) *Suppose* U *is a bounded connected open set of* $\mathbb{R}^2$, *and* $(0, 0) \in U$, *then there exists a connected component* Γ' *of the boundary of* U, *and each one sided ray* l *through the origin satisfies* $l \cap \Gamma' \neq \emptyset$ *(see also Liu* [138]).

Theorem 8.2.2 (Dancer and Zhang [71]) *Suppose that* (H_1), (H_2) *are satisfied,* (8.22), (8.23) *(or* (8.24), (8.25)*) are satisfied, and* $a_1, d_1 < \lambda_1$, $(a_2, d_2) \in \tilde{S}$, $(a_2, d_2) \in \Sigma$. *Then* (8.21) *has at least one sign-changing solution, one positive solution, one negative solution.*

Proof Under these hypotheses (8.22), (8.23) (or (8.24), (8.25)), by Lemma 8.2.3, we know J has the retracting property for any $a \in \mathbb{R}$ in X.

First we prove that θ is a strictly local minimum of J. (This is well known, we only give this for completeness.)

We only give the proof for $\Omega \subset \mathbb{R}^N$, $N \geq 3$ (the proof is similar for $N \leq 2$). By (H_1) and $f(x, 0) = 0$, $a_1, d_1 < \lambda_1$, we know that $\exists 0 < \varepsilon_0 < \lambda_1$ and $\delta > 0$ such that $|f(x, t)| < (\lambda_1 - \varepsilon_0)|t|$, $\forall |t| < \delta$, thus

$$\left| F(x, u) \right| \leq \int_0^u \left| f(x, t) \right| dt < \frac{1}{2}(\lambda_1 - \varepsilon_0)u^2, \quad \text{as } 0 < |u| < \delta. \tag{8.32}$$

By (H_1), (H_2), we have $\exists b_1 > 0$ such that

$$\left| F(x, u) \right| \leq \int_0^u \left| f(x, t) \right| dt < \frac{1}{2}(\lambda_1 - \varepsilon_0)u^2 + b_1 |u|^{\frac{2n}{n-2}},$$

$$\text{for } -\infty < u < \infty. \tag{8.33}$$

Therefore, by the Poincaré inequality, we have

$$\left| \int_\Omega F(x, u(x)) \, dx \right| \leq \int_\Omega \left| F(x, u(x)) \right| dx$$

$$< \frac{1}{2}(\lambda_1 - \varepsilon_0) \int_\Omega u^2(x) \, dx + b_1 \int_\Omega \left| u(x) \right|^{\frac{2n}{n-2}} dx$$

$$\leq \frac{1}{2}\|u\|^2 - \frac{1}{2}\varepsilon_0 C^2(\Omega)\|u\|^2 + b_1 \int_\Omega \left| u(x) \right|^{\frac{2n}{n-2}} dx, \tag{8.34}$$

and by the definition of J, we have

$$J(u) > \frac{1}{2}\varepsilon_0 C^2(\Omega)\|u\|^2 - b_1 \int_\Omega |u(x)|^{\frac{2n}{n-2}}\,dx, \quad \forall u \in H. \tag{8.35}$$

Since $H \hookrightarrow L_{\frac{2n}{n-2}}(\Omega)$, we see that $\exists b_2 > 0$ such that

$$b_1 \int_\Omega |u|^{\frac{2n}{n-2}}(x)\,dx \le b_2 \|u\|^{\frac{2n}{n-2}}.$$

Thus, we know

$$J(u) > \frac{1}{2}\varepsilon_0 C^2(\Omega)\|u\|^2 - b_2\|u\|^{\frac{2n}{n-2}}.$$

Therefore, there exist $b_3 > 0$ and an open neighborhood U_{δ_0} of θ such that

$$J(u) > b_3\|u\|^2, \quad \forall u \in U_{\delta_0}, \quad \text{and} \quad f(0) < f(u), \quad \forall u \in U_{\delta_0}, u \ne 0, \tag{8.36}$$

i.e., θ is a strictly local minima of J, where $U_{\delta_0} = \{u \in H, \|u\| < \delta_0\}$.

Now we prove there exists a neighborhood of θ in X, which is an invariant set of J.

Let $U_{1/n} = \{u \in U_{\delta_0} \cap X \,|\, J(u) < \frac{1}{n}\}$, $n = 1, 2, 3, \ldots$, then $U_{1/n}$ is an open neighborhood of θ in X for each n. Therefore, for sufficiently large n_0 such that

$$\frac{1}{n_0 b_3} < \delta_0^2,$$

and for $u_0 \in U_{1/n_0}$, the solution $u(t, u_0)$ of (8.29) satisfies the following formula:

$$b_3\|u(t, u_0)\|^2 < J\big(u(t, u_0)\big) \le J(u_0) < \frac{1}{n_0}.$$

Thus we find that U_{1/n_0} is an open, invariant set of descent flow of J.

Let $U_0 = U_{1/n_0}$.

Set

$$U_1 = \big\{u_1 \in X \,|\, \exists t' > 0 \text{ such that } u(t', u_1) \in U_0\big\}.$$

It is easy to know that U_1 is an open invariant set of descent flow of J in X. Since $u(t, u_1)$ has continuous dependence on the initial value u_1, we may prove that ∂U_1 is also an invariant set of descent flow of J in X if $\partial U_1 \ne \emptyset$. It is clear that $J(u) > 0$ on ∂U_1, and hence $J(u(t, u_0)) > 0$ for $u_0 \in \partial U_1$ by invariance.

Since $(a_2, d_2) \in \tilde{S}$, there exists (a_0, d_0) such that $(a_0, d_0) \in \Gamma$ and $a_2 > a_0$, $d_2 > d_0$. By Lemma 8.2.1, we see that there exists a path L_0 in $\Delta(a_2, d_2)$ connecting $\varphi_1, -\varphi_1$. By the proof of Lemma 2.3 of [69],

$$L_0 = \big\{t\varphi_0^+ + (1-t)\varphi_1 : 0 \le t \le 1\big\} \cup \big\{t\varphi_0^+ + (1-t)\varphi_0^- : 0 \le t \le 1\big\}$$

$$\cup \big\{t\varphi_0^- + (1-t)(-\varphi_1) : 0 \le t \le 1\big\}, \tag{8.37}$$

where φ_0 is one nontrivial solution of (8.15) with $(a,d) = (a_0, d_0)$. Since L_0 is compact, there is an ε neighborhood L_ε in H, such that $L_\varepsilon \subset \Delta(a_2, d_2)$. In this neighborhood, we can choose a path L in X, such that $L \subset \Delta(a_2, d_2)$. In fact, we can choose $\varphi^1 \in X$, $\varphi^2 \in X$ such that

$$\left\| \varphi^1 - \varphi_0^+ \right\| < \frac{\varepsilon}{2}, \qquad \left\| \varphi^2 - \varphi_0^- \right\| < \frac{\varepsilon}{2}.$$

Define

$$L = \left\{ t\varphi^1 + (1-t)\varphi_1 : 0 \le t \le 1 \right\} \cup \left\{ t\varphi^1 + (1-t)\varphi^2 : 0 \le t \le 1 \right\}$$

$$\cup \left\{ t\varphi^2 + (1-t)(-\varphi_1) : 0 \le t \le 1 \right\}.$$

We improve L by replacing it by $\frac{u}{\|u\|_X}$ for $u \in L$, and finally we can shrink the curve to remove intersections and get L with no self intersections on ∂B_1, where B_1 is the unit ball in X. (L can be chosen arc connected by Theorem 4.1 on p. 27 of Whyburn [192].)

(H_2) implies that for $\delta > 0$, $\delta\lambda_1^{-1} \sup\{\|u\| : u \in L\} < \inf\{|J_{(a_2,d_2)}(u)| : u \in L\}$, there exists $t_0 > 0$ such that

$$\left| f(x,t) - a_2 t \right| < \delta t, \quad \forall t > t_0,$$
$$\left| f(x,t) - d_2 t \right| < \delta|t|, \quad \forall t < -t_0.$$

Thus

$$\left| F(x,t) - \frac{a_2}{2} t^2 \right| \le \frac{\delta}{2} t^2 + Ct, \quad \forall t \in \mathbb{R}^+,$$

$$\left| F(x,t) - \frac{d_2}{2} t^2 \right| \le \frac{\delta}{2} t^2 + C|t|, \quad \forall t \in \mathbb{R}^-.$$

Therefore, for $u \in H$,

$$\int_\Omega \left| F(x, u^+) - \frac{a_2}{2}(u^+)^2 \right| dx + \int_\Omega \left| F(x, u^-) - \frac{d_2}{2}(u^-)^2 \right| dx$$

$$\le \delta \int_\Omega u^2 \, dx + C \int_\Omega |u| \, dx \le \delta\lambda_1^{-1} \|u\|^2 + C\|u\|.$$

Since

$$J(u) = J_{(a_2,d_2)}(u) - \int_\Omega \left(F(x, u^+) - \frac{a_2}{2}(u^+)^2 \right) dx - \int_\Omega \left(F(x, u^-) - \frac{d_2}{2}(u^-)^2 \right) dx$$

for $u \in H$ and since for $u \in L$ and $t \ge 0$, $J_{(a_2,d_2)}(tu) = t^2 J_{(a_2,d_2)}(u)$, we have

$$J(tu) \le t^2 J_{(a_2,d_2)}(u) + \delta\lambda_1^{-1} t^2 \|u\| + tC\|u\|,$$

$$\le t^2 \sup\left\{ J_{(a_2,d_2)}(u) : u \in L \right\} + \delta\lambda_1^{-1} t^2 \sup\left\{ \|u\| : u \in L \right\}$$

$$+ tC \sup\{\|u\| : u \in L\}. \tag{8.38}$$

Thus

$$J(tu) \to -\infty \ (t \to +\infty) \quad \text{uniformly for } u \in L \ \left(\|u\|_X = 1\right). \tag{8.39}$$

Define the surface $Z := \{tu : t \geq 0, u \in L\}$. It is easily seen to be homeomorphic to a closed half space $\mathbb{R}^{2^+} := \{(x, y) \in \mathbb{R}^2 | y \geq 0\}$ in $\mathbb{R}^2$ by our careful choice of L.

It is easy to know $U_1 \cap Z$ is a bounded and relatively open set in Z, $\partial U_1 \cap Z \neq \emptyset$. We may assume that $U_1 \cap Z$ is connected. (Otherwise, we consider the connected component $U_1' \subset Z$ of $U_1 \cap Z$, with $(0, 0) \in U_1'$, instead of $U_1 \cap Z$.) By Lemma 8.2.4 there exists at least one connected component $\Lambda \subset (\partial U_1 \cap Z)$ such that $\Lambda \cap \overset{\circ}{P} \neq \emptyset$, $\Lambda \cap (-\overset{\circ}{P}) \neq \emptyset$, $\Lambda \cap [X \backslash (-P \cup P)] \neq \emptyset$. (Note that to apply Lemma 8.2.4, we consider the homeomorphism of U_1 in the half space $\mathbb{R}^{2^+}$, and then add its reflection in the other half space to obtain an open set in $\mathbb{R}^2$.)

Thus without loss of generality, we assume that $\partial U_1 \cap Z$ is connected, and ∂U_1 is connected.

Let

$$V_1 = \left\{h \in \partial U_1 | u' = -J'(u), u(0) = h, \exists t_0 \geq 0 \text{ such that } u(t_0, h) \in \overset{\circ}{P}\right\}, \tag{8.40}$$

$$V_2 = \left\{h \in \partial U_1 | u' = -J'(u), u(0) = h, \exists t_0 \geq 0 \text{ such that } u(t_0, h) \in -\overset{\circ}{P}\right\}. \tag{8.41}$$

Noticing (8.30) and $f(x, u)u \geq 0$, by the strongly order preserving property of K we know that $P \cap \partial U_1$, $(-P) \cap \partial U_1$ are both invariant sets of descent flow of J in X. V_1, V_2 are disjoint relatively open (in ∂U_1) invariant sets of descent flow of J, and V_1, V_2 are unchanged if we replace $\overset{\circ}{P}$ by P in (8.40)–(8.41). By connectedness of ∂U_1, $\partial U_1 \backslash (V_1 \cup V_2)$ is not empty. Thus by the deformation lemma (Lemma 8.2.3), we find that every solution of (8.29) goes to negative energy or approaches a critical point. Regularity implies that it converges in X. So

$$\forall u_0 \in P \cap \partial U_1, \quad u(t, u_0) \to u_1 \in M \cap P \text{ in } X,$$

$$\forall u_0 \in -P \cap \partial U_1, \quad u(t, u_0) \to u_2 \in M \cap (-P) \text{ in } X,$$

$$\forall u_0 \in \partial U_1 \backslash (V_1 \cup V_2), \quad u(t, u_0) \to u_3 \in M \text{ in } X.$$

Thus (8.21) has at least three solutions $u_1 \in P \cap \partial U_1$, $u_2 \in (-P) \cap \partial U_1$, $u_3 \in \partial U_1 \backslash (V_1 \cup V_2)$. By the maximum principle, we know that $u_1 \in \overset{\circ}{P}$, $u_2 \in -\overset{\circ}{P}$. It is clear that u_3 is sign-changing, and $\exists \varepsilon_0 > 0$ such that $J(u_i) \geq \varepsilon_0$, $i = 1, 2, 3$. (If $u(t, u_0)$ approached a point in $P \cup (-P)$, $u_0 \in V_1 \cup V_2$.) $\qquad \square$

Theorem 8.2.3 (Dancer and Zhang [71]) *Suppose that (H_1), (H_2), (H_3) are satisfied, (8.22), (8.23) (or (8.24), (8.25)) are satisfied, $(a_1, d_1) \in D_k$, $k \geq 2$, $(a_2, d_2) \in \tilde{S}$, and $(a_2, d_2) \in \Sigma$. Then (8.21) has at least seven solutions, and three are sign-changing, two are positive, two are negative.*

Proof By a small modification of the proofs of Theorem 1 and Corollary 1 in [70], we find that there exist four nontrivial solutions u_i, $i = 1, 2, 3, 4$ in $[\varphi, \psi]$, u_1 is positive, u_2 is negative, and u_3, u_4 are sign-changing. In the proof of Theorem 8.2.2, we replace U_0 by the interior of $[\varphi, \psi]$ (note the interior of $[\varphi, \psi]$ is connected). This is also an open invariant set of descent flow of J (note (8.29)). Much as before, we see that there exist three critical points u_5, u_6, u_7 outside $[\varphi, \psi]$, and $u_5 \in \partial U_1 \backslash (V_1 \cup V_2)$, $u_6 \in \overset{\circ}{P}$, $u_7 \in - \overset{\circ}{P}$. $\square$

When $a_2 = d_2$, we have

Corollary 8.2.1 *If $f \in C(\bar{\Omega} \times R)$ satisfies*

$$\limsup_{t \to 0} \frac{f(x,t)}{t} < \lambda_1, \qquad \lim_{|t| \to \infty} \frac{f(x,t)}{t} = \lambda_l, \quad l > 2, \tag{8.42}$$

and

$$\lim_{|t| \to \infty} \left(2F(x,t) - tf(x,t) \right) = -\infty, \quad \textit{uniformly in } \Omega; \tag{8.43}$$

or

$$\lim_{|t| \to \infty} \left(2F(x,t) - tf(x,t) \right) = +\infty, \quad \textit{uniformly in } \Omega. \tag{8.44}$$

Then (8.21) has at least one sign-changing solution, one positive solution, one negative solution.

Remark 8.2.3 As in [153], we do not assume $p(x,t) := f(x,t) - a_2 t^+ - d_2 t^-$ grows sublinearly. For example (8.22), (8.23) (or (8.24), (8.25)) are satisfied if $p(x,t) = -t/\ln|t|$ (resp. $t/\ln|t|$) for $|t|$ large, even though there is no $\sigma \in (0, 1)$ such that $|p(x,t)| \leq C(|t|^\sigma + 1)$ in this case.

Remark 8.2.4 (H_3) is satisfied when f satisfies one of the following three conditions:

(F_1) $\limsup_{|t| \to +\infty} \frac{f(x,t)}{t} < \lambda_1$;
(F_2) there exist $t_1 < 0, t_2 > 0$ such that $f(x, t_1) = f(x, t_2) = 0$;
(F_3) there is a number $k > 0$ such that $|f(x,t)| < k$ for $t \in [-c, c]$, where $c = \max_\Omega e(x)$ and $e(x)$ satisfies $-\Delta e = k$ in Ω, $e = 0$ on $\partial \Omega$.

Remark 8.2.5 It is easy to see that $f(x, u)u \geq 0$ is not necessary. Because $f(x, u)$ is asymptotically linear uniformly for $x \in \Omega$, there exists a positive number m such that $g(x, u) = f(x, u) + mu$ and $g(x, u)u \geq 0$, we can consider the equation $-\Delta u + mu = g(x, u), u|_{\partial \Omega} = 0$ instead of (8.21).

Remark 8.2.6 If $f \in C^1(\bar{\Omega} \times \mathbb{R})$, $f'_t(x, 0) < \lambda_1$; $f'_t(x, t) \to \lambda_l$, $l > 2$ as $|t| \to +\infty$, and (8.43) or (8.44) are satisfied, the conclusion of Corollary 8.2.1 is clearly valid.

Remark 8.2.7 From the proof of Theorem 8.2.2, we know that $J(u_i) \geq \varepsilon_0 > 0$, $i = 1, 2, 3$. So in fact for the compact condition, we only use that J satisfies $(\psi)_c$ for any $c > 0$, and J has the retracting property for any $c > 0$ in X. From the proof of Lemma 5.1 of [153], we can use

$$H(x, t) \geq W_1(x) \in L^1(\Omega), \quad \forall t \in \mathbb{R},$$

$$H(x, t) \geq W_2(x) \in L^1(\Omega) \quad \text{for large } |t|, \quad \text{and} \quad \int_\Omega W_2(x)\, dx \geq 0 \tag{8.45}$$

instead of (8.24), (8.25). For the convenience of the reader, we give the proof here.

Proof We prove for any $c > 0$,

$$J(u_k) \to c, \qquad \left(1 + \|u_k\|\right) J'(u_k) \to 0 \tag{8.46}$$

implies that $\{u_k\}$ has a convergent subsequence, i.e., J satisfies $(\psi)_c$ for any $c > 0$.
   By (8.46), we have

$$J_{(a_2,d_2)}(u_k) - \int_\Omega P(x, u_k)\, dx \to c, \tag{8.47}$$

$$\left(J'_{(a_2,d_2)}(u_k), v\right) - \left(p(x, u_k), v\right) \to 0, \quad \forall v \in H, \tag{8.48}$$

and

$$\left|\left(J'(u_k), u_k\right)\right| \leq \|u_k\| \cdot \|J'(u_k)\| \to 0.$$

Thus

$$\left(J'(u_k), u_k\right) = 2J_{(a_2,d_2)}(u_k) - \int_\Omega u_k\, p(x, u_k)\, dx \to 0. \tag{8.49}$$

   Assume that $\rho_k = \|u_k\| \to +\infty$, and let $\tilde{u}_k = u_k/\rho_k$. Then $\|\tilde{u}_k\| = 1$. Thus there is a subsequence such that $\tilde{u}_k \to \tilde{u}$ weakly in H, strongly in $L^2(\Omega)$, and a.e. in Ω. By (8.49), we obtain

$$1 = \|\nabla \tilde{u}_k\|^2_{L^2(\Omega)} = a_2\|\tilde{u}_k^+\|^2_{L^2(\Omega)} + d_2\|\tilde{u}_k^-\|^2_{L^2(\Omega)} + \rho_k^{-2}\int_\Omega u_k\, p(x, u_k)\, dx + \varepsilon_k,$$

where $\varepsilon_k \to 0$ as $k \to +\infty$. Since $\tilde{u}_k \to \tilde{u}$ strongly in $L^2(\Omega)$ and $p(x, t)/t \to 0$, as $|t| \to +\infty$ uniformly in x, we can pass to the limit and find

$$1 = a_2\|\tilde{u}^+\|^2_{L^2(\Omega)} + d_2\|\tilde{u}^-\|^2_{L^2(\Omega)}. \tag{8.50}$$

This shows that $\tilde{u} \not\equiv 0$.
   Moreover, (8.48) implies $J'_{(a_2,d_2)}(\tilde{u}) = 0$.
   In fact, (8.48) implies

$$\left(J'_{(a_2,d_2)}(\tilde{u}_k), v\right) - \left(\frac{p(x, u_k)}{\rho_k}, v\right) \to 0, \quad \forall v \in H.$$

Taking the limit, we find $(J'_{(a_2,d_2)}(\tilde{u}), v) = 0$ for all $v \in H$, and hence $J'_{(a_2,d_2)}(\tilde{u}) = 0$. Multiplying by $\tilde{u}$ gives

$$\|\tilde{u}\|^2 = a_2 \|\tilde{u}^+\|^2_{L^2(\Omega)} + d_2 \|\tilde{u}^-\|^2_{L^2(\Omega)}.$$

Hence by (8.50), $\|\tilde{u}\| = 1$. This show that $\tilde{u}_k \to \tilde{u}$ strongly in H. Combining (8.47) and (8.49), we obtain

$$\frac{1}{2} \int_\Omega H(x, u_k)\, dx \to -c < 0. \tag{8.51}$$

Let $\Omega_0 = \{x \in \Omega : \tilde{u}(x) = 0\}, \Omega_1 = \Omega \backslash \Omega_0$. Then Ω_0 has measure zero, $|u_k(x)| \to \infty$ for almost all $x \in \Omega_1$. By (8.45) and Fatou's Lemma, we have

$$\liminf_{k\to\infty} \int_\Omega H\big(x, u_k(x)\big)\, dx \geq \int_{\Omega_0} W_1(x)\, dx + \liminf_{k\to\infty} \int_{\Omega_1} H\big(x, u_k(x)\big)\, dx$$

$$\geq \int_{\Omega_1} \liminf_{k\to\infty} H\big(x, u_k(x)\big)\, dx$$

$$\geq \int_{\Omega_1} W_2(x)\, dx = \int_\Omega W_2(x)\, dx$$

$$\geq 0, \tag{8.52}$$

contradicting (8.51). Hence, the sequence $\{u_k\}$ is bounded in H. A standard argument now shows that it has a convergent subsequence.

Under (8.22)–(8.23), or (8.24)–(8.25), we can obtain

$$\int_\Omega H(x, u_k)\, dx \leq \int_{\Omega_0} W(x)\, dx + \int_{\Omega_1} H(x, u_k)\, dx \to -\infty, \tag{8.53}$$

or

$$\int_\Omega H(x, u_k)\, dx \geq \int_{\Omega_0} W(x)\, dx + \int_{\Omega_1} H(x, u_k)\, dx \to +\infty. \tag{8.54}$$

Both (8.53) and (8.54) contradict that $\frac{1}{2} \int_\Omega H(x, u_k)\, dx \to -c$ for all $c \in \mathbb{R}$. $\qquad\square$

8.2.3 Sign-Changing Solutions of p-Laplacian Equations

We consider the existence of multiple and sign-changing solutions of the problem

$$\begin{cases} -\Delta_p u = h(u) & \text{in } \Omega, \\ u(x) = 0 & \text{on } \partial\Omega, \end{cases} \tag{8.55}$$

where $-\Delta_p u = -\operatorname{div}(|\nabla u|^{p-2}\nabla u)$ is the p-Laplacian, $1 < p < +\infty$, Ω is a smooth bounded domain in $\mathbb{R}^N$ $(N \geq 1)$, $h(u)$ is local Lipschitz continuous,

$h(0) = 0$, $h(u)u \geq 0$ and $h(u)$ has "jumping" nonlinearities at zero or infinity.

$$(H_1) \quad \lim_{u \to 0^+} \frac{h(u)}{|u|^{p-2}u} = a_0, \qquad \lim_{u \to 0^-} \frac{h(u)}{|u|^{p-2}u} = d_0;$$

$$(H_2) \quad \lim_{u \to +\infty} \frac{h(u)}{|u|^{p-2}u} = a_1, \qquad \lim_{u \to -\infty} \frac{h(u)}{|u|^{p-2}u} = d_1.$$

Zhang–Li [205] seems to be the first to consider multiple and sign-changing solutions about p-Laplacian in the case of $N < p < \infty$.

We construct carefully a pseudo-gradient vector field (in short, p.g.v.f.) in $C_0^1(\bar{\Omega})$, by which we obtain the positive and negative cones of $C_0^1(\bar{\Omega})$ are all the invariant sets of the descent flow of the corresponding functional, then we use differential equations theory in Banach space to obtain sign-changing and multiple solutions of (8.55) (see [209]).

Before giving our main result, let us recall a related homogeneous "jumping" nonlinear problem

$$\begin{cases} -\Delta_p u = a \cdot (u^+)^{p-1} - d \cdot (u^-)^{p-1} & \text{in } \Omega, \\ u|_{\partial\Omega} = 0, \end{cases} \tag{8.56}$$

where $u^+ = \max\{u, 0\}$, $u^- = \max\{-u, 0\}$. By [62], the Fučik spectrum of p-Laplacian on $W_0^{1,p}(\Omega)$ is defined as the set of Σ_p of those $(a, d) \in \mathbb{R}^2$ such that (8.56) has a nontrivial solution u. Also in [62], Cuesta et al. construct the first nontrivial curve Γ in Σ_p and there exists a continuous function $\eta(t)$ defined on $(\lambda_1, \lambda_2]$ in ad-plane, such that

(a) (8.56) has a nontrivial solution for $(a, d) \in \Gamma = \{(a, \eta(a)); \lambda_1 < a \leq \lambda_2\}$
$\cup\{(\eta(d), d); \lambda_1 < d \leq \lambda_2\}$;

(b) (8.56) has no nontrivial solution for $\lambda_1 < d < \eta(a)$, $a \in (\lambda_1, \lambda_2]$ or $\lambda_1 < a < \eta(d), d \in (\lambda_1, \lambda_2]$,

where λ_1 is the first eigenvalue of $-\Delta_p$ in $W_0^{1,p}(\Omega)$ with Dirichlet boundary condition, λ_1 is simple with eigenfunction $\phi_1(x) > 0$ in Ω and $\phi_1(x) \in C_0^1(\bar{\Omega})$ and λ_2 is defined as

$$\lambda_2 = \inf\{\lambda > \lambda_1; \lambda \text{ is an eigenvalue of } -\Delta_p \text{ in } W_0^{1,p}(\Omega)\}.$$

Now we denote by S the set of points (a, d) which are above Γ in the ad-plane, $\tilde{S} = S \cup \{+\infty\}$, and our main result reads as follows.

Theorem 8.2.4 (Zhang, Chen and Li [209]) *Suppose that $(H_1), (H_2)$ hold, a_1, $d_1 < \lambda_1$, $(a_0, d_0) \in \tilde{S}$, and*

(H_3) *$h(u)$ is nondecreasing in u and for any $u \in \mathbb{R}$ any $0 < t < 1$, there exists a continuous function $\eta(u, t) > 0$ such that $h(tu) \geq (1 + \eta)t^{p-1}h(u)$, $\forall u \geq 0$; $h(tu) \leq (1 - \eta)t^{p-1}h(u)$, $\forall u \leq 0$.*

Then (8.55) *has at least three solutions, including one unique positive, one unique negative, and one sign-changing solution.*

We can give an example here: $h(u) = a(u)|u|^{p-2}, a(u)u \geq 0$. $a(u)$ is locally Lipschitz continuous,

$$\frac{a(u)}{u} \to a_0 \quad (u \to 0^+), \quad \text{and} \quad \frac{a(u)}{u} \to d_0 \quad (u \to 0^-);$$

$$\frac{a(u)}{u} \to a_1 \quad (u \to +\infty), \quad \text{and} \quad \frac{a(u)}{u} \to d_1 \quad (u \to -\infty);$$

and $\forall u \in \mathbb{R}$, $0 < t < 1$, $\exists \eta(u, t) > 0$ such that

$$a(tu) \geq (1 + \eta)t^{p-1}a(u), \forall u \geq 0; \quad \text{and} \quad a(tu) \leq (1 - \eta)t^{p-1}a(u), \quad \forall u \leq 0;$$

(i.e. $a(u)$ is sublinear).

We omit the proof here, if the readers are interested, please see [209].

Remark 8.2.8 A useful inequality (see [50]) and (S_+) condition for the operator $-\Delta_p$. According to the elementary inequalities: For any $\xi, \eta \in \mathbb{R}^N$,

$$\left(|\xi|^{p-2}\xi - |\eta|^{p-2}\eta\right) \cdot (\xi - \eta) \geq c_p \begin{cases} |\xi - \eta|^p, & p \geq 2, \\ (1 + |\xi| + |\eta|)^{p-2}|\xi - \eta|^2, & 1 < p < 2 \end{cases}$$

(where $c_p > 0$ is a constant) and the Hölder inequality, we have

$$\left\langle -\Delta_p u - (-\Delta_p v), u - v \right\rangle$$

$$\geq c_p \begin{cases} \int_\Omega |\nabla u - \nabla v|^p & \text{if } p \geq 2, \\ \left(\int_\Omega (1 + |\nabla u| + |\nabla v|)^p\right)^{1-\frac{2}{p}} \left(\int_\Omega |\nabla u - \nabla v|^p\right)^{\frac{2}{p}} & \text{if } 1 < p < 2. \end{cases}$$

The operator $-\Delta_p$ satisfies the (S_+) condition (see [86]): if $u_n \rightharpoonup u$ (weakly in $W_0^{1,p}(\Omega)$) and $\limsup_{n\to\infty}\langle -\Delta_p u_n, u_n - u\rangle \leq 0$, then $u_n \to u$ (strongly in $W_0^{1,p}(\Omega)$).

8.2.4 Sign-Changing Solutions of Schrödinger Equations

With the results we established in Sect. 7.3 we can readily extend from bounded domains to the entire space many known existence and multiplicity results on nonlinear elliptic problems involving Fučik spectrum, though this would not need much new technical ideas. We mention a couple of such results with the proofs omitted. Let us first consider the existence of solutions of

$$-\Delta u + V(x)u = f(u) \quad \text{in } \mathbb{R}^N \tag{8.57}$$

which satisfy $u(x) \to 0$ as $|x| \to \infty$. This type of equations arise from study of standing wave solutions of time-dependent nonlinear Schrödinger equations. We consider the compact first, i.e., we assume (V_0). For a coercive case, let λ_1 be the first eigenvalue of $-\Delta + V$. We assume that $f(s)$ satisfies

$(f1)$ $f \in C(\mathbb{R}, \mathbb{R})$, $f(s)s \geq 0$;

$(f2)$ $\limsup_{|s| \to \infty} \frac{f(s)}{s} < \lambda_1$;

$(f3)$ $\lim_{s \to 0^+} \frac{f(s)}{s} = \alpha$, $\lim_{s \to 0^-} \frac{f(s)}{s} = \beta$, (α, β) is above the curve Θ in $\mathbb{R}^2$ associated with the operator (7.14).

For another case we assume

$(f4)$ $\limsup_{|s| \to 0} \frac{f(s)}{s} < \lambda_1$;

$(f5)$ For some $(\alpha, \beta) \notin \Sigma$, (α, β) is above the curve Θ in $\mathbb{R}^2$ associated with (7.14), $\lim_{s \to +\infty} \frac{f(s)}{s} = \alpha$, $\lim_{s \to -\infty} \frac{f(s)}{s} = \beta$.

Theorem 8.2.5 (Bartsch, Wang and Zhang [31])

(a) *Under assumptions $(f1)$–$(f3)$, (8.57) has at least three non-trivial solutions including one positive and one negative.*

(b) *Under assumptions $(f1)$, $(f4)$, $(f5)$, (8.57) has at least one positive, one negative, and one sign-changing solution.*

For the proof of the above results we can adapt the arguments from bounded domains case (e.g., [127, 143, 205]). We leave the details to the interested readers.

For the steep potential case when we assume (V_1)–(V_3) in Sect. 7.3, we can also consider the existence and multiplicity of solutions for nonlinear Schrödinger equations of the form:

$$-\Delta u + V_\lambda(x)u = f(u) \quad \text{in } \mathbb{R}^N. \tag{8.58}$$

Let μ_1 be the first eigenvalue of $-\Delta + a$ in Ω with Dirichlet boundary condition and $\tilde{\Theta}$ be the first nontrivial Fučik spectrum curve. We assume

$(\tilde{f}2)$ $\limsup_{|s| \to \infty} \frac{f(x,s)}{s} < \mu_1$;

$(\tilde{f}3)$ $\lim_{s \to 0^+} \frac{f(s)}{s} = \alpha$, $\lim_{s \to 0^-} \frac{f(s)}{s} = \beta$, (α, β) is above the curve $\tilde{\Theta}$ in $\mathbb{R}^2$;

$(\tilde{f}4)$ $\limsup_{|s| \to 0} \frac{f(s)}{s} < \mu_1$;

$(\tilde{f}5)$ For some $(\alpha, \beta) \notin \Sigma$, (α, β) is above the curve $\tilde{\Theta}$ in $\mathbb{R}^2$, $\lim_{s \to +\infty} \frac{f(s)}{s} = \alpha$, $\lim_{s \to -\infty} \frac{f(s)}{s} = \beta$.

Theorem 8.2.6 (Bartsch, Wang and Zhang [31])

(a) *Under assumptions $(f1)$, $(\tilde{f}2)$, $(\tilde{f}3)$, there is $\Lambda > 0$ such that for $\lambda > \Lambda$, (8.58) has at least three non-trivial solutions including one positive and one negative.*

(b) *Under assumptions $(f1)$, $(\tilde{f}4)$, $(\tilde{f}5)$, there is $\Lambda > 0$ such that for $\lambda > \Lambda$, (8.58) has at least one positive, one negative, and one sign-changing solution.*

For Theorem 8.2.6 we sketch the proof as follows. First for the corresponding energy functional in case (a), the (PS) condition is satisfied for λ large and this can be done by the arguments in [143]. In case (b), a modified argument of [143] works also. By Remarks 7.3.1 and 7.3.3 for λ large $(\tilde{f}2)$ and $(\tilde{f}4)$ are satisfied with μ_1 being replaced by μ_1^λ. By Remarks 7.3.2 and 7.3.3 for λ large again, we have $(\tilde{f}3)$ and $(\tilde{f}5)$ are satisfied with (α, β) being above the curve Θ_λ and $(\alpha, \beta) \notin \Sigma_\lambda$. From here the arguments for the compact case can be carried over with little modifications.

Chapter 9
Extension of Brezis–Nirenberg's Results and Quasilinear Problems

9.1 Introduction

Let $\Omega \subset \mathbb{R}^N$ ($N \geq 3$) be a bounded smooth domain and $p > 1$. Consider the functional

$$J(u) = \frac{1}{p} \int_\Omega |Du|^p \, dx - \int_\Omega F(u) \, dx, \quad u \in W_0^{1,p}(\Omega), \tag{9.1}$$

where $F(u) = \int_0^u f(s) \, ds$ and $f \in C^1(-\infty, \infty)$ satisfies

(F) $|f(s)| \leq \alpha_2 + \alpha_1|s|^\beta$, $|f'(s)| \leq \alpha_4 + \alpha_3|s|^{\beta-1}$, where $\alpha_1, \alpha_3 \in [0, 1]$, $\alpha_2, \alpha_4 \in (0, +\infty)$, $p - 1 < \beta < \frac{Np}{N-p} - 1$ as $1 < p < N$ and $|f(s)| \leq \alpha_6 + \alpha_5|s|^{\beta_1}$, $|f'(s)| \leq \alpha_8 + \alpha_7|s|^{\beta_1-1}$, where $\alpha_5, \alpha_7 \in [0, 1]$, $\alpha_6, \alpha_8 \in (0, +\infty)$, $p - 1 < \beta_1 < +\infty$ as $p \geq N$.

In many cases we need to know the answer of the following question:

(Q) If $u_0 \in W_0^{1,p}(\Omega)$ is a local minimizer of J in the C^1-topology, is it still a local minimizer of J in $W_0^{1,p}(\Omega)$?

For $p = 2$, H. Brezis and L. Nirenberg [36] gave a positive answer. For $p > 1$ and $p \neq 2$, their method does not apply since we have a nonlinear operator. We first give a positive answer to the above question for $p > 2$. Then using this positive answer we study the structure of solutions of the quasilinear elliptic problems

$$\begin{cases} -\Delta_p u = f_\lambda(u) & \text{in } \Omega, \\ u = 0 & \text{on } \partial\Omega \end{cases} \tag{9.2}$$

where $p > 2$; $f_\lambda : \mathbb{R} \to \mathbb{R}$ and $\lambda > 0$ is a real parameter.

The background of (9.2) can be found in [113]. The existence and uniqueness of, possibly multiple, solutions of (9.2) have been studied by Zongming Guo et al. in previous papers (see [90, 103–105, 107–109, 112–114, 195]). Such problems have also been treated by many other authors, see, for example, [18–20, 63, 82, 83, 87, 96, 97, 118, 119, 124, 127, 132, 161, 175, 204, 205, 208] and the references therein.

Z. Zhang, *Variational, Topological, and Partial Order Methods with Their Applications*, Developments in Mathematics 29,
DOI 10.1007/978-3-642-30709-6_9, © Springer-Verlag Berlin Heidelberg 2013

When $f_\lambda = \lambda g(s)$, it was shown in [103] that there exist at least two positive solutions of (9.2) when λ is sufficiently large provided that g is strictly increasing on $\mathbb{R}^+$, $g(0) = 0$, $\lim_{s \to 0^+}(g(s)/s^{p-1}) = 0$ and $g(s) \leq \alpha_1 + \alpha_2 s^\theta$, $0 < \theta < p - 1$. On the other hand, the structure of positive solutions of (9.2) with $f_\lambda(s) = \lambda g(s)$ and g changing sign has also been studied in [107–109, 113].

Definition 9.1.1 A solution of (9.2) is a pair of $(\lambda, u) \in \mathbb{R}^+ \times (W_0^{1,p}(\Omega) \cap C_0^1(\overline{\Omega}))$ which satisfies (9.2) in the weak sense.

When $f_\lambda(u) = \lambda u^q$, $0 < q < p - 1$, the sub- and supersolution argument as in [112] easily provides the existence of a unique positive solution of (9.2) for all $\lambda > 0$.

As an application of the answer of our question (Q), we shall study problem (9.2) when f_λ is, roughly, the sum of two terms: one has the growth less than $p - 1$, another one has the growth larger than or same as $p - 1$. We consider two types of condition on f_λ:

(i) $f_\lambda(s) = \lambda s^q + s^\omega$ for $s > 0$, here $0 < q < p - 1 < \omega$.
(ii) $f_\lambda(s) = \lambda |s|^{q-1}s + g(s)$ for $s \in (-\infty, +\infty)$; $q \in (0, p-1)$; $g \in C^1(-\infty, +\infty)$
     satisfies

> (H_1) $g'(s) \geq 0$ for $s \in (-\infty, +\infty)$, $g(s)s \geq 0$ for any $s \in (-\infty, +\infty)$ and $\lim_{|s| \to 0} g(s)/|s|^{p-1} = 0$.

In Theorem 9.3.2 below we show that there exists a positive constant $\Lambda > 0$ such that if $\lambda \in (0, \Lambda)$, there exist at least two solutions of the problem

$$\begin{cases} -\Delta_p u = \lambda u^q + u^\omega, & u > 0 \quad \text{in } \Omega, \\ u = 0 \quad \text{on } \partial\Omega. \end{cases} \tag{9.3}$$

Such kind of problems has been studied in [176] by variational method and genus. We shall obtain Theorem 9.3.2 by different ideas and provide more information on the solutions. Using the assumption that ω is an arbitrary positive number, we only get one positive solution of (9.3). To get the second positive solution, we need a subcritical growth assumption on ω. When $p = 2$, this problem has been treated in [12], but their methods cannot be easily used to deal with (9.2) here, since the linearization of the operator in (9.2) is difficult to handle. To overcome the difficulty arising from our operator, we use a scale argument instead.

In Theorem 9.3.3 below, we study structure of solutions of the problem

$$\begin{cases} -\Delta_p u = \lambda |u|^{q-1}u + g(u) & \text{in } \Omega, \\ u = 0 \quad \text{on } \partial\Omega. \end{cases} \tag{9.4}$$

Problem (9.4) with $1 < p < N$ was discussed in [20] for g with critical exponent.

Theorem 9.2.1 and Theorem 9.3.2 have been also obtained in [21] with different proofs. In [21], the authors considered the problem,

$$-\Delta_p u = f_\lambda(u) \quad \text{in } \Omega, \qquad u = 0 \quad \text{on } \partial\Omega,$$

where $f_\lambda(u) = |u|^{r-2}u + \lambda|u|^{q-2}u$ with $1 < q < p < r < p^*$, $p^* = \frac{Np}{N-p}$ $(p < N)$, $p^* = \infty$ $(p \geq N)$.

The main result of [21] is Theorem 1.1 (for $p > 1$): If $u_0 \in W_0^{1,p}(\Omega)$ is a local minimizer of J in $C^1(\Omega)$, then u_0 is a local minimizer in $W_0^{1,p}(\Omega)$. Then they use Theorem 1.1 and the Mountain Pass theorem to prove the above equation has at least two positive solutions for all $\lambda \in (0, \Lambda)$; there is no positive solution for $\lambda > \Lambda$; there exists at least one positive solution for $\lambda = \Lambda$ (Theorem 1.3). From the proof of Theorem 1.1 of [21], we know that the main work is to prove the uniform $C^{1,\alpha}$ estimate of v_ϵ: $\|v_\epsilon\|_{C^{1,\alpha}} \leq C$ (Theorem 1.2, v_ϵ is the same as that in the proof of Theorem 9.2.1 here).

We should point out that our proof for the principal result (Theorem 9.2.1) is much simpler than that of Theorem 1.1 in [21], although we only prove for $p > 2$ and there are some hypotheses on f' (see hypothesis (F)). Theorem 9.3.2 here is similar to Theorem 1.3 of [21]. Theorem 9.3.1 (ω can be arbitrary for the existence of one solution) and the results of the final section are new.

9.2 $W_0^{1,p}(\Omega)$ Versus $C_0^1(\bar{\Omega})$ Local Minimizers

In this section we shall give a positive answer to our question (Q) for $p > 2$. The main result in this section is

Theorem 9.2.1 (Guo and Zhang [115]) *Assume that $p > 2$ and (F) holds. Assume that $u_0 \in W_0^{1,p}(\Omega) \cap C_0^1(\bar{\Omega})$ is a local minimizer of J in the C^1-topology; this means that there is some $r > 0$ such that*

$$J(u_0) \leq J(u_0 + v) \quad \forall v \in C_0^1(\bar{\Omega}) \quad \text{with } \|v\|_{C_0^1(\bar{\Omega})} \leq r. \tag{9.5}$$

Then u_0 is local minimizer of J in $W_0^{1,p}(\Omega)$, i.e. there exists $\kappa > 0$ such that

$$J(u_0) \leq J(u_0 + v) \quad \forall v \in W_0^{1,p}(\Omega) \quad \text{with } \|v\|_{W_0^{1,p}(\Omega)} \leq \kappa. \tag{9.6}$$

Proof We first consider the case of $2 < p < N$. Recall that u_0 satisfies in the weak sense the problem

$$-\Delta_p u_0 = f(u_0) \quad \text{in } \Omega, \qquad u_0 = 0 \quad \text{on } \partial\Omega. \tag{9.7}$$

Then it follows from the condition (F) and the regularity results in [124] and [180, 181] (see the proof of Proposition 2.2 of [103]) that $u_0 \in C^{1,\sigma}(\bar{\Omega})$ $(0 < \sigma < 1)$. Here we use the growth conditions on f.

Suppose the conclusion does not hold. Then

$$\forall \epsilon > 0, \ \exists v_\epsilon \in B_\epsilon \quad \text{such that} \quad J(u_0 + v_\epsilon) < J(u_0) \tag{9.8}$$

where $B_\epsilon = \{v \in W_0^{1,p}(\Omega) : \|v\|_{W_0^{1,p}(\Omega)} \leq \epsilon\}$. It is easily known that J is lower semi-continuous on the convex set B_ϵ.

Notice that B_ϵ is weak sequence compact and weakly closed in $W_0^{1,p}(\Omega)$. By the embedding of $W_0^{1,p}(\Omega)$ to $L^\gamma(\Omega)$ with $p - 1 < \gamma \leq \frac{Np}{N-p} - 1$ and a standard lower semi-continuity argument, we know that J is bounded from below on B_ϵ and $\exists v_\epsilon \in B_\epsilon$ such that

$$J(u_0 + v_\epsilon) = \inf_{v \in B_\epsilon} J(u_0 + v).$$

We shall prove that $v_\epsilon \to 0$ in C^1 as $\epsilon \to 0$, but then (9.5) and (9.8) are contradictory. The corresponding Euler equation for v_ϵ involves a Lagrange multiplier $\mu_\epsilon \leq 0$ (by Theorem 26.1 of [81]), namely, v_ϵ satisfies

$$J_v'(u_0 + v_\epsilon)(h) = \mu_\epsilon \int_\Omega \left[|Dv_\epsilon|^{p-2} Dv_\epsilon Dh \right] dx, \quad \forall h \in W_0^{1,p}(\Omega),$$

i.e.

$$-\Delta_p(u_0 + v_\epsilon) - f(u_0 + v_\epsilon) = -\mu_\epsilon \Delta_p v_\epsilon. \tag{9.9}$$

Thus,

$$-\Delta_p u_0 - \left[\Delta_p(u_0 + v_\epsilon) - \Delta_p u_0\right] = f(u_0 + v_\epsilon) - \mu_\epsilon \Delta_p v_\epsilon$$

and

$$-\left[\Delta_p(u_0 + v_\epsilon) - \Delta_p u_0\right] + \mu_\epsilon \Delta_p v_\epsilon = f(u_0 + v_\epsilon) - f(u_0). \tag{9.10}$$

Writing (9.10) to the form

$$\begin{aligned}
-\operatorname{div}\bigl(A(v_\epsilon)\bigr) :=\ & -\operatorname{div}\bigl(|D(u_0 + v_\epsilon)|^{p-2} D(u_0 + v_\epsilon) \\
& \quad - |Du_0|^{p-2} Du_0 - \mu_\epsilon |Dv_\epsilon|^{p-2} Dv_\epsilon \bigr) \\
=\ & f(u_0 + v_\epsilon) - f(u_0) \\
=\ & f'(\xi)v_\epsilon,
\end{aligned}$$

where $\xi \in (\min\{u_0, u_0 + v_\epsilon\}, \max\{u_0, u_0 + v_\epsilon\})$. We know from Lemma 2.1 of [63] that for $p > 2$ there exists $\rho > 0$ independent of u_0 and v_ϵ such that

$$\left[|D(u_0 + v_\epsilon)|^{p-2} D(u_0 + v_\epsilon) - |Du_0|^{p-2} Du_0 \right] \cdot Dv_\epsilon \geq \rho |Dv_\epsilon|^p.$$

Thus,

$$A(v_\epsilon) \cdot Dv_\epsilon \geq (\rho - \mu_\epsilon) |Dv_\epsilon|^p \geq \rho |Dv_\epsilon|^p,$$

since $\mu_\epsilon \leq 0$. On the other hand, using the growth condition (F) on $f'(s)$, we have

$$\left| f'(\xi) \right| \leq \alpha_4 + \alpha_3 |\xi|^{\beta - 1} \leq \alpha_4 + C\left[|u_0|^{\beta - 1} + |v_\epsilon|^{\beta - 1} \right]$$

since $\beta - 1 > p - 2 > 0$. Thus, by the regularity results obtained in [124] (see Theorem 7.1 in [124], pp. 286–287, and Theorem 1.1 in [124], p. 251) we see that for some $0 < \sigma < 1$, there exists $C > 0$ independent of ϵ such that

$$\|v_\epsilon\|_{C^\sigma(\Omega)} \le C\left(\|v_\epsilon\|_{W_0^{1,p}(\Omega)}\right) \le C$$

By the regularity results in [132] (see also [85]), we also have

$$\|v_\epsilon\|_{C_0^{1,\sigma}(\Omega)} \le C^*$$

where C^* is determined by C. This implies that $v_\epsilon \to v_0$ in C^1 as $\epsilon \to 0$. Since $\|v_\epsilon\|_{W_0^{1,p}(\Omega)} \to 0$ as $\epsilon \to 0$, we have $v_0 \equiv 0$. This completes the proof of the case $2 < p < N$.

The proof of the case of $p \ge N$ is similar. Note that in this case the embedding $W_0^{1,p}(\Omega) \to C^0(\overline{\Omega})$ holds. By Theorem 1 of [132], we know that

$$\|v_\epsilon\|_{C^{1,\sigma}(\Omega)} \le C$$

where C is independent of ϵ. This completes the proof. $\qquad\square$

9.3 Multiplicity Results for the Quasilinear Problems

We assume that Ω has a good property as in [105, 112, 113], that is, Ω is a bounded smooth domain in $\mathbb{R}^N$, for example, $\Omega \in C^3$. We denote by $d(x)$ the distance from $x \in \Omega$ to the boundary of $\partial\Omega$, and by $s(x)$ the point of $\partial\Omega$ which is closest to x (which is uniquely defined if x is close enough to $\partial\Omega$). The boundary strip $\Omega_\delta :=\{x \in \Omega : 0 < d(x) < \delta\}$ (where $\delta > 0$ is small enough) is defined as in [105, 112], which is covered (only covered) by the straight lines in the inner normal direction $-n_{s(x)}$ and emanating from $s(x)$.

We consider below the problem of finding solutions of the boundary value problem (9.3). To emphasize the dependence on λ, the problem (9.3) often referred to as problem $(9.3)_\lambda$ (the subscript λ is omitted if no confusion arises).

Our first result is

Theorem 9.3.1 (Guo and Zhang [115]) *Let $p > 1$. For all $0 < q < p - 1 < \omega$ there exists $\Lambda > 0$ such that for $\lambda \in (0, \Lambda)$, the problem $(9.3)_\lambda$ has a minimal solution u_λ, which is increasing with respect to λ and $\|u_\lambda\|_\infty \to 0$ as $\lambda \to 0$. For all $\lambda > \Lambda$, the problem $(9.3)_\lambda$ has no solution. Moreover, for $\lambda = \Lambda$, the problem $(9.3)_\lambda$ has at least one weak solution $u_\lambda \in W_0^{1,p}(\Omega) \cap L^{\omega+1}(\Omega)$.*

Remark 9.3.1 The existence of a solution of $(9.3)_\lambda$ with $\lambda > 0$ sufficiently small has been obtained in [19, 20].

To prove Theorem 9.3.1 we need the following lemmas.

Lemma 9.3.1 (Guo and Zhang [115]) *Let $\Lambda = \sup\{\lambda > 0 : (9.3)_\lambda \text{ has a solution}\}$. Then $0 < \Lambda < \infty$.*

Proof Let e be the unique positive solution of

$$-\Delta_p e = 1 \quad \text{in } \Omega, \qquad e = 0 \quad \text{on } \partial\Omega.$$

Since $0 < q < p - 1 < \omega$, we can find $\lambda_0 > 0$ such that for all $0 < \lambda \le \lambda_0$ there exists $M = M(\lambda) > 0$ satisfying

$$M^{p-1} \ge \lambda M^q \|e\|_\infty^q + M^\omega \|e\|_\infty^\omega.$$

As a consequence, the function Me satisfies

$$-\Delta_p(Me) = M^{p-1} \ge \lambda M^q \|e\|_\infty^q + M^\omega \|e\|_\infty^\omega$$

and hence it is a supersolution of $(9.3)_\lambda$. Moreover, let ϕ_1 with $\|\phi_1\|_\infty = 1$ be the first eigenfunction corresponding to the first eigenvalue $\lambda_1 > 0$ of the problem

$$-\Delta_p u = \lambda |u|^{p-2} u \quad \text{in } \Omega, \qquad u = 0 \quad \text{on } \partial\Omega.$$

Then, any $\epsilon \phi_1$ is a subsolution of $(9.3)_\lambda$ provided

$$-\Delta_p(\epsilon \phi_1) = \epsilon^{p-1} \lambda_1 \phi_1^{p-1} \le \lambda \epsilon^q \phi_1^q + \epsilon^\omega \phi_1^\omega,$$

which is satisfied for all $\epsilon > 0$ small enough and any fixed λ. Taking ϵ possibly smaller, we also have

$$\epsilon \phi_1 < Me.$$

It follows from the sub- and supersolution argument as in [112, 113] that $(9.3)_\lambda$ has a solution $\epsilon \phi_1 \le u \le Me$ (here we use the monotonicity of the function $f_\lambda(s) = \lambda s^q + s^\omega$) whenever $\lambda \le \lambda_0$ and thus $\Lambda \ge \lambda_0$. Next, let $\overline{\lambda}$ be such that

$$\overline{\lambda} t^q + t^\omega > \lambda_1 t^{p-1} \quad \text{for } t > 0. \tag{9.11}$$

For fixed $\lambda \ge \overline{\lambda}$, if there exists a positive solution u of (9.3) (we omit the subscript here and below), then

$$-\Delta_p u = \lambda u^q + u^\omega > \lambda_1 u^{p-1}.$$

Then, let

$$\beta = \sup\{\mu \in \mathbb{R} : u - \mu \phi_1 > 0 \text{ in } \Omega\}.$$

We have

$$u \ge \beta \phi_1 \quad \text{in } \Omega.$$

We claim that $0 < \beta < \infty$. It follows from Lemma 2.3 of [112] (i.e., Lemma 9.4.1 below) that there exist $\ell_i > 0$ $(i = 1, 2, 3, 4)$ such that

$$\ell_1 d(x) \le u(x) \le \ell_2 d(x),$$

$$\ell_3 d(x) \le \phi_1(x) \le \ell_4 d(x)$$

where $d(x) = \mathrm{dist}(x, \partial\Omega)$. These imply that

$$\frac{\ell_1}{\ell_4}\phi_1(x) \le u(x) \le \frac{\ell_2}{\ell_3}\phi_1(x).$$

Thus $\frac{\ell_1}{\ell_4} \le \beta \le \frac{\ell_2}{\ell_3}$. This is our claim. Moreover,

$$-\Delta_p u - \{-\Delta_p(\beta\phi_1)\} > \lambda_1\left[u^{p-1} - (\beta\phi_1)^{p-1}\right] \ge 0.$$

By a scale argument similar to [112] (for convenience of the readers, we shall give the outline of the proof later), we show that there exists $\delta_1 > 0$ such that $u \equiv \beta\phi_1$ in Ω_{δ_1}, where $\Omega_{\delta_1} = \{x \in \Omega : d(x, \partial\Omega) < \delta_1\}$. This clearly implies that

$$-\Delta_p u = -\Delta_p(\beta\phi_1)$$

$$= \lambda_1(\beta\phi_1)^{p-1} = \lambda_1 u^{p-1} \quad \text{in } \Omega_{\delta_1}.$$

This contradicts

$$-\Delta_p u > \lambda_1 u^{p-1} \quad \text{in } \Omega.$$

This also implies that $\lambda < \bar{\lambda}$ and shows that $\Lambda \le \bar{\lambda}$.

Now we prove that there exists $\delta_1 > 0$ such that $u \equiv \beta\phi_1$ in Ω_{δ_1}. We first show that there exists $\eta \in \Omega$ such that $u(\eta) = \beta\phi_1(\eta)$. On the contrary, we have $u > \beta\phi_1$ in Ω. Since $\frac{\partial u}{\partial n_s} < 0$ and $\frac{\partial\phi_1}{\partial n_s} < 0$ on $\partial\Omega$ and $\partial\Omega$ is compact, we know that there exists $\delta_1 > 0$ and $\gamma > 0$ such that

$$\frac{\partial u}{\partial n_{s(x)}} < -\gamma < 0 \quad \text{and} \quad \frac{\partial\phi_1}{\partial n_{s(x)}} < -\gamma < 0 \quad \text{in } \Omega_{\delta_1}.$$

Thus,

$$t\frac{\partial u}{\partial n_{s(x)}}(x) + (1-t)\frac{\partial(\beta\phi_1)}{\partial n_{s(x)}}(x) \le -\gamma \quad \text{for } x \in \Omega_{\delta_1} \text{ and all } t \in [0, 1]. \qquad (9.12)$$

Hence, using the mean value theorem, we obtain

$$0 \le -\Delta_p u - \{-\Delta_p(\beta\phi_1)\}$$

$$= -\sum_{i,j}\frac{\partial}{\partial x_i}\left[a^{ij}(x)\frac{\partial(u - \beta\phi_1)}{\partial x_j}\right] \quad \text{in } \Omega_{\delta_1},$$

where $a^{ij}(x) = \int_0^1 \frac{\partial a^i}{\partial q_j}[t\,Du + (1-t)D(\beta\phi_1)]\,dt$ and $a^i = |q|^{p-2}q_i$ $(i = 1, 2, \ldots, N)$ for $q = (q_1, q_2, \ldots, q_N) \in \mathbb{R}^N$. Put

$$L\cdot = \sum_{i,j} \frac{\partial}{\partial x_i}\left[a^{ij}(x)\frac{\partial}{\partial x_j}\cdot \right].$$

Using (9.12), we see that L is a uniformly elliptic operator on Ω_{δ_1}. Consequently, we have

$$-L(u - \beta\phi_1) \geq 0 \quad \text{in } \Omega_{\delta_1}, \tag{9.13}$$

$$u(x) > \beta\phi_1(x) \quad \text{in } \Omega_{\delta_1}, \quad \text{and} \quad u - \beta\phi_1 = 0 \quad \text{on } \partial\Omega \text{ (part of } \partial\Omega_{\delta_1}). \tag{9.14}$$

By Hopf's boundary point lemma ([95], Lemma 3.4) we obtain $\frac{\partial(u-\beta\phi_1)}{\partial n_s} < 0$ on $\partial\Omega$. By arguments similar to those in [112], we see that there exists $\theta > 0$ such that

$$u(x) \geq (\beta + \theta)\phi_1(x) \quad \text{for } x \in \Omega. \tag{9.15}$$

This contradicts the definition of β. By the same argument as that in the proof of Theorem 3.1 in [112], we also see that there exists a point $z \in \Omega_{\delta_1}$ where $u - \beta\phi_1$ vanishes and therefore

$$u \equiv \beta\phi_1 \quad \text{in } \Omega_{\delta_1}.$$

This completes the proof. $\square$

Lemma 9.3.2 (Guo and Zhang [115]) *For all $0 < \lambda < \Lambda$, the problem $(9.3)_\lambda$ has a solution.*

Proof Given $\lambda < \Lambda$, let u_μ be a solution of $(9.3)_\mu$ with $\lambda < \mu < \Lambda$. Plainly, such a u_μ is a supersolution of $(9.3)_\lambda$. Since $\epsilon\phi_1 < u_\mu$ provided $\epsilon > 0$ is sufficiently small, it follows that $(9.3)_\lambda$ has a solution. This completes the proof. $\square$

We next prove that $(9.3)_\lambda$ possesses a minimal solution. To this end we need the following lemma.

Lemma 9.3.3 (Guo and Zhang [115]) *Assume that f is a non-decreasing C^1 function with $f(0) = 0$ such that $s^{1-p}f(s)$ is strictly decreasing for $s > 0$. Let $v, w \in W_0^{1,p} \cap C^1(\overline{\Omega})$ satisfy*

$$-\Delta_p v \leq f(v), \quad v > 0 \text{ in } \Omega, \quad v = 0 \text{ on } \partial\Omega \tag{9.16}$$

and

$$-\Delta_p w \geq f(w), \quad w > 0 \text{ in } \Omega, \quad w = 0 \text{ on } \partial\Omega. \tag{9.17}$$

Moreover, $\frac{\partial v}{\partial n} < 0$ on $\partial\Omega$, where n is the outward norm vector of $\partial\Omega$. Then $w \geq v$ in Ω.

Proof We also use the scale argument to prove this lemma. Let

$$\beta = \sup\{\mu \in \mathbb{R} : w - \mu v > 0 \text{ in } \Omega\}.$$

Then $0 < \beta < \infty$ and $w \geq \beta v$ in Ω. We shall prove that $\beta \geq 1$. Suppose that $\beta < 1$, then

$$-\Delta_p w - \left\{-\Delta_p(\beta v)\right\} \geq f(w) - \beta^{p-1} f(v)$$
$$> f(w) - f(\beta v) \geq 0 \quad \text{in } \Omega, \tag{9.18}$$

where we use the facts that f is non-decreasing and that $s^{1-p} f(s)$ is strictly decreasing for $s > 0$. Using arguments similar to those in the proof of Lemma 9.3.1 we see that there exists $\delta_2 > 0$ such that

$$w \equiv \beta v \quad \text{in } \Omega_{\delta_2}.$$

This clearly contradicts (9.18). This completes the proof. $\square$

It should be noted that the scale argument used in the proof of Lemma 9.3.1 works with any positive β because the function involved is s^{p-1}; while the same argument used in the proof of Lemma 9.3.3 works only for $0 < \beta < 1$ because the involved function $f(s)$ satisfies: $s^{1-p} f(s)$ is strictly decreasing for $s > 0$.

Lemma 9.3.4 (Guo and Zhang [115]) *For all $0 < \lambda < \Lambda$, the problem $(9.3)_\lambda$ has a minimal solution y_λ and $\|y_\lambda\|_\infty \to 0$ as $\lambda \to 0$.*

Proof Let $z_\lambda \in C_0^1(\overline{\Omega})$ be the unique positive solution of

$$-\Delta_p z = \lambda z^q \quad \text{in } \Omega, \qquad z = 0 \quad \text{on } \partial\Omega.$$

We already know that there exists a solution $u_\lambda > 0$ of $(9.3)_\lambda$ for every $\lambda \in (0, \Lambda)$. Since $-\Delta_p u_\lambda \geq \lambda u_\lambda^q$, we use Lemma 9.3.3 with $w = u_\lambda$ and $v = z_\lambda$ to deduce that any solution of $(9.3)_\lambda$ must satisfy $u_\lambda \geq z_\lambda$. (It follows from Lemma 2.2 of [103] that $\frac{\partial v_\lambda}{\partial n} < 0$ on $\partial\Omega$.) Clearly, z_λ is a subsolution of $(9.3)_\lambda$. The monotone iteration

$$-\Delta_p u_{n+1} = \lambda u_n^q + u_n^\omega, \quad u_0 = z_\lambda$$

and the maximum principle [103] imply that $u_n \uparrow y_\lambda$, with y_λ a solution of $(9.3)_\lambda$. It is easy to check that y_λ is a minimal solution of $(9.3)_\lambda$. Indeed, for any solution u_λ of $(9.3)_\lambda$, we have $u_\lambda \geq z_\lambda$. Then the weak comparison principle implies that $u_n \leq u_\lambda$ for any n and thus $y_\lambda \leq u_\lambda$. Since $M(\lambda) \to 0$ as $\lambda \to 0$ (see the proof of Lemma 9.3.1), it follows that $\|y_\lambda\|_\infty \to 0$ as $\lambda \to 0$.

Now we show the existence of a positive solution of $(9.3)_\lambda$ for $\lambda = \Lambda$. Let

$$f_\lambda(s) = \begin{cases} \lambda s^q + s^\omega, & s \geq 0, \\ 0, & s < 0 \end{cases}$$

and

$$F_\lambda(u) = \int_0^u f_\lambda(s)\, ds,$$

we may define the functional $\overline{I}_\lambda : W_0^{1,p}(\Omega) \to \mathbb{R}$ by setting

$$\overline{I}_\lambda(u) = \frac{1}{p}\|u\|^p_{W^{1,p}(\Omega)} - \int_\Omega F_\lambda(u)\, dx.$$

It is well known that the critical points of $\overline{I}_\lambda$ correspond to the solutions of $(9.3)_\lambda$. $\square$

Now we have the following lemma. (In the sequel λ is fixed.)

Lemma 9.3.5 (Guo and Zhang [115]) *For all $\lambda \in (0, \Lambda)$, the problem $(9.3)_\lambda$ has a solution u_λ which is in addition a local minimizer of $\overline{I}_\lambda$ in the C^1-topology. Moreover, there exists $C = C(\Lambda) > 0$ such that*

$$\overline{I}_\lambda(u_\lambda) < 0$$

and

$$\|u_\lambda\|^p_{W_0^{1,p}(\Omega)} \le C, \qquad \|u_\lambda\|^{\omega+1}_{L^{\omega+1}(\Omega)} \le C.$$

Proof We fix $\hat{\lambda}_1 < \lambda < \hat{\lambda}_2 < \Lambda$ and consider the minimal solutions $u_1 := y_{\hat{\lambda}_1}$ and $u_2 = y_{\hat{\lambda}_2}$ defined in Lemma 9.3.4. We first show that $u_1 \le u_2$. In fact, we can show that for any $\lambda^* \in (0, \Lambda)$

$$y_\lambda \le y_{\lambda^*} \quad \text{whenever} \quad \lambda \le \lambda^*.$$

Indeed, if $\lambda < \lambda^*$ then y_{λ^*} is a supersolution of $(9.3)_\lambda$. Since, for $\epsilon > 0$ small, $\epsilon\phi_1$ is a subsolution of $(9.3)_\lambda$ and $\epsilon\phi_1 < y_{\lambda^*}$, then $(9.3)_\lambda$ possesses a solution v_λ, with

$$(\epsilon\phi_1 \le)\ v_\lambda \le y_{\lambda^*}.$$

Since y_λ is the minimal solution of $(9.3)_\lambda$, we infer that $y_\lambda \le v_\lambda \le y_{\lambda^*}$. Now we show actually

$$u_1 < u_2 \quad \text{in } \Omega, \qquad \frac{\partial(u_2 - u_1)}{\partial n} < 0 \quad \text{on } \partial\Omega$$

where n is the outward normal vector on $\partial\Omega$. Clearly, u_1, respectively, u_2, is a subsolution, respectively, supersolution, of $(9.3)_\lambda$. Moreover,

$$-\Delta_p u_2 - \{-\Delta_p u_1\} = \hat{\lambda}_2 u_2^q + u_2^\omega - \left(\hat{\lambda}_1 u_1^q + u_1^\omega\right)$$

$$\ge \hat{\lambda}_1 u_2^q + u_2^\omega - \left(\hat{\lambda}_1 u_1^q + u_1^\omega\right) \ge 0 \quad \text{in } \Omega.$$

Since $u_1 \not\equiv u_2$ (because $\hat\lambda_1 < \hat\lambda_2$), by arguments similar to those in the proof of Lemma 9.3.1, we find that there exists $\delta_3 > 0$ such that

$$0 \le -\Delta_p u_2 - \{-\Delta_p u_1\} = -L(u_2 - u_1) \quad \text{in } \Omega_{\delta_3}$$

where L is a uniformly elliptic operator in Ω_{δ_3}. Since $u_2 \ge u_1$ in Ω, if there exists an $\eta \in \Omega$ such that $u_2(\eta) = u_1(\eta)$, we can prove that there exists a $\xi \in \Omega_{\delta_3}$ such that $u_2(\xi) = u_1(\xi)$. On the contrary, we can find $\Omega_1 \Subset \Omega$ such that $\eta \in \Omega_1$ and $\partial\Omega_1 \subset \Omega_{\delta_3}$ and $u_2 \ge u_1 + \zeta$ on $\partial\Omega_1$ with $\zeta > 0$. Let $u_3 = u_1 + \zeta$. Then

$$-\Delta_p u_2 - \{-\Delta_p u_3\} \ge 0 \quad \text{in } \Omega_1$$

and

$$u_2 \ge u_3 \quad \text{on } \partial\Omega_1.$$

The weak comparison principle implies that

$$u_2 \ge u_3 \quad \text{in } \Omega_1.$$

This contradicts $u_2(\eta) = u_1(\eta)$. Since L is uniformly elliptic in Ω_{δ_3} and $\xi \in \Omega_{\delta_3}$, we have $u_1 \equiv u_2$ in Ω_{δ_3}. This contradicts $\hat\lambda_1 < \hat\lambda_2$. Thus, $u_2 > u_1$ in Ω. Since L is uniformly elliptic in Ω_{δ_3} and $-L(u_2 - u_1) \ge 0$ in Ω_{δ_3} with $u_2 - u_1 = 0$ on $\partial\Omega$, the Hopf boundary point lemma yields

$$\frac{\partial}{\partial n}(u_2 - u_1) < 0 \quad \text{on } \partial\Omega.$$

Now we set

$$\tilde f_\lambda(x, s) = \begin{cases} f_\lambda(u_1(x)), & s \le u_1, \\ f_\lambda(s), & u_1 < s < u_2, \\ f_\lambda(u_2(x)), & s \ge u_2, \end{cases}$$

$$\tilde F_\lambda(x, u) = \int_0^u \tilde f_\lambda(x, s)\, ds$$

and

$$\tilde I_\lambda(u) = \frac{1}{p} \int_\Omega |Du|^p\, dx - \int_\Omega \tilde F_\lambda(x, u)\, dx.$$

By the standard way, one can prove that $\tilde I_\lambda$ achieves its (global) minimum at some $u_\lambda \in W_0^{1,p}(\Omega)$. Moreover,

$$-\Delta_p u_\lambda = \tilde f_\lambda(x, u_\lambda) \quad \text{for } x \in \Omega.$$

We also know from Proposition 2.2 of [103] that $u_\lambda \in C_0^1(\overline\Omega)$. It is clear that $\tilde f_\lambda(x, u_\lambda) \ge f_\lambda(x, u_1)$ in Ω. Using the scale argument as above, we obtain

$$u_1 < u_\lambda < u_2 \quad \text{in } \Omega, \tag{9.19}$$

$$\frac{\partial}{\partial n}(u_\lambda - u_1) < 0, \qquad \frac{\partial}{\partial n}(u_\lambda - u_2) > 0 \quad \text{on } \partial\Omega. \tag{9.20}$$

These imply that u_λ is a solution of $(9.3)_\lambda$. From (9.19)–(9.20) it follows that if

$$\|v - u_\lambda\|_{C^1} = \epsilon$$

with ϵ small, then $u_1 \le v \le u_2$. Moreover $\overline{I}_\lambda(v) - \tilde{I}_\lambda(v)$ is constant for $u_1 \le v \le u_2$ and therefore u_λ is also a local minimizer for $\overline{I}_\lambda$ in the C^1-topology. Let

$$J_\lambda(\epsilon) = \overline{I}_\lambda(u_\lambda + \epsilon h),$$

for any $h \in C_0^1(\overline{\Omega})$ and $h > 0$. Then J_λ attains a local minimum at $\epsilon = 0$. Thus,

$$J_\lambda''(0) = \left(\overline{I}_\lambda''(u_\lambda)h, h\right) \ge 0.$$

Setting $h = u_\lambda$, we have

$$\int_\Omega |Du_\lambda|^p\, dx - \left(\lambda q/(p-1)\right) \int_\Omega u_\lambda^{q+1}\, dx - \left(\omega/(p-1)\right) \int_\Omega u_\lambda^{\omega+1}\, dx \ge 0.$$

This together with

$$\overline{I}_\lambda(u_\lambda) = \frac{1}{p} \int_\Omega |Du_\lambda|^p\, dx - \frac{\lambda}{q+1} \int_\Omega u_\lambda^{q+1}\, dx - \frac{1}{\omega+1} \int_\Omega u_\lambda^{\omega+1}\, dx$$

and

$$\int_\Omega |Du_\lambda|^p\, dx = \lambda \int_\Omega u_\lambda^{q+1}\, dx + \int_\Omega u_\lambda^{\omega+1}\, dx$$

imply that

$$\overline{I}_\lambda(u_\lambda) < 0$$

and there exists $C = C(\Lambda)$ such that

$$\|u_\lambda\|_{W^{1,p}(\Omega)}^p \le C, \tag{9.21}$$

$$\|u_\lambda\|_{L^{\omega+1}(\Omega)}^{\omega+1} \le C. \tag{9.22}$$

This completes the proof. $\qquad\qquad\qquad\qquad\qquad\qquad\qquad\qquad\qquad\qquad\quad\square$

Lemma 9.3.6 (Guo and Zhang [115]) *There exists a solution* $u^* \in W_0^{1,p}(\Omega) \cap L^{\omega+1}(\Omega)$ *of* $(9.3)_\lambda$ *for* $\lambda = \Lambda$.

Proof In fact, let $\{\lambda_n\}$ be a sequence such that $\lambda_n \uparrow \Lambda$. By Lemma 9.3.5, there exists a solution $u_n \in W^{1,p}(\Omega) \cap C_0^1(\overline{\Omega})$ of $(9.3)_{\lambda_n}$ such that $\overline{I}_{\lambda_n}(u_n) < 0$ and (9.21)–(9.22) hold. Then there exists $u^* \in W_0^{1,p}(\Omega) \cap L^{\omega+1}(\Omega)$ such that $u_n \to u^*$ a.e in Ω, weakly in $W_0^{1,p}(\Omega)$ and $L^{\omega+1}(\Omega)$. Such an u^* is thus a weak solution of $(9.3)_\lambda$ for $\lambda = \Lambda$. When $p - 1 < \omega < [Np/(N - p)] - 1$ for $1 < p < N$ and $\omega > p - 1$ for $p \geq N$, we know from the proof of Theorem 9.2.1 that $u^* \in C_0^{1,\alpha}(\overline{\Omega})$. This completes the proof. $\square$

Proof of Theorem 9.3.1 The proof of Theorem 9.3.1 can be obtained directly from the lemmas above.

Now we are looking for a second positive solution of $(9.3)_\lambda$. Denote y_λ the solution obtained in Lemma 9.3.4. We have the following theorem. $\square$

Theorem 9.3.2 (Guo and Zhang [115]) *Let* $0 < q < (p - 1) < \omega < [Np/(N - p)] - 1$ *for* $2 < p < N$, $0 < q < (p - 1) < \omega < +\infty$ *for* $p \geq N$. *Then for all* $\lambda \in (0, \Lambda)$, *the problem* $(9.3)_\lambda$ *has another solution* w_λ *with* $w_\lambda \not\equiv y_\lambda$ *and* $w_\lambda > y_\lambda$.

Proof Let

$$\tilde{f}_\lambda(x, s) = \begin{cases} f_\lambda(s), & \text{if } s \geq y_\lambda(x), \\ f_\lambda(y_\lambda(x)), & \text{if } s < y_\lambda(x). \end{cases}$$

Then $\tilde{f}_\lambda$ is continuous on x and s. By arguments similar to those in the proof of Lemma 9.3.5, we see that the functional

$$\tilde{I}_\lambda(u) = \frac{1}{p} \int_\Omega |Du|^p \, dx - \int_\Omega \tilde{F}_\lambda(x, u) \, dx$$

with $\tilde{F}_\lambda(x, s) = \int_0^s \tilde{f}_\lambda(x, \xi) \, d\xi$ restricted to $E = C_0^1(\overline{\Omega})$ has a local minimizer $\tilde{u}_\lambda$ in the interval $[\tilde{u}_1, \tilde{u}_2]$ in E, where $\tilde{u}_1 = y_{\lambda-\epsilon}$ and $\tilde{u}_2 = y_{\lambda+\epsilon}$ for $\epsilon > 0$ sufficiently small. Theorem 9.2.1 implies that this local minimizer is also a local minimizer of $\tilde{I}_\lambda$ in $W_0^{1,p}(\Omega)$. We assume that y_λ is the unique solution of the problem

$$-\Delta_p u = \tilde{f}_\lambda(x, u) \quad \text{in } \Omega, \qquad u = 0 \quad \text{on } \partial\Omega$$

in $[\tilde{u}_1, \tilde{u}_2]_E$, otherwise we have obtained our conclusion. Thus, $\tilde{u}_\lambda \equiv y_\lambda$ in Ω and y_λ is the only local minimizer of $\tilde{I}_\lambda$ in $W_0^{1,p}(\Omega)$, this implies that y_λ is a strictly local minimizer of $\tilde{I}_\lambda$ in $W_0^{1,p}(\Omega)$. One easily checks that $\tilde{I}_\lambda(t\phi_1) \to -\infty$ as $t \to +\infty$. Moreover, if $\tilde{I}_\lambda$ satisfies (PS) condition, then it is easy to show that for some $\epsilon > 0$ small, $\inf\{\tilde{I}_\lambda(u) : \|u - y_\lambda\|_{W_0^{1,p}(\Omega)} = \epsilon\} > \tilde{I}_\lambda(y_\lambda)$. Hence one can use the Mountain Pass lemma to obtain a critical point $w_\lambda \in W_0^{1,p}(\Omega)$ of $\tilde{I}_\lambda$ such that $w_\lambda \not\equiv y_\lambda$. We also know that $w_\lambda \in C_0^1(\overline{\Omega})$ and thus, $w_\lambda \geq y_\lambda$. We can show $w_\lambda > y_\lambda$ by the scale argument as above. This implies that w_λ is another solution of $(9.3)_\lambda$.

We still have to show that $\tilde{I}_\lambda$ satisfies (PS) condition. This can be done by arguments similar to those in the proof of Theorem 4.6 of [105]. (We need to use the embedding $W_0^{1,p}(\Omega) \to C^0(\Omega)$ for $p \geq N$.) This completes the proof. $\square$

Remark 9.3.2 When $\omega = [Np/(N - p)] - 1$ for $1 < p < N$, some existence results of the problem $(9.3)_\lambda$ have been obtained in [20].

On Multiplicity Results for the Problem (9.4)   We next study problem $(9.4)_\lambda$ with f_λ satisfying (ii) and g satisfying (H_1). It is clear that the nonlinearity of problem (9.3) is a special case of the f_λ discussed here. To obtain our multiplicity results for $(9.4)_\lambda$, we first use sub- and supersolution arguments like above to obtain a minimal positive solution and a maximal negative solution for $(9.4)_\lambda$. Using Theorem 9.2.1, we easily know that the minimal positive solution and the maximal negative solution are strictly local minimizers of a corresponding functional of $(9.4)_\lambda$ with some extra conditions on g. Finally, using the Mountain Pass Lemma we can obtain another positive solution and another negative solution of $(9.4)_\lambda$ by arguments similar to those above. Moreover, we can also provide a sign-changing solution for $(9.4)_\lambda$.

In the following, $\lambda_1 > 0$ denotes the first eigenvalue of the problem

$$\begin{cases} -\Delta_p u = \lambda |u|^{p-2} u & \text{in } \Omega, \\ u = 0 & \text{on } \partial\Omega. \end{cases} \tag{9.23}$$

It is well-known that λ_1 is simple.

Theorem 9.3.3 (Guo and Zhang [115]) *Suppose $p > 2$, g satisfies (H_1) and either*

(H_2) $\lim_{|u|\to\infty} g(u)/|u|^{p-2}u = a > \lambda_1$, *or*
(H_3) $\lim_{|u|\to\infty} g(u)/|u|^{\gamma-1}u = b > 0$, *where $p - 1 < \gamma < [Np/(N-p)] - 1$ for $2 < p < N$; $\gamma > p - 1$ for $p \geq N$.*

Then there exist Λ^+, $\Lambda^- > 0$ such that

(i) *for $\lambda > \Lambda^+$ (resp. Λ^-), $(9.4)_\lambda$ has no positive (resp. negative) solution;*
(ii) *for $0 < \lambda < \Lambda^+$ (resp. Λ^-), $(9.4)_\lambda$ has at least two positive (resp. negative) solutions;*
(iii) *for $\lambda = \Lambda^+$ (resp. Λ^-), $(9.4)_\lambda$ has at least one positive (resp. negative) solution;*
(iv) *when g satisfies (H_2), for $0 < \lambda < \min\{\Lambda^-, \Lambda^+\}$, $(9.4)_\lambda$ has at least one sign-changing solution.*
(v) *when g satisfies (H_3) with $2 < p < N$, $p - 1 < \gamma < [Np/(N-p)] - 1$ and Ω is an N-ball, for $0 < \lambda < \min\{\Lambda^-, \Lambda^+\}$, $(9.4)_\lambda$ has at least one sign-changing radial solution.*

Remark 9.3.3 We expect that (v) is true for any bounded smooth domain Ω and $p > 2$. The key point here is how to get the upper bound of the solutions of $(9.4)_\lambda$. Normally, we use a blow up argument to obtain this (see [113]). We need to know the structure of the corresponding equation in $\mathbb{R}^N$. When Ω is a ball and $p - 1 < \gamma < [Np/(N-p)] - 1$ for $1 < p < N$, we know from [96] that there is no bounded positive radial solution for the equation $-\Delta_p u = u^\gamma$ in $\mathbb{R}^N$. Then we can find the upper bound of the positive radial solutions of $(9.4)_\lambda$ in this case by a blow up argument.

The proof of Theorem 9.3.3 can be obtained from the following lemmas and theorems.

Lemma 9.3.7 (Guo and Zhang [115]) *Suppose g satisfies (H_1) and (H_2). Then there exist Λ^+, $\Lambda^- \in (0, \infty)$ such that*

(i) *for $\lambda > \Lambda^+$ (resp. Λ^-), $(9.4)_\lambda$ has no positive (resp. negative) solution;*
(ii) *for $0 < \lambda < \Lambda^+$ (resp. Λ^-), $(9.4)_\lambda$ has at least one positive (resp. negative) solution.*

Proof (i) Define

$$\Lambda^+ = \sup\{\lambda > 0\colon (9.4)_\lambda \text{ has a positive solution}\},$$

$$\Lambda^- = \sup\{\lambda > 0\colon (9.4)_\lambda \text{ has a negative solution}\}.$$

Then the conclusion follows from a simple variation of the proof of Lemma 9.3.1.

(ii) The same sub- and supersolution arguments as in the proofs of Lemmas 9.3.1–9.3.3 show that $(9.4)_\lambda$ has a minimal positive solution y_λ for $0 < \lambda < \Lambda^+$, and a maximal negative solution $\overline{y}_\lambda$ for $0 < \lambda < \Lambda^-$. This completes the proof. $\square$

Theorem 9.3.4 (Guo and Zhang [115]) *Suppose that g satisfies (H_1) and (H_2), Λ^+, Λ^- are as in Lemma 9.3.7. Then*

(i) *for $\lambda = \Lambda^+$ (resp. Λ^-), $(9.4)_\lambda$ has at least one positive (resp. negative) solution;*
(ii) *for $0 < \lambda < \Lambda^+$ (resp. Λ^-), $(9.4)_\lambda$ has at least two positive (resp. negative) solutions;*
(iii) *for $0 < \lambda < \min\{\Lambda^-, \Lambda^+\}$, $(9.4)_\lambda$ has at least one sign-changing solution.*

Proof (i) We carry out the proof for $\lambda = \Lambda^+$ only; another case can be proved analogously. By Lemma 9.3.4, $(9.4)_\lambda$ has a minimal positive solution y_λ for any $\lambda \in (0, \Lambda^+)$. Since

$$\lim_{u \to +\infty} \left(\lambda u^q + g(u)\right)/u^{p-1} = a > \lambda_1$$

uniformly for $\lambda \in (0, \Lambda^+)$, we shall prove that there exists $C > 0$ independent of λ such that

$$\|y_\lambda\|_\infty \leq C$$

for all $\lambda \in (0, \Lambda^+)$. Indeed, suppose that there exists a sequence $\{\lambda_n\} \subset (0, \Lambda^+)$ such that $\{y_n\} \equiv \{y_{\lambda_n}\}$ with $\|y_n\|_\infty \to \infty$ as $n \to \infty$. Then let $v_n = y_n/\|y_n\|_\infty$, we have

$$-\Delta_p v_n = \lambda_n \|y_n\|_\infty^{-(p-q-1)} v_n^q + g\big(\|y_n\|_\infty v_n\big)/\|y_n\|_\infty^{p-1} \quad \text{in } \Omega,$$

$$v_n = 0 \quad \text{on } \partial\Omega.$$

The regularity of $-\Delta_p \cdot$ (see [103]) implies that $v_n \to v$ in $C_0^1(\overline{\Omega})$ with $v \geq 0$, $\|v\|_\infty = 1$ and v satisfies

$$-\Delta_p v = a v^{p-1} \quad \text{in } \Omega, \qquad v = 0 \quad \text{on } \partial\Omega.$$

The scale argument as in the proof of Lemma 9.3.1 implies that there exists $\delta_4 > 0$ such that

$$v \equiv \beta\phi_1 \quad \text{in } \Omega_{\delta_4}.$$

This contradicts $a > \lambda_1$. This shows that $\|y_\lambda\|_\infty \leq C$. It follows from the regularity of the p-Laplacian that $\{y_\lambda : 0 < \lambda < \Lambda^+\}$ is precompact in C^1. Hence for some sequence $\lambda_n \to \Lambda^+$, y_{λ_n} converges to a solution y_{Λ^+} of $(9.4)_{\Lambda^+}$. By arguments similar to those in the proof of Lemma 9.3.4, $u_{\Lambda^+} \geq z_{\Lambda^+}$, where z_{Λ^+} is the unique positive solution of

$$-\Delta_p z = \lambda z^q \quad \text{in } \Omega, \qquad z = 0 \quad \text{on } \partial\Omega$$

which is always a subsolution to $(9.4)_\lambda$. Therefore, y_{Λ^+} must be a positive solution of $(9.4)_{\Lambda^+}$ and $y_{\Lambda^+} \geq z_{\Lambda^+}$. This implies that $(9.4)_\lambda$ has a minimal positive solution y_λ for $\lambda = \Lambda^+$ as well.

(ii) Again we consider the case $0 < \lambda < \Lambda^+$ only. Another case is similar. Choose $\lambda' \in (\lambda, \Lambda^+)$ and define $\bar{u} = y_{\lambda'}$. Then $\bar{u}$ is an supersolution to $(9.4)_\lambda$. Let u^* be the minimal solution of $(9.4)_\lambda$ between z_λ and $\bar{u}$. Then the proof of Theorem 9.3.2 shows that $(9.4)_\lambda$ has a second solution $u \geq u^*$.

(iii) We may assume that $\Lambda^+ < \Lambda^-$. The proofs for other cases are similar. We need only to show that $(9.4)_\lambda$ has a sign-changing solution between the maximal negative solution $\bar{y}_\lambda$ and the minimal positive solution y_λ for any $0 < \lambda < \Lambda :=$ $\min\{\Lambda^+, \Lambda^-\} = \Lambda^+$. By truncating $f_\lambda(u) = \lambda|u|^{q-1}u + g(u)$ as

$$\bar{f}_\lambda(x, s) = \begin{cases} f_\lambda(\bar{y}_\lambda(x)), & \text{if } s < \bar{y}_\lambda(x), \\ f_\lambda(s), & \text{if } \bar{y}_\lambda(x) \leq s \leq y_\lambda(x), \\ f_\lambda(y_\lambda(x)), & \text{if } s > y_\lambda(x) \end{cases}$$

we get the functional

$$\bar{J}_\lambda(u) = \frac{1}{p}\int_\Omega |Du|^p \, dx - \int_\Omega \bar{F}_\lambda(x, u) \, dx$$

with $\bar{F}_\lambda(x, s) = \int_0^s \bar{f}_\lambda(x, \xi)\, d\xi$ which satisfies (PS) condition in $W_0^{1,p}(\Omega)$.
   Define

$$\mathcal{D} = [\bar{y}_\lambda, y_\lambda] := \left\{ w \in C_0^1(\overline{\Omega}) : \bar{y}_\lambda \leq w \leq y_\lambda \right\}.$$

The proof of Lemma 9.3.5 implies that $\bar{J}_\lambda$ has a positive and a negative minimizers in $W_0^{1,p}(\Omega)$ and they are y_λ and $\bar{y}_\lambda$, respectively. We only show that y_λ is the positive minimizer. The proof of $\bar{y}_\lambda$ is similar. Setting $\tilde{u}_1 = y_{\lambda'}$ with $0 < \lambda' < \lambda$, $\tilde{u}_2 = y_\lambda$ and

$$\tilde{f}_\lambda(x, s) = \begin{cases} f_\lambda(\tilde{u}_2(x)), & \text{if } s > \tilde{u}_2(x), \\ f_\lambda(s), & \text{if } \tilde{u}_1(x) \leq s \leq \tilde{u}_2(x), \\ f_\lambda(\tilde{u}_1(x)), & \text{if } s < \tilde{u}_1(x) \end{cases}$$

we obtain by arguments similar to those in the proof of Lemma 9.3.5 that the functional

$$\tilde{I}_\lambda(u) = \frac{1}{p} \int_\Omega |Du|^p \, dx - \int_\Omega \tilde{F}_\lambda(x, u) \, dx$$

with $\tilde{F}_\lambda(x, s) = \int_0^s \tilde{f}_\lambda(x, \xi) \, d\xi$ has a local minimizer $\tilde{u}_\lambda$ in $W_0^{1,p}(\Omega)$ and $\tilde{u}_\lambda \in [\tilde{u}_1, \tilde{u}_2]_E$. Since y_λ is the minimal positive solution of $(9.4)_\lambda$, we easily know that $\tilde{u}_\lambda \equiv y_\lambda$. Arguments similar to those in the proof of Theorem 9.3.2 imply that y_λ and $\overline{y}_\lambda$ are strictly local minimizers of $\overline{J}_\lambda$ in $W_0^{1,p}(\Omega)$. Therefore, there is a mountain pass critical point w of $\overline{J}_\lambda$ in $W_0^{1,p}(\Omega)$. It is clear that 0 is not a mountain pass critical point of $\overline{J}_\lambda$, so $w \neq 0$. Now we show that w must be a sign-changing critical point. On the contrary, we have $w \geq 0$ or $w \leq 0$, but $w \neq 0$. We shall only derive a contradiction for the first case. The proof of the second case is similar. Since the regularity of $-\Delta_p$ implies $w \in C_0^1(\overline{\Omega})$, by the fact $g(w) \geq 0$, we easily know $-\Delta_p w \geq 0$. The strong maximum principle in [103] implies $w > 0$. Now we show $w \in [0, y_\lambda]_E$ and thus w is a positive solution of $(9.4)_\lambda$, which is a contradiction since y_λ is the minimal positive solution of $(9.4)_\lambda$. Indeed, we know that

$$-\Delta_p w = \overline{f}_\lambda(w) \leq f_\lambda(y_\lambda) = -\Delta_p y_\lambda$$

where we use the monotonicity of g. Then the weak comparison principle of $-\Delta_p$ implies

$$w \leq y_\lambda$$

and thus $w \in [0, y_\lambda]_E$. The analysis above shows that w must be a sign-changing critical point. We also need to show that

$$w \in \mathcal{D}.$$

We only show that $w \leq y_\lambda$. On the contrary, suppose that there exists $\Omega_1 \subset \Omega$ such that $w > y_\lambda$ on Ω_1. Then

$$-\Delta_p w = \overline{f}_\lambda(w) \equiv f_\lambda(y_\lambda) = -\Delta_p y_\lambda \quad \text{in } \Omega_1.$$

Defining $\psi = (w - y_\lambda)^+$, we have

$$\int_{\Omega_1} \left(|Dw|^{p-2} Dw - |Dy_\lambda|^{p-2} Dy_\lambda \right)(Dw - Dy_\lambda) \, dx = 0.$$

On the other hand, we know from [63] that there exists $\gamma_0 > 0$ independent of p such that

$$\int_{\Omega_1} \left(|Dw|^{p-2} Dw - |Dy_\lambda|^{p-2} Dy_\lambda \right)(Dw - Dy_\lambda) \, dx$$

$$\geq \gamma_0 \begin{cases} \int_{\Omega_1} (1 + |Dw| + |Dy_\lambda|)^{p-2} |Dw - Dy_\lambda|^2 \, dx & \text{if } 1 < p < 2, \\ \int_{\Omega_1} |Dw - Dy_\lambda|^p \, dx & \text{if } p \geq 2 \end{cases}$$

and hence we derive a contradiction. Therefore, $w \in \mathcal{D}$ and hence w is a sign-changing solution of $(9.4)_\lambda$. This completes the proof. $\qquad\square$

Theorem 9.3.5 (Guo and Zhang [115]) *Suppose $p > 2$ and g satisfies (H_1) and*

(H_3) $\lim_{|u| \to \infty} g(u)/|u|^{\gamma-1} u = b > 0$, *where $p - 1 < \gamma < [Np/(N - p)] - 1$ for $2 < p < N$ and $\gamma > p - 1$ for $p \geq N$.*

Then there exist $\Lambda^+, \Lambda^- > 0$ such that

  (i) *for $\lambda > \Lambda^+$ (resp. Λ^-), $(9.4)_\lambda$ has no positive (resp. negative) solution;*
  (ii) *for $\lambda = \Lambda^+$ (resp. Λ^-), $(9.4)_\lambda$ has at least one positive (resp. negative) solution;*
  (iii) *for $0 < \lambda < \Lambda^+$ (resp. Λ^-), $(9.4)_\lambda$ has at least two positive (resp. negative) solutions;*
  (iv) *when Ω is an N-ball and g satisfies (H_3) for $2 < p < N$, $p - 1 < \gamma < [Np/(N - p)] - 1$, $(9.4)_\lambda$ has at least one sign-changing radial solution for $0 < \lambda < \min\{\Lambda^+, \Lambda^-\}$.*

Proof (i) can be obtained by the idea similar to that in the proof of Lemma 9.3.7.

  (ii) We consider the case $\lambda = \Lambda^+$ only. By the idea similar to that in the proof of Lemma 9.3.4, $(9.4)_\lambda$ has a minimal positive solution y_λ for any $\lambda \in (0, \Lambda^+)$. As above, one easily sees that $y_{\lambda'} \leq y_{\lambda''}$ if $0 < \lambda' \leq \lambda'' < \Lambda^+$. Also, if z_λ is defined as in the proof of Lemma 9.3.4, then $z_\lambda \leq y_\lambda$ and z_λ is a subsolution of $(9.4)_\lambda$ for any $\lambda \in (0, \Lambda^+)$. Thus, as in the proof of Lemma 9.3.4, one concludes that $(9.4)_\lambda$ has a solution $\tilde{u}_\lambda$ between z_λ and $y_{\lambda'}$ for $\lambda' \in (\lambda, \Lambda^+)$ which minimizes J_λ on $[z_\lambda, y_{\lambda'}]_{C^1}$, where

$$J_\lambda(u) = \frac{1}{p} \int_\Omega |Du|^p \, dx - \frac{1}{q+1} \int_\Omega |u|^{q+1} \, dx - \int_\Omega G(u) \, dx,$$

$$G(u) = \int_0^u g(s) \, ds.$$

In particular, $J_\lambda(\tilde{u}_\lambda) \leq J_\lambda(z_\lambda)$. This implies that

$$J_\lambda(\tilde{u}_\lambda) \leq M \equiv \max_{\Lambda^+/2 \leq \xi \leq \Lambda^+} J_\xi(z_\xi) \quad \text{for any } \lambda \in \left(\Lambda^+/2, \Lambda^+\right).$$

Since $\tilde{u}_\lambda$ is also a local minimizer of J_λ in $W_0^{1,p}(\Omega)$, then (H_3) (actually, we see that J_λ satisfies the (PS) condition) implies that $\sup\{\|\tilde{u}_\lambda\|_{W_0^{1,p}(\Omega)} : \Lambda^+/2 < \lambda < \Lambda^+\} < \infty$. The proof of Proposition 2.2 of [103] implies that $\tilde{u}_\lambda \in C_0^{1,\alpha}(\Omega)$ and $\|\tilde{u}_\lambda\|_{C^{1,\alpha}(\Omega)} \leq C$ where C is independent of λ. Now choosing a sequence $\lambda_n \to \Lambda^+$ such that $\tilde{u}_{\lambda_n} \to u_{\Lambda^+}$, and passing to the limit in $(9.4)_\lambda$ with $(\lambda, u) = (\lambda_n, \tilde{u}_n)$, using (H_1), we conclude that u_{Λ^+} is a positive solution of $(9.4)_\lambda$ with $\lambda = \Lambda^+$. u_{Λ^+} must be a positive solution since $u_{\Lambda^+} \geq z_{\Lambda^+}$. This implies that $(9.4)_\lambda$ has a minimal positive solution at $\lambda = \Lambda^+$.

(iii) can be obtained by a simple variation of the proof of Theorem 9.3.4. Note that (H_1) and (H_3) guarantee that we still have the (PS) condition and the mountain pass lemma applies as before.

(iv) can be obtained by arguments similar to those in the proof of Theorem 9.3.4. Now since we only consider the radial solutions of $(9.4)_\lambda$, we can write $(9.4)_\lambda$ to the form

$$-\left(r^{N-1}|u'|^{p-2}u'\right)' = \lambda r^{N-1}|u|^{q-1}u + r^{N-1}g(u)$$

under the corresponding boundary conditions. Writing the operator A in Theorem 9.3.4 in the radial form, we know the compactness of A from [104]. The existence of sub- and supersolutions of $(9.4)_\lambda$ can also be obtained by arguments similar to those in the proof of Theorem 9.3.4. Now we only need to prove that all the positive and negative radial solutions of $(1.4)_\lambda$ are uniformly bounded in $C^0(\Omega)$. We only find the boundedness of the positive solutions of $(9.4)_\lambda$. This can be obtained by a blow up argument as in [94, 113] since we know from [96] that there is no bounded positive radial solution of the equation

$$-\Delta_p u = u^\gamma \quad \text{in } \mathbb{R}^N$$

where γ is as in the assumption of (iv). $\square$

9.4 Uniqueness Results

Assume that Ω is a bounded smooth domain in $\mathbb{R}^N$, $N \geq 2$.

We study uniqueness of positive solution:

$$\begin{cases} -\Delta_p u = \lambda f(u(x)) & \text{in } \Omega, \\ u(x) = 0 & \text{on } \partial\Omega, \end{cases} \tag{9.24}$$

where $f(s) > 0$ for $s > 0$, $p > 1$ and $\lambda > 0$.

Lemma 9.4.1 (Zongming Guo, see [105, 112]) *Assume that $\rho \geq 0, \rho \not\equiv 0$ in Ω. Suppose that $u \in C_0^1(\bar{\Omega})$ is the unique positive solution of the problem*

$$\begin{cases} -\Delta_p u = \rho(x) & \text{in } \Omega, \\ u(x) = 0 & \text{on } \partial\Omega. \end{cases} \tag{9.25}$$

Then, there exist constants $l_1 \geq k_1 > 0$ such that $k_1 d(x) \leq u(x) \leq l_1 d(x)$ on $\bar{\Omega}$.

Lemma 9.4.2 (Zongming Guo see [105, 112]) *For any $p > 1$, the problem*

$$\begin{cases} -\Delta_p u = 1 & \text{in } \Omega, \\ u(x) = 0 & \text{on } \partial\Omega, \end{cases} \tag{9.26}$$

has a unique positive solution $v_0(x) \in C_0^{1,\alpha}(\bar{\Omega})$, $0 < \alpha < 1$. When Ω is an N-ball $B_R(0)$ or an annulus, v_0 is an radial solution.

Uniqueness of large positive solutions

Lemma 9.4.3 (Zongming Guo, see [105, 112]) *There exists a unique positive solution $v_\beta(x)$ of the problem*

$$\begin{cases} -\Delta_p u = u^\beta(x) & in\ \Omega, \\ u(x) = 0 & on\ \partial\Omega, \end{cases} \tag{9.27}$$

where $0 < \beta < p - 1$. When Ω is an N-ball $B_R(0)$ or an annulus, v_β is a radial solution.

Definition 9.4.1 We say u_λ is a large solution of (9.24) if $u_\lambda(x) \geq v_0(x)$ in Ω. We say u_λ is a small solution if $\lim_{\lambda \to \infty} u_\lambda = 0$ in Ω.

Theorem 9.4.1 (Zongming Guo [105]) *Suppose that f satisfies*

(H_1) $f \in C^1((0, \infty)) \cap C^\gamma([0, \infty))$ *for some $\gamma \in (0, 1]$, f is non-decreasing for $s > 0$,*
(H_2) $s^{-\beta} f(s) \to \mu > 0$, *as $s \to \infty$,*
(H_3) $f(s) \sim s^\alpha$ *as $s \to 0^+$, where $p - 1 < \alpha < [Np/(N - p)] - 1$ if $p < N$; $N - 1 < \alpha$ if $p = N$ (without loss of generality, we assume $(H_3)'$ $\lim_{s \to 0^+} \frac{f(s)}{s^\alpha} = 1$),*
(H_4) *there exist $T, Y > 0$ with $Y \geq T$ such that*

$$\left(\frac{f(s)}{s^{p-1}} \right)' > 0 \quad for\ s \in (0, T)$$

and

$$\left(\frac{f(s)}{s^{p-1}} \right)' < 0 \quad for\ s > Y.$$

Then, there exists $\lambda^* > 0$ such that for $\lambda \geq \lambda^*$, (9.24) has a unique positive solution $u_\lambda(x) \geq v_0(x)$ and for u_λ,

$$\lim_{\lambda \to \infty} u_\lambda \lambda^{-1/(p-1-\beta)} = \mu^{1/(p-1-\beta)} v_\beta(x) \quad in\ C_0^1(\bar{\Omega}). \tag{9.28}$$

Theorem 9.4.2 (Zongming Guo [105]) *Suppose that $\Omega = \{x : r_1 < |x| < r_2, r_2 > r_1 > 0\}$ is an annulus in $\mathbb{R}^N$, f satisfies the conditions of Theorem 9.4.1. Then the large solution of (9.24) is a radial solution.*

For the proofs, please see [105], we omit them here.
Next we introduce two results about flat core of positive solution.

Theorem 9.4.3 (Zongming Guo, see [108]) *Assume f satisfies*

(F_1) $f \in C^1((0, \infty) \setminus \{a\}) \cap C^0([0, \infty))$, *where $a > 0$, $f(0) = f(a) = 0$, $f(s) > 0$ for $s \in (0, a)$, $f(s) < 0$ for $s \in (a, \infty)$, $\liminf_{s \to 0^+} f(s)/s^{p-1} = a^* > 0$, $f'(s) \geq 0$ for $s \in (0, s_0]$ and $0 < s_0 < a/2$.*

(F_2) *There exists $\delta > 0$ sufficiently small such that $f'(s) < 0$ for $s \in (a - \delta, a)$ and there exists $M > 0$ such that*

$$f(s) \leq M(a - s)^{p-1} \quad \text{for } 0 < s < a. \tag{9.29}$$

Then for λ sufficiently large, (9.24) possesses exactly one positive solution u_λ such that $\sup_\Omega u_\lambda < a$ and

$$\lim_{\lambda \to \infty} \sup_\Omega u_\lambda = a.$$

Moreover, for any compact set $K \Subset \Omega$, $u_\lambda \to a$ on K as $\lambda \to \infty$.

When $p = 2$ Theorem 9.4.3 has been obtained in [65]. The problem becomes complicated for $p \neq 2$ since the operator $-\Delta_p$ is degenerate and the nonlinearity is not monotone. The technique of the proof is the well-known sub- and supersolution method, but some technical difficulties arising from the degeneracy and the lack of monotonicity are overcome in [108].

Theorem 9.4.4 (Zongming Guo, see [108]) *Assume f satisfies (F_1) and*

(F_3) *There exists $\delta > 0$ sufficiently small such that $f'(s) < 0$ for $s \in (a - \delta, a)$, $\lim_{s \to a^-} \frac{f(s)}{(a-s)^\omega} = C, 0 < \omega < p - 1$, for some $C > 0$.*

Then (9.24) has a unique positive solution u_λ in the order interval $[0, a]$ for λ sufficiently large. Moreover, u_λ satisfies $u_\lambda \to a$ in $C^0_{\text{loc}}(\Omega)$ as $\lambda \to \infty$ and

(i) *the flat core $G_\lambda := \{x \in \Omega : u_\lambda(x) = a\} \neq \emptyset$,*
(ii) *if $d(\lambda) = \text{dist}(G_\lambda, \partial\Omega)$, then*

$$\lim_{\lambda \to \infty} \lambda^{1/p} d(\lambda) = \frac{C(F)^{1/p}}{2}$$

where

$$C(F) = \frac{p-1}{p} \left(\int_0^a \frac{2\,ds}{(F(a) - F(s))^{1/p}} \right)^p \quad \text{and} \quad F(s) = \int_0^s f(\zeta)\,d\zeta.$$

Moreover,

$$\lim_{\lambda \to \infty} \lambda^{1/p} \text{dist}(x, G_\lambda) = \frac{C(F)^{1/p}}{2} \quad \text{for any } x \in \partial\Omega.$$

Chapter 10
Nonlocal Kirchhoff Elliptic Problems

10.1 Introduction

We study and obtain existence of solutions for the following problem:

$$
\begin{cases}
-\left(a + b \int_{\Omega} |\nabla u|^2\right) \Delta u = f(x,u) & \text{in } \Omega, \\
u = 0 & \text{on } \partial\Omega,
\end{cases}
\tag{10.1}
$$

where Ω is a smooth bounded domain in $\mathbb{R}^n$, $a, b > 0$, and $f(x,t)$ is locally Lipschitz continuous in $t \in \mathbb{R}$, uniformly in $x \in \overline{\Omega}$, and subcritical:

$$
|f(x,t)| \leq C\left(|t|^{p-1} + 1\right) \quad \text{for some } 2 < p < 2^* =
\begin{cases}
\frac{2n}{n-2}, & n \geq 3, \\
\infty, & n = 1, 2
\end{cases}
\tag{10.2}
$$

where C denotes a generic positive constant.

This problem is related to the stationary analogue of the equation

$$
u_{tt} - \left(a + b \int_{\Omega} |\nabla u|^2\right) \Delta u = g(x,t)
\tag{10.3}
$$

proposed by Kirchhoff [121] as an extension of the classical D'Alembert's wave equation for free vibrations of elastic strings. Kirchhoff's model takes into account the changes in length of the string produced by transverse vibrations. Some early classical studies of Kirchhoff equations were Bernstein [33] and Pohozaev [155]. However, equation (10.3) received much attention only after Lions [134] proposed an abstract framework to the problem. Some interesting results can be found, for example, in [17, 46, 77]. More recently Alves et al. [6] and Ma and Rivera [144] obtained positive solutions of such problems by variational methods. Similar nonlocal problems also model several physical and biological systems where u describes a process which depends on the average of itself, for example the population density, see [4, 14, 53, 54, 183].

Z. Zhang, *Variational, Topological, and Partial Order Methods with Their Applications*, Developments in Mathematics 29,
DOI 10.1007/978-3-642-30709-6_10, © Springer-Verlag Berlin Heidelberg 2013

10.2 Yang Index and Critical Groups to Nonlocal Problems

We next obtain nontrivial solutions of a class of nonlocal quasilinear elliptic boundary value problems using the Yang index and critical groups.

Assume that f is a Carathéodory function on $\Omega \times \mathbb{R}$ such that

$$\lim_{t \to 0} \frac{f(x,t)}{at} = \lambda, \qquad \lim_{|t| \to \infty} \frac{f(x,t)}{bt^3} = \mu \quad \text{uniformly in } x. \tag{10.4}$$

Denote by $0 < \lambda_1 < \lambda_2 \leq \cdots$ the Dirichlet eigenvalues of $-\Delta$ on Ω:

$$\begin{cases} -\Delta u = \lambda u & \text{in } \Omega, \\ u = 0 & \text{on } \partial\Omega. \end{cases} \tag{10.5}$$

We also need the eigenvalue of the following problem:

$$\begin{cases} -\|u\|^2 \Delta u = \mu u^3 & \text{in } \Omega, \\ u = 0 & \text{on } \partial\Omega. \end{cases} \tag{10.6}$$

Yang Index   We briefly recall the definition and some basic properties of the Yang index. Yang [194] considered compact Hausdorff spaces with fixed-point-free continuous involutions and used the Čech homology theory, but for our purposes here it suffices to work with closed symmetric subsets of Banach spaces that do not contain the origin and singular homology groups.

Following [194], we first construct a special homology theory defined on the category of all pairs of closed symmetric subsets of Banach spaces that do not contain the origin and all continuous odd maps of such pairs. Let (X, A), $A \subset X$ be such a pair and $C(X, A)$ its singular chain complex with $\mathbb{Z}_2$ coefficients, and denote by $T_\#$ the chain map of $C(X, A)$ induced by the antipodal map $T(x) = -x$. We say that a q-chain c is symmetric if $T_\#(c) = c$, which holds if and only if $c = c' + T_\#(c')$ for some q-chain c'. The symmetric q-chains form a subgroup $C_q(X, A; T)$ of $C_q(X, A)$, and the boundary operator ∂_q maps $C_q(X, A; T)$ into $C_{q-1}(X, A; T)$, so these subgroups form a subcomplex $C(X, A; T)$. We denote by

$$Z_q(X, A; T) = \left\{ c \in C_q(X, A; T) : \partial_q c = 0 \right\}, \tag{10.7}$$

$$B_q(X, A; T) = \left\{ \partial_{q+1} c : c \in C_{q+1}(X, A; T) \right\}, \tag{10.8}$$

$$H_q(X, A; T) = Z_q(X, A; T) / B_q(X, A; T) \tag{10.9}$$

the corresponding cycles, boundaries, and homology groups. A continuous odd map $f : (X, A) \to (Y, B)$ of pairs as above induces a chain map $f_\# : C(X, A; T) \to C(Y, B; T)$ and hence homomorphisms

$$f_* : H_q(X, A; T) \to H_q(Y, B; T). \tag{10.10}$$

Example 10.2.1 (See [194]) For the m-sphere,

$$H_q\left(S^m; T\right) = \begin{cases} \mathbb{Z}_2 & \text{for } 0 \le q \le m, \\ 0 & \text{for } q > m. \end{cases} \tag{10.11}$$

Let X be as above, and define homomorphisms $v : Z_q(X; T) \to \mathbb{Z}_2$ inductively by

$$v(z) = \begin{cases} \text{In}(c) & \text{for } q = 0, \\ v(\partial c) & \text{for } q > 0 \end{cases} \tag{10.12}$$

if $z = c + T_\#(c)$, where the index of a 0-chain $c = \sum_i n_i \sigma_i$ is defined by $\text{In}(c) = \sum_i n_i$. As in [194], v is well defined and $v B_q(X; T) = 0$, so we define the *index homomorphism* $v_* : H_q(X; T) \to \mathbb{Z}_2$ by $v_*([z]) = v(z)$.

Proposition 10.2.1 (See [194]) *If F is a closed subset of X such that $F \cup T(F) = X$ and $A = F \cap T(F)$, then there is a homomorphism $\Delta : H_q(X; T) \to H_{q-1}(A; T)$ such that $v_*(\Delta[z]) = v_*([z])$.*

Taking $F = X$ we see that if $v_* H_m(X; T) = \mathbb{Z}_2$, then $v_* H_q(X; T) = \mathbb{Z}_2$ for $0 \le q \le m$. We define the *Yang index* of X by

$$i(X) = \inf\{m \ge -1 : v_* H_{m+1}(X; T) = 0\}, \tag{10.13}$$

taking $\inf \emptyset = \infty$. Clearly, $v_* H_0(X; T) = \mathbb{Z}_2$ if $X \ne \emptyset$, so $i(X) = -1$ if and only if $X = \emptyset$.

Example 10.2.2 (See [194]) $i(S^m) = m$.

Proposition 10.2.2 (See [194]) *If $f : X \to Y$ is as above, then $v_*(f_*([z])) = v_*([z])$ for $[z] \in H_q(X; T)$, and hence $i(X) \le i(Y)$. In particular, this inequality holds if $X \subset Y$.*

Recall that the *Krasnosel'skii Genus* of X is defined by

$$\gamma(X) = \inf\{m \ge 0 : \exists \text{ a continuous odd map } f : X \to S^{m-1}\}. \tag{10.14}$$

By Example 10.2.2 and Proposition 10.2.2, we have

Proposition 10.2.3 $\gamma(X) \ge i(X) + 1$.

Proposition 10.2.4 (Perera and Zhang [154]) *If $i(X) = m \ge 0$, then the reduced group $\widetilde{H}_m(X) \ne 0$.*

Proof By (10.13),

$$v_* H_q(X; T) = \begin{cases} \mathbb{Z}_2 & \text{for } 0 \le q \le m, \\ 0 & \text{for } q > m. \end{cases} \tag{10.15}$$

We show that if $[z] \in H_m(X; T)$ is such that $\nu_*([z]) \neq 0$, then $[z] \neq 0$ in $\widetilde{H}_m(X)$. Arguing indirectly, assume that $z \in B_m(X)$, say, $z = \partial c$. Since $z \in B_m(X; T)$, $T_\#(z) = z$. Let $c' = c + T_\#(c)$. Then $c' \in Z_{m+1}(X; T)$ since $\partial c' = z + T_\#(z) = 2z = 0 \mod 2$, and $\nu_*([c']) = \nu(c') = \nu(\partial c) = \nu(z) \neq 0$, contradicting $\nu_* \, H_{m+1}(X; T) = 0$. $\qquad\square$

Variational Eigenvalues of (10.6)   The eigenvalues of (10.6) are the critical values of the functional

$$I(u) = \|u\|^4, \quad u \in S := \left\{ u \in H = H_0^1(\Omega) : \int_\Omega u^4 = 1 \right\}. \tag{10.16}$$

Let $H = H_0^1(\Omega)$ be the usual Sobolev space, normed by

$$\|u\| = \left(\int_\Omega |\nabla u|^2 \right)^{1/2}. \tag{10.17}$$

By the Lagrange multiplier rule, $u \in S$ is a critical point of I if and only if

$$\|u\|^2 \int_\Omega \nabla u \cdot \nabla v = \mu \int_\Omega u^3 v, \quad \forall v \in H \tag{10.18}$$

for some $\mu \in \mathbb{R}$, i.e., u is a weak solution of (10.6). Taking $v = u$ we see that the Lagrange multiplier μ equals the corresponding critical value $I(u)$. So the eigenvalues of (10.6) are precisely the critical values of the functional I. We use the customary notation

$$I^a = \left\{ u \in S : I(u) \le a \right\} \tag{10.19}$$

for the sublevel sets of I.

Lemma 10.2.1 (Perera and Zhang [154]) *I satisfies the Palais–Smale condition* (PS), *i.e., every sequence $\{u_j\}$ in H such that $I(u_j)$ is bounded and $I'(u_j) \to 0$, called a* (PS) *sequence, has a convergent subsequence.*

Proof Since $\|u_j\|$ is bounded, for a subsequence, u_j converges to some u weakly in H and strongly in $L^4(\Omega)$. Denoting by

$$P_j v = v - \left(\int_\Omega u_j^3 v \right) u_j \tag{10.20}$$

the projection of $v \in H$ onto the tangent space to S at u_j, we have

$$\|u_j\|^2 \int_\Omega \nabla u_j \cdot \nabla(u_j - u) = I(u_j) \int_\Omega u_j^3 (u_j - u)$$

$$+ \frac{1}{4} \big(I'(u_j), P_j(u_j - u) \big) \to 0, \tag{10.21}$$

so, passing to a subsequence, $u_j \to 0$ or u. $\qquad\square$

The first eigenvalue $\mu_1 > 0$ obtained by minimizing I, i.e. $\mu_1 := \inf_{u \in S} I(u)$. If ψ is a minimizer, then so is $|\psi|$, so we may assume that $\psi \geq 0$. Since ψ is a nontrivial solution of (10.6), $\psi > 0$ in Ω and the interior normal derivative $\frac{\partial \psi}{\partial \nu} > 0$ on $\partial \Omega$ by the strong maximum principle.

Denote by $\mathcal{A}$ the class of closed symmetric subsets of S, let

$$\mathcal{F}_m = \big\{ A \in \mathcal{A} : i(A) \geq m - 1 \big\}, \tag{10.22}$$

and set

$$\mu_m := \inf_{A \in \mathcal{F}_m} \max_{u \in A} I(u). \tag{10.23}$$

Proposition 10.2.5 (Perera and Zhang [154]) μ_m *is an eigenvalue of* (10.6) *and* $\mu_m \nearrow \infty$.

Proof If μ_m is not a critical value of I, then there is an $\varepsilon > 0$ and an odd homeomorphism η of S such that $\eta(I^{\mu_m + \varepsilon}) \subset I^{\mu_m - \varepsilon}$ by the first deformation lemma. Taking $A \in \mathcal{F}_m$ with $\max I(A) \leq \mu_m + \varepsilon$, we have $A' = \eta(A) \in \mathcal{F}_m$ by Proposition 10.2.2, but $\max I(A') \leq \mu_m - \varepsilon$, contradicting (10.23).

Clearly, $\mu_{m+1} \geq \mu_m$. Since the sequence of Ljusternik–Schnirelmann eigenvalues $\widetilde{\mu}_m$ of (10.6) defined using the genus γ is unbounded (see, e.g., Struwe [168]) and $\mu_m \geq \widetilde{\mu}_m$ by Proposition 10.2.3, $\mu_m \to \infty$. $\qquad \square$

When μ is not an eigenvalue of (10.6), 0 is the only critical point of the associated variational functional

$$I_\mu(u) = \|u\|^4 - \mu \int_\Omega u^4, \quad u \in H \tag{10.24}$$

and hence its critical groups at 0 are defined and given by

$$C_q(I_\mu, 0) = H_q\big(I_\mu^0, I_\mu^0 \setminus \{0\}\big) \tag{10.25}$$

(see, e.g., Chang [49]).

Proposition 10.2.6 (Perera and Zhang [154]) *If* $\mu \in (\mu_m, \mu_{m+1})$ *is not an eigenvalue of* (10.6), *then*

$$C_m(I_\mu, 0) \neq 0. \tag{10.26}$$

Proof Since I_μ is positive homogeneous, I_μ^0 is radially contractible to 0 and $I_\mu^0 \setminus \{0\}$ is homotopic to $I_\mu^0 \cap S$ via the radial projection onto S, so it follows from the long exact sequence of reduced homology groups of the pair $(I_\mu^0, I_\mu^0 \setminus \{0\})$ that

$$C_m(I_\mu, 0) = H_m\big(I_\mu^0, I_\mu^0 \setminus \{0\}\big) \cong \widetilde{H}_{m-1}\big(I_\mu^0 \cap S\big) = \widetilde{H}_{m-1}\big(I^\mu\big) \tag{10.27}$$

where the last equality follows from $I_\mu|_S = I - \mu$. Since $\mu > \mu_m$, there is an $A \in \mathcal{F}_m$ such that $A \subset I^\mu$, so $i(I^\mu) \geq i(A) \geq m - 1$ by Proposition 10.2.2. On the other hand, $I^\mu \notin \mathcal{F}_{m+1}$ since $\mu < \mu_{m+1}$, so $i(I^\mu) < m$. Hence $i(I^\mu) = m - 1$, and $\widetilde{H}_{m-1}(I^\mu) \neq 0$ by Proposition 10.2.4. $\qquad\square$

Theorem 10.2.1 (Perera and Zhang [154]) *If $\lambda \in (\lambda_l, \lambda_{l+1})$ and $\mu \in (\mu_m, \mu_{m+1})$ is not an eigenvalue of (10.6), with $l \neq m$, then problem (10.1) has a nontrivial solution.*

Proof Recall that a function $u \in H$ is called a weak solution of (10.1) if

$$\left(a + b\|u\|^2\right) \int_\Omega \nabla u \cdot \nabla v = \int_\Omega f(x, u)v \quad \forall v \in H, \tag{10.28}$$

Weak solutions are the critical points of the C^1 functional

$$\Phi(u) = \frac{a}{2}\|u\|^2 + \frac{b}{4}\|u\|^4 - \int_\Omega F(x, u), \quad u \in H, \tag{10.29}$$

where $F(x, t) = \int_0^t f(x, s)\,ds$. They are also classical solutions if f is locally Lipschitz on $\overline{\Omega} \times \mathbb{R}$. $\qquad\square$

Lemma 10.2.2 (Perera and Zhang [154]) *If μ is not an eigenvalue of (10.6), Φ satisfies* (PS).

Proof As usual, it suffices to show that every (PS) sequence $\{u_j\}$ of Φ is bounded (see, e.g., Alves et al. [6], Lemma 1). Suppose that $\rho_j = \|u_j\| \to \infty$ for a subsequence. Setting $v_j = u_j/\rho_j$ and passing to a further subsequence, v_j converges to some v weakly in H, strongly in $L^4(\Omega)$, and a.e. in Ω. Passing to the limit in

$$\int_\Omega \nabla v_j \cdot \nabla w - \int_\Omega \frac{f(x, u_j)}{b u_j^3} \frac{v_j^3}{1 + a/b\rho_j^2} w = \frac{(\Phi'(u_j), w)}{(a + b\rho_j^2)\rho_j} \tag{10.30}$$

gives

$$\int_\Omega \nabla v \cdot \nabla w - \mu v^3 w = 0 \tag{10.31}$$

for each $w \in H$, and passing to the limit with $w = v_j - v$ shows that $\|v\| = 1$, so μ is an eigenvalue of (10.6), contrary to assumption. $\qquad\square$

Let

$$\Phi_0(u) = \frac{a}{2}\left(\|u\|^2 - \lambda \int_\Omega u^2\right), \qquad \Phi_\infty(u) = \frac{b}{4}\left(\|u\|^4 - \mu \int_\Omega u^4\right). \tag{10.32}$$

Proposition 10.2.7 (Perera and Zhang [154]) *If λ and μ are not eigenvalues of* (10.5) *and* (10.6), *respectively, then for all sufficiently small $\rho > 0$ and sufficiently large $R > 4\rho$ there is a functional $\widetilde{\Phi} \in C^1(H, \mathbb{R})$ such that*

(i)

$$\widetilde{\Phi}(u) = \begin{cases} \Phi_0(u), & \|u\| \le \rho, \\ \Phi(u), & 2\rho \le \|u\| \le R/2, \\ \Phi_\infty(u), & \|u\| \ge R, \end{cases}$$

in particular,

$$C_q(\widetilde{\Phi}, 0) = C_q(\Phi_0, 0), \qquad C_q(\widetilde{\Phi}, \infty) = C_q(\Phi_\infty, 0), \tag{10.33}$$

(ii) *$u = 0$ is the only critical point of Φ and $\widetilde{\Phi}$ with $\|u\| \le 2\rho$ or $\|u\| \ge R/2$, in particular, critical points of $\widetilde{\Phi}$ are the solutions of* (10.1),

(iii) *$\widetilde{\Phi}$ satisfies* (PS).

Proof Since λ and μ are not eigenvalues of (10.45) and (10.6), respectively, Φ_0 and Φ_∞ satisfy (PS) and have no critical points with $\|u\| = 1$, so

$$\delta_0 := \inf_{\|u\|=1} \left\| \Phi_0'(u) \right\| > 0, \qquad \delta_\infty := \inf_{\|u\|=1} \left\| \Phi_\infty'(u) \right\| > 0, \tag{10.34}$$

and

$$\inf_{\|u\|=\rho} \left\| \Phi_0'(u) \right\| = \rho \delta_0, \qquad \inf_{\|u\|=R} \left\| \Phi_\infty'(u) \right\| = R^3 \delta_\infty \tag{10.35}$$

by homogeneity. Let

$$\begin{aligned}
\Psi_0(u) &= \frac{b}{4} \|u\|^4 + \int_\Omega \frac{a\lambda}{2} u^2 - F(x, u), \\
\Psi_\infty(u) &= \frac{a}{2} \|u\|^2 + \int_\Omega \frac{b\mu}{4} u^4 - F(x, u).
\end{aligned} \tag{10.36}$$

By (10.2),

$$\sup_{\|u\|=\rho} \left| \Psi_0(u) \right| = o(\rho^2), \qquad \sup_{\|u\|=R} \left| \Psi_\infty(u) \right| = o(R^4) \tag{10.37}$$

and

$$\sup_{\|u\|=\rho} \left\| \Psi_0'(u) \right\| = o(\rho), \qquad \sup_{\|u\|=R} \left\| \Psi_\infty'(u) \right\| = o(R^3) \tag{10.38}$$

as $\rho \to 0$ and $R \to \infty$. Since $\Phi = \Phi_0 + \Psi_0 = \Phi_\infty + \Psi_\infty$, it follows from (10.35) and (10.38) that

$$\inf_{\|u\|=\rho} \left\| \Phi'(u) \right\| = \rho(\delta_0 + o(1)), \qquad \inf_{\|u\|=R} \left\| \Phi'(u) \right\| = R^3(\delta_\infty + o(1)). \tag{10.39}$$

Take smooth functions $\varphi_0, \varphi_\infty : [0, \infty) \to [0, 1]$ such that

$$\varphi_0(t) = \begin{cases} 1, & t \le 1, \\ 0, & t \ge 2, \end{cases} \qquad \varphi_\infty(t) = \begin{cases} 0, & t \le 1/2, \\ 1, & t \ge 1 \end{cases} \tag{10.40}$$

and set

$$\widetilde{\Phi}(u) = \Phi(u) - \varphi_0\big(\|u\|/\rho\big)\Psi_0(u) - \varphi_\infty\big(\|u\|/R\big)\Psi_\infty(u). \tag{10.41}$$

Since

$$\big\|d\big(\varphi_0(\|u\|/\rho)\big)\big\| = O\big(\rho^{-1}\big), \qquad \big\|d\big(\varphi_\infty(\|u\|/R)\big)\big\| = O\big(R^{-1}\big), \tag{10.42}$$

(10.39) holds with Φ replaced by $\widetilde{\Phi}$ also, and (i) and (ii) follow.

As for (iii), $\|\widetilde{\Phi}'\|$ is bounded away from 0 for $\rho \le \|u\| \le 2\rho$ and for $\|u\| \ge R/2$ by construction, so every (PS) sequence for $\widetilde{\Phi}$ has a subsequence in $\|u\| < \rho$ or in $2\rho < \|u\| < R/2$, which is then a (PS) sequence for Φ_0 or for Φ, respectively. $\square$

We are now ready to prove Theorem 10.2.1. Since $\lambda \in (\lambda_l, \lambda_{l+1})$ and $l \ne m$

$$C_m(\widetilde{\Phi}, 0) = C_m(\Phi_0, 0) = 0, \tag{10.43}$$

but

$$C_m(\widetilde{\Phi}, \infty) = C_m(\Phi_\infty, 0) = C_m(I_\mu, 0) \ne 0 \tag{10.44}$$

by Proposition 10.2.6, so $\widetilde{\Phi}$ must have a nontrivial critical point.

10.3 Variational Methods and Invariant Sets of Descent Flow

We assume that

$$tf(x, t) \ge 0 \tag{10.45}$$

and consider three cases:

  (i) 4-sublinear case: $p < 4$,
  (ii) asymptotically 4-linear case:

$$\lim_{|t| \to \infty} \frac{f(x, t)}{b\,t^3} = \mu \quad \text{uniformly in } x, \tag{10.46}$$

(iii) 4-superlinear case:

$$\exists \nu > 4: \quad \nu F(x, t) \le tf(x, t), \quad |t| \text{ large} \tag{10.47}$$

where $F(x, t) = \int_0^t f(x, s)\, ds$, which implies

$$F(x, t) \ge C\big(|t|^\nu - 1\big). \tag{10.48}$$

By (10.2) and (10.46) (resp. (10.48)),

$$4 \le p < 2^* \quad \left(\text{resp. } 4 < \nu \le p < 2^*\right), \tag{10.49}$$

so $n = 1, 2$, or 3 in (ii) and (iii).

Case (ii) leads to the nonlinear eigenvalue problem (10.6), whose eigenvalues are the critical values of the functional

$$I(u) = \|u\|^4, \quad u \in S := \left\{ u \in H : \int_\Omega u^4 = 1 \right\}. \tag{10.50}$$

We see in Lemma 10.2.1 that I satisfies the Palais–Smale condition (PS) and that the first eigenvalue $\mu_1 > 0$ obtained by minimizing I i.e. $\mu_1 := \inf_{u \in S} I(u)$, (10.6) has an eigenfunction $\psi > 0$. We define a second eigenvalue $\ge \mu_1$ by

$$\mu_2 := \inf_{\gamma \in \Gamma} \max_{u \in \gamma([0,1])} I(u) \tag{10.51}$$

where Γ is the class of paths $\gamma \in C([0, 1], S)$ joining $\pm \psi$ such that $\gamma \cup (-\gamma)$ is non-self-intersecting.

Theorem 10.3.1 (Zhang and Perera [206]) *Problem (10.1) has a positive solution, a negative solution, and a sign-changing solution in the following cases:*

(i) $p < 4$ *and*

$$\exists \lambda > \lambda_2: \quad F(x, t) \ge \frac{a\lambda}{2} t^2, \quad |t| \text{ small}, \tag{10.52}$$

(ii) (10.46) *and* (10.52) *hold and* $\mu < \mu_1$,

(iii) (10.46) *holds,* $\mu > \mu_2$ *is not an eigenvalue of* (10.6), *and*

$$F(x, t) \le \frac{a\lambda_1}{2} t^2, \quad |t| \text{ small}, \tag{10.53}$$

(iv) (10.47) *and* (10.53) *hold.*

Recall that a function $u \in H$ is called a weak solution of (10.1) if

$$\left(a + b\|u\|^2\right) \int_\Omega \nabla u \cdot \nabla v = \int_\Omega f(x, u)v \quad \forall v \in H. \tag{10.54}$$

Weak solutions are the critical points of the C^{2-0} functional

$$\Phi(u) = \frac{a}{2}\|u\|^2 + \frac{b}{4}\|u\|^4 - \int_\Omega F(x, u), \quad u \in H. \tag{10.55}$$

They are also classical solutions if f is locally Lipschitz on $\overline{\Omega} \times \mathbb{R}$.

Let $X = C_0^1(\overline{\Omega})$, with the usual norm

$$\|u\|_X = \max_{0 \le |\alpha| \le 1} \sup_{x \in \overline{\Omega}} \left|D^\alpha u(x)\right|, \tag{10.56}$$

which is densely embedded in H. By the elliptic regularity theory, the critical point set $K := \{u \in H : \Phi'(u) = 0\} \subset X$. Let $\widetilde{\Phi} = \Phi|_X$. A standard argument shows that $\widetilde{\Phi}$ has the retracting property (see, e.g., [71], proof of Lemma 1).

Consider the initial value problem

$$\begin{cases} \dfrac{du}{dt} = -W(u), & t > 0 \\[2mm] u(0) = u_0 \in X \end{cases} \tag{10.57}$$

where

$$W(u) = \frac{\widetilde{\Phi}'(u)}{a + b\|u\|^2} = u - Au, \tag{10.58}$$

$A = KG$, $G : X \to L^\infty(\Omega)$, $u \mapsto \frac{f(x,u(x))}{a+b\|u\|^2}$, and $K = (-\Delta)^{-1} : L^\infty(\Omega) \to X$. Since $f(x,t)$ is locally Lipschitz in t, uniformly in $\overline{\Omega}$, G is locally Lipschitz, and K is a bounded linear operator, so W is locally Lipschitz on X. So a unique solution $u(t, u_0)$ of (10.57) satisfying

$$u(t, u_0) = e^{-t}\left(u_0 + \int_0^t e^s \, Au(s, u_0)\, ds\right) \tag{10.59}$$

exists in some maximal existence interval $[0, T(u_0))$, $T(u_0) \le \infty$. Since

$$\frac{d}{dt}\big(\widetilde{\Phi}\big(u(t, u_0)\big)\big) = \left(\widetilde{\Phi}'(u), \frac{du}{dt}\right) = -\frac{\|\widetilde{\Phi}'(u)\|^2}{a + b\|u\|^2} \le 0, \tag{10.60}$$

$\widetilde{\Phi}$ is non-increasing along the orbits.

Therefore, we can use the similar proof of Theorem 8.2.2 to prove Theorem 10.2.1.

Now, by the new information concerning functionals without Palais–Smale condition, we study Kirchhof-type problems with 4-superlinear growth as $|u| \to +\infty$ and with weak conditions than (10.47). We know that

$$tf(x,t) \ge 0$$

which implies that

$$F(x,t) = \int_0^t f(x,s)\, ds \ge 0.$$

Let $\tilde{F}(x, u) := \frac{1}{4} f(x, u)u - F(x, u)$. We make the following assumptions:

(S1) $\frac{F(x,u)}{u^4} \to +\infty$ as $|u| \to +\infty$ uniformly in $x \in \Omega$;

(S2) $f(x, u) = o(|u|)$ as $|u| \to 0$ uniformly in $x \in \Omega$;

(S3) $\tilde{F}(x, u) \to +\infty$ as $|u| \to +\infty$ uniformly in $x \in \Omega$, and there exists $\sigma > \max\{1, N/2\}$ such that $|f(x, u)|^\sigma \le C\tilde{F}(x, u)|u|^\sigma$ for $|u|$ large.

We obtain the following.

Theorem 10.3.2 (Mao and Zhang [145]) *Assume* $(S1)$–$(S3)$ *hold. Then problem* (10.1) *has at least one nontrivial solution.*

Furthermore, we assume

$(S4)$ $f(x, t)$ is locally Lipschitz continuous in $t \in \mathbb{R}$, uniformly in $x \in \overline{\Omega}$.

Theorem 10.3.3 (Mao and Zhang [145]) *Assume* $(S1)$–$(S4)$ *hold. Then problem* (10.1) *has at least one positive solution, one negative solution, and one sign-changing solution.*

The proofs of our results are based on combining the critical point theory, invariant set of descent flow and the minimax methods using the Cerami condition (see [145]).

Examples satisfying the 4-superlinear conditions $(S1)$–$(S3)$ are the following functions:

- *Example 1.* $f(x, u) = u^3 \ln(1 + |u|)$;
- *Example 2.* $F(x, u) = |u|^\mu + (\mu - 4)|u|^{\mu-\epsilon} \sin^2(\frac{|u|^\epsilon}{\epsilon})$ where $N = 3$, $2^* > \mu > 4$ and $0 < \epsilon < \mu + N - \mu N/2$.

Remark 10.3.1 For a class of nonlocal eigenvalue problems (see Agarwal, Kanishka and Zhang [2]),

$$-\|u\|^{p-q}\Delta_q u = \lambda \|u\|_r^{p-r}|u|^{r-2}u, \quad u \in W_0^{1,q}(\Omega) \tag{10.61}$$

where $q \in (1, \infty)$, $p \in [q, \infty)$, and $r \in (1, q^*) \cap (1, p]$. Ω is a bounded domain in $\mathbb{R}^n$, $n \geq 1$, $W_0^{1,q}(\Omega)$ is the usual Sobolev space with the norm

$$\|u\| = \left(\int_\Omega |\nabla u|^q\right)^{1/q},$$

and

$$q^* = \begin{cases} nq/(n-q), & q < n, \\ \infty, & q \geq n. \end{cases}$$

Both sides are positive homogeneous of degree $p - 1$, similar to the p-Laplacian. In fact, (10.61) reduces to the familiar eigenvalue problem for the p-Laplacian

$$-\Delta_p u = \lambda |u|^{p-2}u, \quad u \in W_0^{1,p}(\Omega) \tag{10.62}$$

when $p = q = r$, so we may view it as a nonlocal generalization of (10.62).

We will call the nonlocal operator $\Delta_{p,q}$ defined by

$$\Delta_{p,q} u = \|u\|^{p-q}\, \Delta_q u = \left(\int_\Omega |\nabla u|^q \right)^{(p-q)/q} \operatorname{div}\left(|\nabla u|^{q-2} \nabla u \right),$$

$$q \in (1, \infty), \quad p \in [q, \infty)$$

the (p, q)-Laplacian, noting that $\Delta_{p,p} = \Delta_p$. The spectrum of the pair $(-\Delta_{p,q}, r)$ where $r \in (1, q^*) \cap (1, p]$, denoted by $\sigma(-\Delta_{p,q}, r)$, is the set of all $\lambda \in \mathbb{R}$ for which the eigenvalue problem

$$-\Delta_{p,q} u = \lambda \|u\|_u^{p-r} |u|^{r-2} u, \quad u \in W_0^{1,q}(\Omega) \tag{10.63}$$

has a solution $u \neq 0$. Again we note that $\sigma(-\Delta_{p,p}, p) = \sigma(-\Delta_p)$.

Solutions of (10.63) are the critical points of the C^1-functional

$$\Phi_\lambda(u) = \frac{1}{p}\|u\|^p - \frac{\lambda}{p}\|u\|_r^p = \frac{1}{p}\left(\int_\Omega |\nabla u|^q \right)^{p/q} - \frac{\lambda}{p}\left(\int_\Omega |u|^r \right)^{p/r},$$

$$u \in W = W_0^{1,q}(\Omega). \tag{10.64}$$

We construct a nondecreasing and unbounded sequence of eigenvalues for (10.63) that yields nontrivial critical groups for the associated variational functional using a nonstandard minimax scheme that involves the $\mathbb{Z}_2$-cohomological index. As an application we prove a multiplicity result for a class of nonlocal boundary value problems using Morse theory.

10.4 Uniqueness of Solution for a Class of Kirchhoff-Type Equations

Let Ω be the open ball in $\mathbb{R}^3$ centered at 0 with radius $R > 0$ and let $q \in (3, 5)$. We consider the function $f : \mathbb{R} \to \mathbb{R}$ defined by $f(t) = t^q$ if $t \geq 0$, $f(t) = 0$ if $t < 0$. We have the following result:

Theorem 10.4.1 (Giovanni Anello [15]) *Let a, b, λ be three positive real numbers. The problem*

$$\begin{cases} -\left(a + b \int_\Omega |\nabla u|^2 \, dx \right) \Delta u = \lambda f(u) & \text{in } \Omega, \\[2mm] u = 0 & \text{on } \partial\Omega, \end{cases} \tag{10.65}$$

admits a unique nonzero weak solution.

Proof First of all, we note that by the Strong Maximum Principle every nonzero solution of problem (10.65) must be positive. The weak solutions of problem (10.65) are exactly the critical points of the functional

$$I(u) = \frac{a}{2} \int_\Omega |\nabla u|^2 \, dx + \frac{b}{4} \left(\int_\Omega |\nabla u|^2 \, dx \right)^2 - \lambda \int_\Omega F(u) \, dx, \quad \forall u \in H_0^1(\Omega),$$

where $F(t) = \int_0^t f(s) \, ds$. We put $\|u\| = (\int_\Omega |\nabla u|^2 \, dx)^{\frac{1}{2}}$, $\forall u \in H_0^1(\Omega)$.

Denoting by C the best embedding constant of $H_0^1(\Omega)$ in $L^{q+1}(\Omega)$, for every $r > 0$ we have

$$\inf_{\|u\|=r} I(u) = \frac{1}{2} \left(ar^2 + \frac{b}{2} r^4 \right) - \lambda C^q r^{q+1}.$$

Hence, being $q + 1 > 4$,

$$\liminf_{\|u\| \to +\infty} I(u) = -\infty$$

and there exists $r_0 > 0$ such that

$$\inf_{\|u\|=r_0} I(u) > 0.$$

Thus, I has a Mountain Pass geometry. Moreover, by standard arguments, we see that I satisfies the Palais–Smale condition. Therefore, I has a nonzero critical point $u_1 \in H_0^1(\Omega)$. Let us show that u_1 is unique. Assume that $u_2 \in H_0^1(\Omega)$ is another nonzero critical point. Note that u_i $(i = 1, 2)$ is a weak solution of the problem

$$\begin{cases} -\Delta u = \dfrac{\lambda}{(a + b\|u_i\|^2)} f(u) & \text{in } \Omega, \\ u = 0 & \text{on } \partial\Omega. \end{cases}$$

Now, in [1] it is proved that the problem

$$\begin{cases} -\Delta u = f(u) & \text{in } \Omega, \\ u = 0 & \text{on } \partial\Omega, \end{cases}$$

has a unique nonzero (and positive) solution $v_1 \in H_0^1(\Omega)$. This implies that for all $\lambda > 0$, the function $v_\lambda = \lambda^{\frac{1}{1-q}} v_1$ is the unique nonzero solution of the problem

$$\begin{cases} -\Delta u = \lambda f(u) & \text{in } \Omega, \\ u = 0 & \text{on } \partial\Omega. \end{cases} \tag{P_λ}$$

In particular, for $\lambda_1, \lambda_2 > 0$, v_{λ_1} and v_{λ_2} are related by

$$v_{\lambda_1} = \left(\frac{\lambda_1}{\lambda_2} \right)^{\frac{1}{1-q}} v_{\lambda_2}.$$

Therefore, since u_i is the nonzero solutions of problem (P_{λ_i}) with

$$\lambda_i = \frac{\lambda}{a + b\|u_i\|^2},$$

we have

$$u_1 = \left(\frac{a + b\|u_2\|^2}{a + b\|u_1\|^2}\right)^{\frac{1}{1-q}} u_2 \tag{10.66}$$

from which we get

$$\|u_1\| = \left(\frac{a + b\|u_2\|^2}{a + b\|u_1\|^2}\right)^{\frac{1}{1-q}} \|u_2\|$$

equivalent to

$$a\|u_1\|^{1-q} + b\|u_1\|^{3-q} = a\|u_2\|^{1-q} + b\|u_2\|^{3-q}. \tag{10.67}$$

Since $q > 3$, the function $\varphi(t) = at^{1-q} + bt^{3-q}$ is strictly decreasing for $t > 0$. Consequently, from (10.67) it follows that $\|u_1\| = \|u_2\|$ and so, from (10.66), $u_1 = u_2$. $\qquad\square$

Chapter 11
Free Boundary Problems, System of Equations for Bose–Einstein Condensate and Competing Species

11.1 Competing System with Many Species

In recent years, people have shown a lot of interest in strongly competing systems with many species, that is, the system (or its elliptic case)

$$\frac{\partial u_i}{\partial t} - \Delta u_i = -\kappa u_i \sum_{j \neq i} b_{ij} u_j,$$

where κ is sufficiently large (or its limit at $\kappa = +\infty$). Conti, Terracini and Verzini [58, 59], Caffarelli, Karakhanyan and Lin [41, 42], etc., established the regularity of the singular limit (and the partial regularity of its free boundary) as $\kappa \to +\infty$ and the uniform regularity for all $\kappa > 0$.

We prove the existence and uniqueness result of the following Dirichlet boundary value problem of elliptic systems in a smooth domain Ω in $\mathbb{R}^n$ for $n \geq 1$:

$$\begin{cases} \Delta u_i = \kappa u_i \sum_{j \neq i} b_{ij} u_j, & \text{in } \Omega, \\ u_i = \varphi_i, & \text{on } \partial\Omega. \end{cases} \tag{11.1}$$

Here $b_{ij} > 0$ are constants and satisfy $b_{ij} = b_{ji}$, φ_i are given Lipschitz continuous functions on $\partial\Omega$, which satisfy $\varphi_i \geq 0$ with disjoint support, namely $\varphi_i \cdot \varphi_j = 0$ for $i \neq j$ almost everywhere on $\partial\Omega$, $1 \leq i, j \leq M$, $M \geq 2$. Here we will simply take $b_{ij} = 1$, without loss of generality.

We show that for the singular limit species are spatially segregated and they satisfy a remarkable system of differential inequalities as $\kappa \to +\infty$.

11.1.1 Existence and Uniqueness of Positive Solution

In [59], Conti et al. proved the existence of the positive solution of (11.1), using Leray–Schauder degree theory. However, the uniqueness of the solution was un-

Z. Zhang, *Variational, Topological, and Partial Order Methods with Their Applications*,
Developments in Mathematics 29,
DOI 10.1007/978-3-642-30709-6_11, © Springer-Verlag Berlin Heidelberg 2013

known. Here, we will use the sub- and sup-solution method, iteration and the comparison principle to show existence and uniqueness of the solution. That is, we have the following.

Theorem 11.1.1 (Wang and Zhang [189]) $\forall \kappa \geq 0$, *there exists a unique positive solution* $(u_1, \ldots, u_M)$ *to* (11.1), *where* $u_i \in C^2(\Omega) \cap C^0(\bar{\Omega})$.

We use the following iteration scheme to prove the uniqueness of solutions for (11.1). First, we know the following harmonic extension is possible:

$$\begin{cases} \Delta u_{i,0} = 0, & \text{in } \Omega, \\ u_{i,0} = \varphi_i, & \text{on } \partial\Omega, \end{cases} \tag{11.2}$$

that is, this equation has a unique positive solution $u_{i,0} \in C^2(\Omega) \cap C^0(\bar{\Omega})$ by Theorem 4.3 of [95].

Then the iteration can be defined as

$$\begin{cases} \Delta u_{i,m+1} = \kappa u_{i,m+1} \sum_{j \neq i} u_{j,m}, & \text{in } \Omega, \\ u_{i,m+1} = \varphi_i & \text{on } \partial\Omega, \end{cases} \tag{11.3}$$

this is a linear equation, and it satisfies the Maximum Principle, so the existence and uniqueness of the solution $u_{i,m+1} \in C^2(\Omega) \cap C^0(\bar{\Omega})$ is clear by Theorem 6.13 of [95].

Now concerning these $u_{i,m}$ we have the following result:

Proposition 11.1.1 *In* Ω

$$u_{i,0}(x) > u_{i,2}(x) > \cdots > u_{i,2m}(x) > \cdots > u_{i,2m+1}(x) > \cdots > u_{i,3}(x) > u_{i,1}(x).$$

Proof We divide the proof into several claims.

Claim 1 $\forall i, m, u_{i,m} > 0$ *in* Ω.

Because $\sum_{j \neq i} u_{j,0} > 0$ in Ω, (11.3) satisfies the maximum principle. From the boundary condition $\varphi_i \geq 0$, then we have $u_{i,1} > 0$ in Ω. By induction, we see the claim is right for all $u_{i,m}$.

Claim 2 $u_{i,1} < u_{i,0}$ *in* Ω.

From the equation, now we have

$$\begin{cases} \Delta u_{i,1} \geq 0, & \text{in } \Omega, \\ u_{i,1} = u_{i,0} & \text{on } \partial\Omega, \end{cases} \tag{11.4}$$

so we have $u_{i,1} < u_{i,0}$ from the comparison principle.

In the following we assume the conclusion of the proposition is valid until $2m + 1$, that is, in Ω

$$u_{i,0} > \cdots > u_{i,2m} > u_{i,2m+1} > u_{i,2m-1} > \cdots > u_{i,1}.$$

Then we have

Claim 3 $u_{i,2m+1} \le u_{i,2m+2}.$

By (11.3) we have

$$\Delta u_{i,2m+2} \le \kappa u_{i,2m+2} \sum_{j \ne i} u_{j,2m}, \tag{11.5}$$

$$\Delta u_{i,2m+1} = \kappa u_{i,2m+1} \sum_{j \ne i} u_{j,2m}. \tag{11.6}$$

Because $u_{i,2m+1}$ and $u_{i,2m+2}$ have the same boundary value, comparing (11.5) and (11.6), by the comparison principle again we obtain $u_{i,2m+1} \le u_{i,2m+2}.$

Claim 4 $u_{i,2m+2} \le u_{i,2m}.$

This can be seen by comparing the equations they satisfy

$$\begin{cases} \Delta u_{i,2m+2} = \kappa u_{i,2m+2} \sum_{j \ne i} u_{j,2m+1}, \\ \Delta u_{i,2m} = \kappa u_{i,2m} \sum_{j \ne i} u_{j,2m-1}. \end{cases} \tag{11.7}$$

By assumption we have $u_{j,2m+1} \ge u_{j,2m-1}$, so the claim follows from the comparison principle again.

Claim 5 $u_{i,2m+3} \ge u_{i,2m+1}.$

This can be seen by comparing the equations they satisfy

$$\begin{cases} \Delta u_{i,2m+3} = \kappa u_{i,2m+2} \sum_{j \ne i} u_{j,2m+2}, \\ \Delta u_{i,2m+1} = \kappa u_{i,2m+1} \sum_{j \ne i} u_{j,2m}. \end{cases} \tag{11.8}$$

By Claim 4 we have $u_{j,2m} \ge u_{j,2m+2}$, so the claim follows from the comparison principle again.

Now we know that there exist two family of functions u_i and v_i, such that

$$\lim_{m \to \infty} u_{i,2m}(x) = u_i(x)$$

and

$$\lim_{m\to\infty} u_{i,2m+1}(x) = v_i(x), \quad \forall x \in \Omega.$$

Moreover, from the elliptic estimate, we know this limit is smooth in Ω, i.e., $u_i \in C^2(\Omega) \cap C^0(\bar{\Omega})$, $v_i \in C^2(\Omega) \cap C^0(\bar{\Omega})$, and the convergence is uniformly on $\bar{\Omega}$. So by taking the limit in (11.3) we obtain the following equations:

$$\begin{cases} \Delta u_i = \kappa u_i \sum_{j\neq i} v_j, \\ \Delta v_i = \kappa v_i \sum_{j\neq i} u_j. \end{cases} \tag{11.9}$$

Because $u_{i,2m+1} \leq u_{j,2m}$, by taking limit we also have

$$v_i \leq u_i. \tag{11.10}$$

Now summing (11.9) we have

$$\begin{cases} \Delta\left(\sum_i u_i\right) = \kappa \sum_i \left(u_i \sum_{j\neq i} v_j\right), \\ \Delta\left(\sum_i v_i\right) = \kappa \sum_i \left(v_i \sum_{j\neq i} u_j\right). \end{cases} \tag{11.11}$$

It is easily seen that

$$\sum_i \left(u_i \sum_{j\neq i} v_j\right) = \sum_i v_i \left(\sum_{j\neq i} u_j\right),$$

so we must have $\sum_i u_i \equiv \sum_i v_i$ because they have the same boundary value. This means, by (11.10), $u_i \equiv v_i$. In particular, they satisfy (11.1). $\qquad\square$

Proposition 11.1.2 *If there exists another positive solution w_i of (11.1), we must have $u_i \equiv w_i$.*

Proof We will prove $u_{i,2m} \geq w_i \geq u_{j,2m+1}, \forall m$, then the proposition follows immediately. We divide the proof into several claims.

Claim 1 $w_i \leq u_{i,0}.$

This is because

$$\begin{cases} \Delta w_i \geq 0, & \text{in } \Omega, \\ w_i = u_{i,0} & \text{on } \partial\Omega. \end{cases} \tag{11.12}$$

Claim 2 $w_i \geq u_{i,1}.$

This is because

$$\begin{cases} \Delta w_i = \kappa w_i \sum_{j \neq i} w_j, \\ \Delta u_{i,1} = \kappa u_{i,1} \sum_{j \neq i} u_{j,0}. \end{cases} \tag{11.13}$$

Noting that we have $w_j < u_{j,0}$, so the comparison principle applies.

In the following we assume that our claim is valid until $2m + 1$, that is,

$$u_{i,2m} \geq w_i \geq u_{i,2m+1}.$$

Then we have

Claim 3 $u_{i,2m+2} \geq w_i$.

This can be seen by comparing the equations they satisfy

$$\begin{cases} \Delta w_i = \kappa w_i \sum_{j \neq i} w_j, \\ \Delta u_{i,2m+2} = \kappa u_{i,2m+3} \sum_{j \neq i} u_{j,2m+1}. \end{cases} \tag{11.14}$$

By assumption we have $u_{j,2m+1} \leq w_j$, so the claim follows from the comparison principle again.

Claim 4 $u_{i,2m+3} \leq w_i$.

This can be seen by comparing the equations they satisfy

$$\begin{cases} \Delta w_i = \kappa w_i \sum_{j \neq i} w_j, \\ \Delta u_{i,2m+3} = \kappa u_{i,2m+3} \sum_{j \neq i} u_{j,2m+2}. \end{cases} \tag{11.15}$$

By Claim 3 we have $u_{j,2m+2} \geq w_j$, so the claim follows from the comparison principle again.

Remark 11.1.1 From our proof, we know that the uniqueness result still holds for equations of more general form:

$$\begin{cases} \Delta u_i = u_i \sum_{j \neq i} b_{ij}(x) u_j, & \text{in } \Omega \\ u_i = \varphi_i & \text{on } \partial\Omega, \end{cases} \tag{11.16}$$

where $b_{ij}(x)$ are positive (and smooth enough) functions defined in $\overline{\Omega}$, which satisfy $b_{ij} \equiv b_{ji}$.

11.1.2 The Limit Spatial Segregation System of Competing Systems

Next we consider the uniqueness of the singular limit of (11.1) as $\kappa \to +\infty$. We know that, as $\kappa \to +\infty$, solutions of (11.1) converge to some $(u_1, \ldots, u_M)$ which satisfy the following conditions:

$$
\begin{cases}
\Delta u_i \geq 0, & \text{in } \Omega, \\[2mm]
\Delta\left(u_i - \sum_{j \neq i} u_j\right) \leq 0, & \text{in } \Omega, \\[2mm]
u_i = \varphi_i & \text{on } \partial\Omega, \\
u_i u_j = 0 & \text{in } \Omega.
\end{cases}
\tag{11.17}
$$

That is

Theorem 11.1.2 (Conti, Terracini and Verzini [59]) *Let $(u_{i,\kappa})$ be a positive solution of* (11.1). *Then, there exists $(u_i) \in S$ such that, up to subsequences, $\|u_{i.k} - u_i\|_{H^1} \to 0$ as $\kappa \to \infty$. Here*

$$
S = \left\{ (u_1, \ldots, u_k) \in \left(H^1(\Omega)\right)^M :
\begin{array}{l}
u_i \geq 0, \quad u_i|_{\partial\Omega} = 0, \\[2mm]
u_i \cdot u_j = 0 \quad \text{if } i \neq j, \\[2mm]
-\Delta u_i \leq 0, \\[2mm]
-\Delta \widehat{u}_i = -\Delta\left(u_i - \sum_{j \neq i} u_j\right) \geq 0
\end{array}
\right\}.
\tag{11.18}
$$

Proof We omit it here, see [59]. $\square$

We establish some results concerning the estimate of the $n - 1$ dimensional Hausdorff measure of the free boundary. From the regularity theory in [41], we know that $\partial\{u_i > 0\}$ and $\partial\{v_i > 0\}$ are smooth hypersurface except a closed set of dimension $n - 2$. What we show is that they have finite $n - 1$ dimension Hausdorff measure in the interior of Ω.

Theorem 11.1.3 (Wang and Zhang [189]) *For any compact set $\Omega' \Subset \Omega$ (i.e., Ω' has compact closure in Ω), we have*

$$
H^{n-1}\left(\Omega' \cap \partial\{u_i > 0\}\right) < +\infty.
$$

Proof See [189], we omit it here. $\square$

Remark 11.1.2 Let X be a metric space. If $S \subset X$ and $d \in [0, \infty)$, the d-dimensional Hausdorff content of S is defined by

$$C_H^d(S) := \inf\left\{\sum_i r_i^d : \text{there is a cover of } S \text{ by balls with radii } r_i > 0\right\}.$$

In other words, $C_H^d(S)$ is the infimum of the set of numbers $\delta \geq 0$ such that there is some (indexed) collection of balls $\{B(x_i, r_i) : i \in I\}$ with $r_i > 0$ for each $i \in I$ which satisfies $\sum_{i \in I} r_i^d < \delta$. (The index set I usually counts the natural numbers $\mathbb{N}$. Here, we use the standard convention that $\inf \emptyset = \infty$.) The Hausdorff dimension of X is defined by

$$\dim_H(X) := \inf\{d \geq 0 : C_H^d(X) = 0\}.$$

Equivalently, $\dim_H(X)$ may be defined as the infimum of the set of $d \in [0, \infty)$ such that the d-dimensional Hausdorff measure of X is zero. This is the same as the supremum of the set of $d \in [0, \infty)$ such that the d-dimensional Hausdorff measure of X is finite(except that when this latter set of numbers d is empty the Hausdorff dimension is zero).

11.2 Optimal Partition Problems

11.2.1 An Optimal Partition Problem Related to Nonlinear Eigenvalues

Let $\Omega \subset \mathbb{R}^N$, $N \geq 2$, be a bounded domain and $f : \mathbb{R} \to \mathbb{R}$ a super-linear function. Set $F(s) = \int_0^s f(t)\,dx$ and let us define, for $u \in H_0^1(\Omega)$, the functional

$$J^*(u) := \int_\Omega \left(\frac{1}{2}|\nabla u|^2 - F\big(u(x)\big)\right) dx.$$

With each open $\omega \subset \Omega$ we associate the first nonlinear eigenvalue [146] as

$$\varphi(\omega) := \inf_{\substack{u \in H_0^1(\omega) \\ u > 0}} \sup_{\lambda > 0} J^*(\lambda u).$$

It is well known that $\varphi(\omega)$ is a critical value of the functional J^* over $H_0^1(\omega)$. Thus it corresponds to (at least) one positive solution u to the boundary value problem

$$\begin{cases} -\Delta u(x) = f\big(u(x)\big), & x \in \omega, \\ u = 0, & x \in \partial\omega. \end{cases}$$

u will be referred to as eigenfunction associated to $\varphi(\omega)$.

Conti–Terracini–Verzini [57] considered the problem of finding a partition of Ω (in open sets) that achieves

$$\inf\left\{\sum_{i=1}^{k}\varphi(\omega_i):\bigcup_{i=1}^{k}\bar{\omega}_i=\bar{\Omega},\ \omega_i\cap\omega_j=\emptyset\ \text{if}\ i\neq j\right\}. \tag{11.19}$$

Many free boundary problems can be formulated in terms of optimal partitions, and they are usually studied in the case $k=2$ components. For instance, we quote [3] for applications to the flow of two liquids in models of jets and cavities. Moreover, optimal partition problems arise in linear eigenvalue theory and various fields of real analysis. For example, the most recent proof of the well-known monotonicity formula relies upon a problem of optimal partition in two subsets related to the first eigenvalue of the Dirichlet operator on the sphere (see [40]).

In this subsection we give their proofs on the existence of a minimal partition to problem (11.19) in a weak sense. In fact, at first we will not find an open partition, but a partition made of sets which are supports of H_0^1-functions, i.e., the eigenfunctions associated to our problem. Throughout all the subsection, Ω will be a bounded domain in $\mathbb{R}^N$, with the additional property, in the results about points on $\partial\Omega$, of being of class C^2.

Let f satisfy the following assumptions:

(f_1) $f\in C^1(\mathbb{R})$, $f(-s)=-f(s)$, and there exist positive constants C, p such that for all $s\in\mathbb{R}$

$$\left|f(s)\right|\leq C\left(1+|s|^{p-1}\right),\quad 2<p<2^*,$$

where $2^*=+\infty$ when $N=2$ and $2^*=2N/(N-2)$ when $N\geq 3$,
(f_2) there exists $\gamma>0$ $(2+\gamma\leq p)$ such that, for all $s\neq 0$,

$$f'(s)s^2-(1+\gamma)f(s)s>0.$$

Remark 11.2.1 It is well known that, when f is not an odd function,

$$\inf_{\substack{u\in H_0^1(\omega)\\ u\neq 0}}\ \sup_{\lambda>0}J^*(\lambda u)$$

is achieved by a one-sign critical point of J^*. If the infimum is restricted to the positive functions and then to the negative ones, it gives two possibly different critical levels and two correspondent critical points, one positive and one negative. To fix the ideas, in this paper we consider only positive critical points, i.e. positive values of s. Thus the assumption $f(-s)=-f(s)$ is not truly necessary: we can extend any other f, without loss of generality, to be an odd function.

Moreover, we can allow f to be x-dependent, although for the sake of simplicity we shall always refer to f as a function of s only.

Observe that, in a standard way, from assumptions (f_1) and (f_2) we can obtain the following properties for the primitive F of f:

$$F(s) \le C\left(1 + |s|^p\right), \qquad f(s)s - (2 + \gamma)F(s) \ge 0.$$

For $u \in H_0^1(\Omega)$ and $U := (u_1, \ldots, u_k) \in (H_0^1(\Omega))^k$ we define the functionals

$$J^*(u) := \int_\Omega \left(\frac{1}{2}|\nabla u|^2 - F\big(u(x)\big)\right) dx,$$

$$J(U) := \sum_{i=1}^k J^*(u_i), \tag{11.20}$$

observing that

$$u_i \cdot u_j = 0 \quad \text{a.e. on } \Omega \text{ for } i \ne j \quad \Longrightarrow \quad J(U) = J^*\left(\sum_{i=1}^k u_i\right).$$

Next we define the Nehari manifolds associated to these functionals:

$$\mathcal{N}(J^*) := \left\{u \in H_0^1(\Omega) : u \ge 0, u \not\equiv 0, \nabla J^*(u) \cdot u = 0\right\},$$

$$\mathcal{N}(J) := \left(\mathcal{N}(J^*)\right)^k,$$

$$\mathcal{N}_0 := \mathcal{N}(J) \cap \{u_i \cdot u_j = 0 \text{ a.e. on } \Omega \text{ for } i \ne j\}.$$

With this notation we introduce the problem we want to study here:

$$c_0 := \inf\left\{\sum_{i=1}^k \varphi(\omega_i) : \bigcup_{i=1}^k \bar\omega_i = \bar\Omega, \omega_i \cap \omega_j = \emptyset \text{ if } i \ne j\right\}$$

$$= \inf\left\{\sum_{i=1}^k \sup_{\lambda > 0} J^*(\lambda_i u_i) : u_i \in H_0^1(\Omega), u_i \ge 0, u_i \not\equiv 0,\right.$$

$$\left. u_i \cdot u_j = 0 \text{ a.e. on } \Omega \text{ for } i \ne j\right\}$$

$$= \inf\left\{J(U) : U \in \mathcal{N}_0\right\}. \tag{11.21}$$

Observe that in the previous equality the first infimum is intended over all the partitions of Ω into subsets which are supports of $H_0^1(\Omega)$-functions. In this sense, (11.21) is a relaxed reformulation of the initial problem (11.19) (for more details about the equivalent characterizations see [56]). A similar characterization, when $k = 2$ was also exploited in [45] when seeking changing-sign solutions to super-linear problems.

To prove the following theorem we need a preliminary lemma:

Lemma 11.2.1 (Conti, Terracini and Verzini [57]) *Let* $u \in \mathcal{N}(J^*)$, $J^*(u) \le c_0 + 1$. *Then there exist positive constants* C_1, C_2 *such that* $\|u\|_{H_0^1(\Omega)} \le C_1$ *and* $\|u\|_{L^p} \ge C_2 > 0$.

Proof By assumptions we have $u \not\equiv 0$ and

$$\int_\Omega \left(|\nabla u|^2 - f(u)u\right) dx = 0,$$

$$\int_\Omega \left(\frac{1}{2}|\nabla u|^2 - F(u)\right) dx \le c_0 + 1.$$

Multiplying the first equation by $2 + \gamma$ (see Assumption (f_2)) and subtracting the second we easily find that $\int_\Omega |\nabla u|^2$ is bounded. On the other hand, using the first equation, the super-linear properties of f and the Poincaré inequality we obtain the bound for $\|u\|_{L^p}$, $p < 2^*$.   □

Our result concerns the existence of the optimal partition:

Theorem 11.2.1 (Conti, Terracini and Verzini [57]) *There exists* (*at least*) *a k-tuple of functions* $U := (u_1, \dots, u_k) \in \mathcal{N}_0$ *such that*

$$-\Delta u_i(x) = f\left(u_i(x)\right), \quad x \in \mathrm{supp}(u_i)$$

and U and their supports achieve c_0.

Proof Let us consider a minimizing sequence $U_n := (u_1^{(n)}, \dots, u_k^{(n)})$ in $\mathcal{N}_0$. This means that $u_i^{(n)} \in \mathcal{N}(J^*)$ and hence the previous lemma applies, providing the existence of $(u_1^{(0)}, \dots, u_k^{(0)})$ both H_0^1-weak limit and L^p-strong limit of a subsequence. Using again the previous Lemma 11.2.1 and L^p-convergence we deduce that $u_i^{(0)} \not\equiv 0$ for every i, and thus we can find positive constants λ_i, $i = 1, \dots, k$, such that

$$\sup_{\lambda > 0} J^*\left(\lambda u_i^{(0)}\right) = J^*\left(\lambda_i u_i^{(0)}\right).$$

Now we observe that, by weak convergence, $\|u_i^{(0)}\| \le \liminf_{n \to \infty} \|u_i^{(n)}\|$, and so

$$\sum_{i=1}^k J^*\left(\lambda_i u_i^{(0)}\right) \le \sum_{i=1}^k J^*\left(\lambda_i u_i^{(n)}\right) + o(1) \le \sum_{i=1}^k J^*\left(u_i^{(n)}\right) + o(1) \le c_0.$$

On the other hand, for every i we have $\lambda_i u_i^{(0)} \in \mathcal{N}(J^*)$, and, by strong L^p-convergence, $u_i^{(0)} \cdot u_j^{(0)} = 0$ almost everywhere on Ω. Thus, by definition of c_0,

$$\sum_{i=1}^k J^*\left(\lambda_i u_i^{(0)}\right) \ge c_0.$$

Hence, we have found a k-tuple of functions that achieves the infimum. Now using the equivalent characterizations of (11.21) and standard critical point techniques the theorem follows. $\square$

Theorem 11.2.2 (Local Lipschitz continuity see [57]) *Let U be as in Theorem 11.2.1. Then U is Lipschitz continuous in the interior of Ω.*

Theorem 11.2.3 (Global Lipschitz continuity see [57]) *Let $\Omega \subset \mathbb{R}^N$ be a regular set of class C^2 and let U be as in Theorem 11.2.1. Then U is Lipschitz continuous in $\bar{\Omega}$*

11.2.2 An Optimal Partition Problem for Eigenvalues

Let us consider the following optimal partition problems for Dirichlet eigenvalues (see Cafferelli and Lin [42]):

(P) Let Ω be a bounded smooth domain in $\mathbb{R}^n$, and let $m \geq 1$ be a positive integer. One seeks for a partition of Ω into m, mutually disjoint subsets, Ω_j, $j = 1, 2, \ldots, m$, such that $\Omega = \bigcup_{j=1}^{m} \Omega_j$ and that is minimizers $\sum_{j=1}^{m} \lambda_1(\Omega_j)$ among all possible partitions of Ω. Here $\lambda_1(A)$ is the first Dirichlet eigenvalue of the Laplacian Δ on A with the zero Dirichlet boundary condition on ∂A.

First we give a few important notions and definitions first.

Let Ω be a bounded, open subset in $\mathbb{R}^n$. The (harmonic) capacity of a subset E of Ω is defined by

$$\mathrm{Cap}(E) = \inf\left\{ \int_{\Omega} |\nabla v|^2 \, dx : v \in U_E \right\}, \qquad (11.22)$$

where U_E is the set of all $v \in H_0^1(\Omega)$ such that $v \geq 1$ almost everywhere in an open neighborhood of E.

A subset $E \subset \Omega$ is called quasi-open, if for every $\varepsilon > 0$, there exists an open subset O of Ω such that $\mathrm{Cap}(E \Delta O) < \varepsilon$, where Δ denotes the symmetric difference of sets (The symmetric difference is equivalent to the union of both relative complements, that is, $E \Delta O = (E \setminus O) \cup (O \setminus E)$.)

Let $\mathcal{A}(\Omega)$ be the class of quasi-open subsets of Ω. One also defines a function $f : \Omega \to \mathbb{R}$ to be quasi-continuous if for every $\varepsilon > 0$ there exists a continuous function f_ε such that $\mathrm{Cap}(f \neq f_\varepsilon) < \varepsilon$. A well-known result due to Ziemer [211] says that Sobolev function $v \in H_0^1(\Omega)$ is quasi-continuous. More precisely, it has a quasi-continuous representative. For conveniences, when we have $v \in H^1(\Omega)$, we always identify it with its quasi-continuous representative.

$\mathcal{A}(\Omega)$ is the admissible subsets of Ω. Let $\Omega = \bigcup_{j=1}^{m} \Omega_j$ be an admissible partition. Each $\lambda_1(\Omega_j)$ is characterized as

$$\lambda_1(\Omega_j) = \min\left\{\int_{\Omega_j} |\nabla v|^2 \, dx : v \in H_0^1(\Omega_j), \int_{\Omega_j} |v|^2 \, dx = 1\right\}$$

$$\text{for } j = 1, 2, \ldots, m. \tag{11.23}$$

Let v'_js be the eigenfunctions:

$$\Delta v_j + \lambda_1(\Omega_j) v_j = 0 \quad \text{in } \Omega_j, \qquad v_j = 0 \quad \text{on } \partial\Omega_j. \tag{11.24}$$

Define a map $v : \Omega \to \Sigma$ as follows:

$$v(x) = \big(v_1(x), v_2(x), \ldots, v_m(x)\big), \quad x \in \Omega, \tag{11.25}$$

where each v_j has been extended to the whole Ω by setting $v_j \equiv 0$ on $\Omega \setminus \Omega_j$, Σ is a singular space in $\mathbb{R}^m$:

$$\Sigma \equiv \big\{y \in \mathbb{R}^m : F(y) = 0\big\}, \qquad F(y) = \sum_{k \neq l} y_k^2 y_l^2. \tag{11.26}$$

Since we may assume each $v_j \geq 0$ in (11.24), we may replace Σ by $\Sigma_+ = \{y \in \Sigma : y_k \geq 0, k = 1, 2, \ldots, m\}$.

We can consider the mapping problem.

$$(P^*) \quad \min\left\{\int_{\Omega} |\nabla v|^2 \, dx : v \in H_0^1(\Omega, \Sigma), \int_{\Omega} v_k^2(x) \, dx = 1, \text{ for } k = 1, 2, \ldots, m\right\}.$$

We observe first that Σ is a global Lipschitz neighborhood retractor of $\mathbb{R}^m$, that is, there is a Lipschitz map $R : \mathbb{R}^m \to \Sigma$ such that $R(y) = y$ for $y \in \Sigma$. Such a map R can be constructed directly. We also note that $\pi_k(\Sigma) = 0$, for $k \geq 1$. Here $\pi_k(\Sigma)$ is the kth fundamental group of the Σ. In particular Σ is simply connected (in fact, contractible to $\underline{0}$). The Sobolev space of mapping $H_0^1(\Omega, \Sigma) = \{v \in H_0^1(\Omega, \mathbb{R}^m) : v(x) \in \Sigma$ for a.e. $x \in \Omega\}$ has all the usual properties. For example, Lipschitz maps are strongly dense in $H_0^1(\Omega, \Sigma)$. The following proposition is trivial.

Proposition 11.2.1 (Cafferelli and Lin [42]) *There is a map $u \in H_0^1(\Omega, \Sigma)$ that solves problem (P^*).*

Proposition 11.2.2 (Cafferelli and Lin [42]) *Let u be a minimizer of (P^*). Then*

$$\int_{\Omega} |\nabla u|^2 \, dx = \varepsilon_m := \inf\left\{\sum_{j=1}^{m} \lambda_1(\Omega_j) : \Omega_j \in \mathcal{A}(\Omega), \Omega_i \cap \Omega_j = \emptyset \text{ for } i \neq j\right\},$$

and $\Omega_j = \{x \in \Omega : u_j(x) > 0\}$, $j = 1, 2, \ldots, m$ gives a solution of (P). The converse is also true.

Proof Assume u is a minimizer of (P^*). We consider the quasi-continuous representative of the map u (still denoted by u), and let

$$\Omega_j = \big\{x \in \Omega : u_j(x) > 0\big\} \quad \text{for } j = 1, 2, \ldots, m.$$

It is obvious that in problem (P^*) one may assume $u_j(x) \geq 0$ on Ω, for each $j = 1, 2, \ldots, m$. Ω_j's thus defined are, of course, quasi-open. Since $u : \Omega \to \Sigma$, we have $\Omega_i \cap \Omega_j = \emptyset$ for $i \neq j$. Moreover, $\int_{\Omega_j} u_j^2(x)\, dx = 1$, $u_j(x) = 0$ quasi everywhere on $\bar{\Omega} \setminus \Omega_j$. Therefore by the definition of $\lambda_1(\Omega_j)$, one has $\lambda_1(\Omega_j) \leq \int_{\Omega_j} |\nabla u_j|^2(x)\, dx$. We thus conclude that the infimum value of

$$\varepsilon_m = \inf\left\{ \sum_{j=1}^m \lambda_1(\Omega_j) : \Omega_j \in \mathcal{A}(\Omega),\ \Omega_i \cap \Omega_j = \emptyset \text{ for } i \neq j \right\}$$

$$\leq \int_{\Omega_j} |\nabla u|^2(x)\, dx. \tag{11.27}$$

On the other hand, let $\Omega_j^n \in \mathcal{A}(\Omega)$, $j = 1, 2, \ldots, m$, $n = 1, 2, \ldots$, be a minimizing sequence of problem (P). Thus $\Omega_j^n \cap \Omega_i^n = \emptyset$ for $i \neq j$, $i, j = 1, \ldots, m$ and $n = 1, 2, \ldots$ and one has

$$\sum_{j=1}^m \lambda_i\left(\Omega_j^n\right) \to \varepsilon_m \quad \text{as } n \to +\infty.$$

Let u_j^n be the corresponding sequence of eigenfunctions (solutions of (11.24) on Ω_j^n) so that $\int_{\Omega_j^n} (u_j^n(x))^2(x)\, dx = 1$. We extend $u_j^n(x) \equiv 0$ on $\bar{\Omega} \setminus \Omega_j^n$. Then the maps $u^n(x) = (u_1^n(x), \ldots, u_n^n(x)) : \Omega \to \Sigma)$ are naturally in the Sobolev space of mappings $H_0^1(\Omega, \Sigma)$ with $\int_\Omega (u_j^n(x))^2\, dx = 1$ for $j = 1, 2, \ldots, m$ and for all n. Therefore,

$$\int_\Omega |\nabla u|^2(x)\, dx \leq \int_\Omega \left|\nabla u^n(x)\right|^2\, dx = \sum_{j=1}^m \lambda_1\left(\Omega_j^n\right) \to \varepsilon_m. \qquad \square$$

We know for a harmonic function in $B_1 \subset \mathbb{R}^n$, i.e.,

$$\Delta u = 0.$$

Define for any $r \in (0, 1)$

$$D(r) = \int_{B_r} |\nabla u|^2, \qquad H(r) = \int_{\partial B_r} u^2$$

and

$$N(r) = \frac{r D(r)}{H(r)}. \tag{11.28}$$

Since $\Delta u^2 = 2|\nabla u|^2$, with Green's formula, we may rewrite $D(r)$ as

$$D(r) = \frac{1}{2} \int_{B_r} \Delta u^2 = \int_{\partial B_r} u u_n, \tag{11.29}$$

where u_n is the normal derivative of u. $N(r)$ is called the frequency of u in B_r, it is a nondecreasing function of $r \in (0, 1)$ [116].

Example 11.2.1 If u is a homogeneous harmonic polynomial of degree k, then $N(r)$ is a constant and $N(r) = k$ (the degree of a term is the sum of the exponents on variables, and that the degree of a polynomial is the largest degree of any one term) (see [116]). To see this, we write $u(x) = r^k \varphi(\theta)$, where $\varphi(\theta)$ is the restriction of u to $\mathbb{S}^{n-1}$. Then we have $u_n = kr^{k-1}\varphi(\theta)$, and hence for any $r > 0$

$$N(r) = \frac{rD(r)}{H(r)} = \frac{r \int_{\partial B_r} uu_n}{\int_{\partial B_r} u^2} = k.$$

Next let $x_0 \in \Omega$, $p = u(x_0)$, $0 < r < d(x_0, \partial\Omega)$, and consider the modified frequency function

$$N(r) = e^{\Lambda r} \frac{rD(r)}{H(r)} \tag{11.30}$$

where $D(r) = \int_{B_r(x_0)} |\nabla u|^2 \, dx$, $H(r) = \int_{\partial B_r(x_0)} d_\Sigma^2(u(x), p)$. Here $d_\Sigma(p, q)$ for $p, q \in \Sigma$, denotes the intrinsic distance on Σ, and $\Lambda := \Lambda(n, \Omega, m)$ only depends on n, Ω, m. $N(r)$ is a monotone increasing function of $r \in (0, d_0)$ for $d_0 = d(x_0, \partial\Omega)$. As a consequence of this monotonicity property, we have

Proposition 11.2.3 (Cafferelli and Lin [42]) *Let u be a minimizer of (P^*). Then u is locally uniformly Lipschitz continuous in Ω. If Ω is sufficiently smooth, say $C^{1,1}$, then u is also global Lipschitz continuous on $\bar{\Omega}$.*

For regularity of free interfaces. Let u be a minimizer of (P^*). We define $\Gamma = \{x \in \Omega : u(x) = 0\}$. Under the assumption $u : B_2 \to \Sigma$ be a stationary (minimizing) map with $\fint_{\partial B_2} |u|^2 = 1$ ($\fint$ denotes the average. Given a set $D \subset \mathbb{R}^n$ and an integrable function f over D, the average value of f over its domain is given by $\frac{1}{\text{meas}(D)} \int_D f(x) \, dx$), then $\Gamma \cap B_1(\underline{0})$ is $C^{2,\alpha}$ for $0 < \alpha < 1$ (see Theorem 2 of [42]). For related results, see [43, 57]. $\square$

11.3 Schrödinger Systems from Bose–Einstein Condensate

For solitary wave solutions of time-dependent coupled nonlinear Schrödinger equations given by

$$\begin{cases} -i\dfrac{\partial}{\partial t}\Phi_1 = \Delta\Phi_1 + \mu_1|\Phi_1|^2\Phi_1 + \beta|\Phi_2|^2\Phi_1 & \text{for } y \in \mathbb{R}^n,\ t > 0, \\[2mm] -i\dfrac{\partial}{\partial t}\Phi_2 = \Delta\Phi_2 + \mu_2|\Phi_2|^2\Phi_2 + \beta|\Phi_1|^2\Phi_2 & \text{for } y \in \mathbb{R}^n,\ t > 0, \\[2mm] \Phi_j = \Phi_j(y,t) \in \mathbb{C}, \quad j = 1,2, \\[2mm] \Phi_j(y,t) \to 0 \quad \text{as } |y| \to \infty,\ t > 0,\ j = 1,2. \end{cases} \tag{11.31}$$

This system has been proposed as mathematical model (for solitary wave solutions) for multispecies Bose–Einstein condensation in coupled different hyperfine spin states.

A Bose–Einstein condensate (BEC) is a state of matter of a dilute gas of weakly interacting bosons confined in an external potential and cooled to temperatures very near absolute zero (0 K or $-273.15\ °\text{C}$). Under such conditions, a large fraction of the bosons occupy the lowest quantum state of the external potential, at which point quantum effects become apparent on a macroscopic scale.

This state of matter was first predicted by Satyendra Nath Bose and Albert Einstein in 1924–1925. Bose first sent a paper to Einstein on the quantum statistics of light quanta (now called photons). Einstein was impressed, translated the paper himself from English to German and submitted it for Bose to the Zeitschrift für Physik, which published it. Einstein then extended Bose's ideas to material particles (or matter) in two other papers.

Seventy years later, the first gaseous condensate was produced by Eric Cornell and Carl Wieman in 1995 at the University of Colorado at Boulder NIST-JILA lab, using a gas of rubidium atoms cooled to 170 nanokelvin (nK) (1.7×10^{-7} K). For their achievements Cornell, Wieman, and Wolfgang Ketterle at MIT received the 2001 Nobel Prize in Physics. In November 2010 the first photon BEC was observed.

This system also appears in many physical fields, e.g. in nonlinear optics. Physically the solution Φ_i denotes the ith component of the beam in Kerr-like photorefractive media, with $\mu_j > 0$, $j = 1,2$ we have self-focusing in both components of the beam. The nonlinear coupling constant β is the interaction between the two components of the beam.

Recently many mathematicians become concerned with this Schrödinger system, see [10, 32, 73, 74, 79, 141, 178] etc. Since there is a limit for this book, we only simply introduce some results.

To obtain solitary wave solution of the system (11.31), we set $\Phi_j(y,t) = e^{i\lambda_j t}u_j(y)$, $j = 1,2$ and we may transform the system (11.31) to a coupled elliptic system given by

$$\begin{cases} -\Delta u_1 + \lambda_1 u_1 = \mu_1 u_1^3 + \beta u_2^2 u_1, & \text{in } \mathbb{R}^n, \\[2mm] -\Delta u_2 + \lambda_2 u_2 = \mu_2 u_2^3 + \beta u_1^2 u_2, & \text{in } \mathbb{R}^n, \\[2mm] u_1, u_2 > 0 & \text{in } \mathbb{R}^n. \end{cases} \tag{11.32}$$

We introduce some results in the following.

11.3.1 Existence of Solutions for Schrödinger Systems

(1) Ambrosetti and Colorado [10] studied the following system:

$$
\begin{cases}
-\Delta u_1 + \lambda_1 u_1 = \mu_1 u_1^3 + \beta u_2^2 u_1 & \text{in } \mathbb{R}^n, \\
-\Delta u_2 + \lambda_2 u_2 = \mu_2 u_2^3 + \beta u_1^2 u_2 & \text{in } \mathbb{R}^n, \\
u_1, u_2 \in W^{1,2}(\mathbb{R}^n),
\end{cases}
\tag{11.33}
$$

where $n = 2, 3$, $\lambda_j, \mu_j > 0$, $j = 1, 2$ and $\beta \in \mathbb{R}$.

They showed there exist $\Lambda' \geq \Lambda > 0$, depending on λ_j, μ_j such that (11.33) has a radially symmetric solution $(u_1, u_2) \in W^{1,2}(\mathbb{R}^n) \times W^{1,2}(\mathbb{R}^n)$, with $u_1, u_2 > 0$, provided $\beta \in (0, \Lambda) \cup (\Lambda', +\infty)$. Moreover, for $\beta > \Lambda'$, these solutions are ground states, in the sense that they have minimal energy and their Morse index is 1. It is worth pointing out that for any β, (11.33) has a pair of semi-trivial solutions with one component equal to zero. These solutions have the form $(U_1, 0)$, $(0, U_2)$, where U_j is the positive radial solution of

$$
-\Delta u + \lambda_j u = \mu_j u^3, \quad u \in W^{1,2}(\mathbb{R}^n).
\tag{11.34}
$$

Of course, we care about the solutions different from the preceding ones. On the other hand, the presence of $(U_1, 0)$, $(0, U_2)$ can be usefully exploited to prove their existence results. Actually, the main idea is to show that the Morse index of $(U_1, 0)$ and $(0, U_2)$ changes with β: for $\beta < \Lambda$ their index is 1, while for $\beta > \Lambda'$ their index is greater than or equal to 2. This fact, jointly an appropriate use of the method of natural constraint, is used to prove the existence of bound and ground states as outlined above.

Let us introduce the following notations:

- $E = W^{1,2}(\mathbb{R}^n)$, the standard Sobolev space, endowed with scalar product and norm

$$
(u|v)_j = \int_{\mathbb{R}^n} [\nabla u \cdot \nabla v + \lambda_j uv]\, dx, \quad \|u\|_j^2 = (u|u)_j, \quad j = 1, 2;
$$

- $\mathbb{E} = E \times E$; the elements in $\mathbb{E}$ will be denoted by $\mathbf{u} = (u_1, u_2)$; as a norm in E we will take $\|\mathbf{u}\|^2 = \|u_1\|_1^2 + \|u_2\|_2^2$;
- we set $\mathbf{0} = (0, 0)$;
- for $\mathbf{u} \in \mathbb{E}$, writing $\mathbf{u} \geq \mathbf{0}$ and $\mathbf{u} > \mathbf{0}$ means that $u_j \geq 0$ and $u_j > 0$, respectively, for all $j = 1, 2$;
- H denotes the space of radially symmetric functions in E;
- $\mathbb{H} = H \times H$.

For $u \in E$ and $\mathbf{u} \in \mathbb{E}$, respectively, we set

$$
I_j(u) = \frac{1}{2} \int_{\mathbb{R}^n} \left(|\nabla u|^2 + \lambda_j u^2 \right) dx - \frac{1}{4} \mu_j \int_{\mathbb{R}^n} u^4\, dx,
$$

$$F(\mathbf{u}) = \frac{1}{4} \int_{\mathbb{R}^n} \left(\mu_1 u_1^4 + \mu_2 u_2^4 \right) dx,$$

$$G(\mathbf{u}) = G(u_1, u_2) = \frac{1}{2} \int_{\mathbb{R}^n} u_1^2 u_2^2 \, dx,$$

$$\Phi(\mathbf{u}) = \Phi(u_1, u_2) = I_1(u_1) + I_2(u_2) - \beta G(u_1, u_2)$$

$$= \frac{1}{2} \|\mathbf{u}\|^2 - F(\mathbf{u}) - \beta G(\mathbf{u}).$$

Let us remark that F and G make sense because $E \hookrightarrow L^4(\mathbb{R}^n)$ for $n = 2, 3$. Any critical point $u \in \mathbb{E}$ of Φ gives rise to a solution of (11.33). If $\mathbf{u} \neq \mathbf{0}$, then we say that such a critical point (solution) is non-trivial. We say that a solution $\mathbf{u}$ of (11.33) is positive if $\mathbf{u} > \mathbf{0}$.

Among non-trivial solutions of (11.33), we shall distinguish between the bound states and the ground states.

Definition 11.3.1 We say that $\mathbf{u} \in \mathbb{E}$ is a non-trivial bound state of (11.33) if $\mathbf{u}$ is a non-trivial critical point of Φ. A positive bound state $\mathbf{u} > \mathbf{0}$ such that its energy is minimal among all the non-trivial bound states, namely

$$\Phi(\mathbf{u}) = \min\left\{ \Phi(\mathbf{v}) : \mathbf{v} \in \mathbb{E} \setminus \{\mathbf{0}\}, \, \Phi'(\mathbf{v}) = 0 \right\} \tag{11.35}$$

is called a ground state of (11.33).

In order to find critical points of Φ, let us set

$$\Psi(\mathbf{u}) = \left(\Phi'(\mathbf{u}) | \mathbf{u} \right) = \|\mathbf{u}\|^2 - 4F(\mathbf{u}) - 4\beta G(\mathbf{u})$$

and introduce the so-called Nehari manifold

$$\mathcal{M} = \left\{ \mathbf{u} \in \mathbb{H} \setminus \{\mathbf{0}\} : \Psi(\mathbf{u}) = 0 \right\}.$$

Plainly, $\mathcal{M}$ contains all the non-trivial critical points of Φ on $\mathbb{H}$. A remarkable advantage of working on the Nehari manifold is that Φ is bounded from below on $\mathcal{M}$. Actually, from $\Psi(u) = 0$ and the definition of $\mathcal{M}$, it follows that

$$\|\mathbf{u}\|^2 = 4F(\mathbf{u}) + 4\beta G(\mathbf{u}). \tag{11.36}$$

Substituting into Φ we get

$$\Phi(u) = \frac{1}{4} \|\mathbf{u}\|^2, \quad \forall \mathbf{u} \in \mathcal{M}. \tag{11.37}$$

Since F, G are homogeneous with degree 4, there is $\rho > 0$ such that

$$\|\mathbf{u}\| \geq \rho, \quad \forall \mathbf{u} \in \mathcal{M}. \tag{11.38}$$

Thus there exists $C > 0$ such that

$$\Phi(\mathbf{u}) \geq C > 0, \quad \forall \mathbf{u} \in \mathcal{M}. \tag{11.39}$$

Lemma 11.3.1 (Ambrosetti and Colorado [10])

(i) *$\mathcal{M}$ is homeomorphic to the unit sphere of $\mathbb{H}$, and there exists $\rho > 0$ such that $\|\mathbf{u}\| \geq \rho$, $\forall \mathbf{u} \in \mathcal{M}$.*

(ii) *$\mathcal{M}$ is a $\mathcal{C}^1$ complete manifold of codimension one in $\mathbb{H}$.*

(iii) *$\mathbf{u} \in \mathbb{H}$ is a non-trivial critical point of Φ if and only if $\mathbf{u}$ is a constrained critical point of the restriction $\Phi|_{\mathcal{M}}$ on $\mathcal{M}$.*

(iv) *$\Phi|_{\mathcal{M}}$ satisfies Palais–Smale condition (Every $\{\mathbf{u}_n\} \in \mathcal{M}$ such that $\Phi(\mathbf{u}_n) \to c$ and $\nabla_{\mathcal{M}}\Phi(\mathbf{u}) \to 0$, has a (strongly) converging subsequence: $\exists \mathbf{u}_0 \in \mathcal{M}$ such that $\mathbf{u}_n \to \mathbf{u}_0$).*

Proof (i) For any $\mathbf{u} \in \mathbb{H} \setminus \{\mathbf{0}\}$, one has

$$t\mathbf{u} \in \mathcal{M} \quad \Longleftrightarrow \quad t^2\|\mathbf{u}\|^2 = t^4\big[4F(\mathbf{u}) + 4\beta G(\mathbf{u})\big]. \tag{11.40}$$

As a consequence, for any $\mathbf{u} \in \mathbb{H} \setminus \{\mathbf{0}\}$ there exists a unique $t(\mathbf{u}) \in \mathbb{R}^+$ such that $t(\mathbf{u})\mathbf{u} \in \mathcal{M}$.

In order to prove the continuity of $t(\mathbf{u})$, we assume that $\mathbf{u}_n \to \mathbf{u}_0$ in $\mathbb{H} \setminus \{\mathbf{0}\}$. It follows from (11.40) that $\{t(\mathbf{u}_n)\}$ is bounded. Passing if necessary to a subsequence, we can assume that $t(\mathbf{u}_n) \to t_0$, then $t_0 = t(\mathbf{u}_0)$ by (11.40) and the uniqueness of $t(\mathbf{u}_0)$. Hence $t(\mathbf{u}_n) \to t(\mathbf{u}_0)$. Moreover, the inverse map of $t(\mathbf{u})|_{S^\infty} : S^\infty \to \mathcal{M}$ can be defined by

$$\mathbf{u} \to \mathbf{u}/\|\mathbf{u}\|,$$

which is also continuous. Therefore, the Nehari manifold $\mathcal{M}$ is homeomorphic to the unit sphere of $\mathbb{H}$.

(ii) Notice that for any $\mathbf{u} \in \mathbb{H} \setminus \{\mathbf{0}\}$

$$\big(\Psi'(\mathbf{u})|\mathbf{u}\big) = 2\|\mathbf{u}\|^2 - 16F(\mathbf{u}) - 16\beta G(\mathbf{u}),$$

and

$$\mathbf{u} \in \mathcal{M} \quad \Longleftrightarrow \quad \|\mathbf{u}\|^2 = 4F(\mathbf{u}) + 4\beta G(\mathbf{u}). \tag{11.41}$$

It follows that

$$\big(\Psi'(\mathbf{u})|\mathbf{u}\big) = -2\|\mathbf{u}\|^2 \leq -2\rho^2 < 0, \quad \forall \mathbf{u} \in \mathcal{M}. \tag{11.42}$$

This jointly with (i) and Implicit Function Theorem shows that $\mathcal{M}$ is a $\mathcal{C}^1$ complete manifold of codimension one in $\mathbb{H}$.

(iii) If $\mathbf{u}$ is a critical point of $\Phi|_{\mathcal{M}}$, then there exists $\omega \in \mathbb{R}$ such that $\Phi'(\mathbf{u}) = \omega\Psi'(\mathbf{u})$. It follows from the fact $\mathbf{u} \in \mathcal{M}$ that

$$\omega\big(\Psi'(\mathbf{u})|\mathbf{u}\big) = \big(\Phi'(\mathbf{u})|\mathbf{u}\big) = \Psi(\mathbf{u}) = 0.$$

Thus we deduced from (11.42) that $\omega = 0$ and $\Phi'(\mathbf{u}) = \mathbf{0}$.

(iv) Assume that $\{\mathbf{u}_n\} \subset \mathcal{M}$ is a $(PS)_c$ sequence of $\Phi|_{\mathcal{M}}$, that is,

$$\Phi(\mathbf{u}_n) \to c \quad \text{and} \quad \nabla_{\mathcal{M}}\Phi(\mathbf{u}_n) \to \mathbf{0}.$$

From (11.39) one has $c > 0$. By (11.37) and $\Phi(\mathbf{u}_n) \to c$ we have $\|\mathbf{u}_n\| \le C < +\infty$, without relabeling, we can assume that $\mathbf{u}_n \rightharpoonup \mathbf{u}_0$. Since H is compactly embedded into $L^4(\mathbb{R}^n)$, $n = 2, 3$ (see Theorem 1.1.5), we infer that

$$F(\mathbf{u}_n) + \beta G(\mathbf{u}_n) \to F(\mathbf{u}_0) + \beta G(\mathbf{u}_0).$$

Moreover, using (11.36) jointly with (11.38), one has $F(\mathbf{u}_n) + \beta G(\mathbf{u}_n) \ge c_0$ and then $\mathbf{u}_0 \neq 0$. There exist $\omega_n \in \mathbb{R}$ such that

$$\nabla_{\mathcal{M}}\Phi(\mathbf{u}_n) = \Phi'(\mathbf{u}_n) - \omega_n \Psi'(\mathbf{u}_n). \tag{11.43}$$

Taking the scalar product with $\mathbf{u}_n$ and recalling that $(\Phi'(\mathbf{u}_n)|\mathbf{u}_n) = \Psi(\mathbf{u}_n) = 0$, we find

$$\omega_n\big(\Psi'(\mathbf{u}_n)|\mathbf{u}_n\big) = \big(\Phi'(\mathbf{u}_n)|\mathbf{u}_n\big) - \big(\nabla\Phi(\mathbf{u}_n)|\mathbf{u}_n\big) \to 0.$$

It follows from (11.42) that

$$\omega_n \to 0. \tag{11.44}$$

Since, in addition $\|\Psi'(\mathbf{u}_n)\| = \|2\mathbf{u}_n - 4(F'(\mathbf{u}_n) + \beta G'(\mathbf{u}_n))\| \le C < +\infty$ (noticing that H is compactly embedded into $L^4(\mathbb{R}^n)$, $n = 2, 3$) we deduce that $\Phi'(\mathbf{u}_n) \to \mathbf{0}$.

We can now conclude that $\mathbf{u}_n \to \mathbf{u}_0$ strongly. Actually, from $\Phi'(\mathbf{u}_n) = \mathbf{u}_n - (F'(\mathbf{u}_n) + \beta G'(\mathbf{u}_n))$, $\Psi'(\mathbf{u}_n) = 2\mathbf{u}_n - 4(F'(\mathbf{u}_n) + \beta G'(\mathbf{u}_n))$ and $\Phi'(\mathbf{u}_n) - \omega_n \Psi'(\mathbf{u}_n) = o(1)$, we obtain

$$(1 - 2\omega_n)\mathbf{u}_n = (1 - 4\omega_n)\big(F'(\mathbf{u}_n) + \beta G'(\mathbf{u}_n)\big) + o(1).$$

Now, F' and G' being the gradients of smooth weakly continuous functionals F and G, respectively, are compact operators. This, the preceding equation and $\omega_n \to 0$ imply that $\mathbf{u}_n \to F'(\mathbf{u}_0) + \beta G'(\mathbf{u}_0)$ strongly. As a consequence, $\mathbf{u}_0 = F'(\mathbf{u}_0) + \beta G'(\mathbf{u}_0)$, whence $\|\mathbf{u}_0\|^2 = 4F(\mathbf{u}_0) + \beta G(\mathbf{u}_0)$, which means that $\mathbf{u}_0 \in \mathcal{M}$. This completes the proof. $\square$

Remark 11.3.1 The above lemma implies that $\inf_{\mathbf{u} \in \mathcal{M}} \Phi(\mathbf{u})$ can be achieved, giving rise to a non-negative solution of problem (11.33). However, such an existence result is useless without any further specification. Actually, for every $\beta \in \mathbb{R}$, (11.33) already possesses two explicit solutions given by $\mathbf{u}_1 = (U_1, 0)$, $\mathbf{u}_2 = (0, U_2)$, where U_j is the unique radial positive solution of $-\Delta u + \lambda_j u = \mu_j u^3$. If we denote U as the unique positive radial solution of $-\Delta u + u = u^3$ (see [123]) then $U_j = \sqrt{\frac{\lambda_j}{\mu_j}} U(\sqrt{\lambda_j}x)$, $j = 1, 2$.

In order to demonstrate that there exists a non-trivial solution of problem (11.33) different from semi-positive ones $\mathbf{u}_1$ and $\mathbf{u}_2$, we set

$$\gamma_1^2 = \inf_{\varphi \in H \setminus \{0\}} \frac{\|\varphi\|_2^2}{\int_{\mathbb{R}^n} U_1^2 \varphi^2}, \qquad \gamma_2^2 = \inf_{\varphi \in H \setminus \{0\}} \frac{\|\varphi\|_1^2}{\int_{\mathbb{R}^n} U_2^2 \varphi^2}$$

and

$$\Lambda = \min\{\gamma_1^2, \gamma_1^2\}, \qquad \Lambda' = \max\{\gamma_1^2, \gamma_1^2\}.$$

The next proposition shows that the nature of $\mathbf{u}_j$ changes in dependence of β, Λ, Λ'.

Proposition 11.3.1 (Ambrosetti and Colorado [10])

 (i) $\forall \beta < \Lambda$, $\mathbf{u}_j$, $j = 1, 2$, *are strict local minima of* Φ *on* $\mathcal{M}$.
(ii) *If* $\beta > \Lambda'$, *then* $\mathbf{u}_j$ *are saddle points of* Φ *on* $\mathcal{M}$. *In particular,* $\inf_{\mathcal{M}} \Phi < \min\{\Phi(\mathbf{u}_1), \Phi(\mathbf{u}_2)\}$.

Let $D^2\Phi|_{\mathcal{M}}(\mathbf{u}_j)$ denote the second derivative of Φ constrained on $\mathcal{M}$. Since $\Phi'(\mathbf{u}_j) = 0$, one has

$$D^2\Phi|_{\mathcal{M}}(\mathbf{u}_1)[\mathbf{h}]^2 = \Phi''(\mathbf{u}_1)[\mathbf{h}]^2, \quad \forall \mathbf{h} \in T_{\mathbf{u}_j}\mathcal{M}. \tag{11.45}$$

Similarly, if $\mathcal{N}_j$ denotes the Nehari manifolds relative to I_j, $j = 1, 2$,

$$\mathcal{N}_j = \left\{u \in H \setminus \{0\} : (I_j'(u)|u) = 0\right\} = \left\{u \in H \setminus \{0\} : \|u\|_j^2 = \mu_j \int_{\mathbb{R}^n} u^4\right\}$$

then, from the fact that $I_j'(U_j) = 0$, it follows that

$$D^2 I_j|_{\mathcal{N}_j}(U_j)[h]^2 = I_j''(U_j)[h]^2, \quad \forall h \in T_{U_j}\mathcal{N}_j. \tag{11.46}$$

Notice that U_j is the minimum of I_j on $\mathcal{N}_j$ and thus, using also (11.46), there exists $c_j > 0$ such that

$$I_j''(U_j)[h_j]^2 \ge c_j \|h_j\|_j^2, \quad j = 1, 2. \tag{11.47}$$

Lemma 11.3.2 (Ambrosetti and Colorado [10]) $\mathbf{h} = (h_1, h_2) \in T_{\mathbf{u}_j}\mathcal{M} \Leftrightarrow h_j \in T_{U_j}\mathcal{N}_j$, $j = 1, 2$.

Proof One has $h_j \in T_{U_j}\mathcal{N}_j$ if and only if $(U_j|\phi)_j = 2\mu_j \int_{\mathbb{R}^n} U_j^3\phi$, while $\mathbf{h} = (h_1, h_2) \in T_{\mathbf{u}}\mathcal{M}$ if and only if

$$(u_1|h_1)_1 + (u_2|h_2)_2 = 2 \int_{\mathbb{R}_n} \left(\mu_1 u_1^3 h_1 + \mu_2 u_2^3 h_2\right) + \beta \int_{\mathbb{R}^n} \left(u_1 h_1 u_2^2 + u_1^2 u_2 h_2\right).$$

Thus $\mathbf{h} = (h_1, h_2) \in T_{\mathbf{u}_j}\mathcal{M}$, if and only if $(U_j|h_j)_j = 2\mu_j \int U_j^3 h_j$. $\qquad \square$

Proof of Proposition 11.3.1 (i) If $\mathbf{u} = (u_1, u_2) \in \mathbb{H}$ and $\mathbf{h} = (h_1, h_2) \in \mathbb{H}$, then one has

$$\Phi'(\mathbf{u})[\mathbf{h}]^2 = I_1''(u_1)[h_1]^2 + I_2''(u_2)[h_2]^2 - \beta \int_{\mathbb{R}^n} \left(u_1^2 h_2^2 + u_2^2 h_1^2 + 4u_1 u_2 h_1 h_2\right).$$

In particular, if $\mathbf{u} = \mathbf{u}_1$, then we get

$$\Phi''(\mathbf{u}_1)[\mathbf{h}]^2 = I_1''(U_1)[h_1]^2 + \|h_2\|_2^2 - \beta \int_{\mathbb{R}_n} U_1^2 h_2^2. \tag{11.48}$$

Now, let us take $\mathbf{h} = (h_1, h_2) \in T_{\mathbf{u}_1}\mathcal{M}$. Then, by Lemma 11.3.2, we have $h_1 \in T_{U_1}\mathcal{N}_1$, and hence (11.47) yields

$$I_1''(U_1)[h_1]^2 \geq c_1 \|h_1\|_1^2.$$

Substituting into (11.48) we infer that

$$\Phi''(\mathbf{u}_1)[\mathbf{h}]^2 \geq c_1 \|h_1\|_1^2 + \|h_2\|_2^2 - \beta \int_{\mathbb{R}^n} U_1^2 h_2^2 \quad \forall \mathbf{h} \in T_{\mathbf{u}_1}\mathcal{M},$$

and this, using the definition of γ_1, yields

$$\Phi''(\mathbf{u}_1)[\mathbf{h}]^2 \geq c_1 \|h_1\|_1^2 + \|h_2\|_2^2 - \frac{\beta}{\gamma_1^2} \|h_2\|_2^2 \quad \forall \mathbf{h} \in T_{\mathbf{u}_1}\mathcal{M}.$$

Therefore, if $\beta < \gamma_1^2$, then there exists $c_2 > 0$ such that

$$\Phi''(\mathbf{u}_1)[\mathbf{h}]^2 \geq c_1 \|h_1\|_1^2 + c_2 \|h_2\|_2^2 \quad \forall \mathbf{h} \in T_{\mathbf{u}_1}\mathcal{M}. \tag{11.49}$$

Taking into account (11.45), we infer that (11.49) implies that $\mathbf{u}_1$ is a local strict minimum of Φ on $\mathcal{M}$. Similarly, if $\beta < \gamma_2^2$, $\mathbf{u}_2$ is a local strict minimum of Φ on $\mathcal{M}$.

(ii) We will evaluate $\Phi''(\mathbf{u}_1)$ on tangent vectors of the form $(0, h_2)$. According to (11.48), one has

$$\Phi''(\mathbf{u}_1)\big[(0, h_2)\big]^2 = \|h_2\|_2^2 - \beta \int_{\mathbb{R}^n} U_1^2 h_2^2.$$

Moreover, Lemma 11.3.2 implies that $(0, h_2) \in T_{\mathbf{u}_1}\mathcal{M}$ for all $h_2 \in H$. If $\beta > \gamma_1^2$, then there exists $\tilde{h}_2 \in H \setminus \{0\}$ such that

$$\gamma_1^2 < \frac{\|\tilde{h}_2\|_2^2}{\int_{\mathbb{R}^n} U_1^2 \tilde{h}_2^2} < \beta,$$

and hence

$$\Phi''(\mathbf{u}_1)\big[(0, \tilde{h}_2)\big]^2 = \|\tilde{h}_2\|_2^2 - \beta \int_{\mathbb{R}^n} U_1^2 \tilde{h}_2^2 < 0.$$

Similarly, if $\beta > \gamma_2^2$, then there exists $\hat{h}_1 \in H \setminus \{0\}$ such that $\Phi''(\mathbf{u}_2)[(\hat{h}_1, 0)]^2 < 0$. $\qquad\square$

Remark 11.3.2 What we have really proved is that $\mathbf{u}_j$ is a minimum and a saddle point of Φ on $\mathcal{M}$, provided $\beta < \gamma_j^2$ and $\beta > \gamma_j^2$, $j = 1, 2$, respectively.

According to Lemma 11.3.1(iii), in order to find a non-trivial solution of (11.33), it suffices to find a critical point of Φ constrained on $\mathcal{M}$. The following lemma is a direct consequence of Proposition 11.3.1 and Lemma 11.3.1(iv).

Lemma 11.3.3 (Ambrosetti and Colorado [10])

(i) *If $\beta < \Lambda$, then Φ has a mountain-pass critical point $\mathbf{u}^*$ on $\mathcal{M}$, and there holds $\Phi(\mathbf{u}^*) > \max\{\Phi(\mathbf{u}_1), \Phi(\mathbf{u}_2)\}$.*
(ii) *If $\beta > \Lambda'$ then Φ has a global minimum $\tilde{\mathbf{u}}$ on $\mathcal{M}$, and $\Phi(\tilde{\mathbf{u}}) < \min\{\Phi(\mathbf{u}_1), \Phi(\mathbf{u}_2)\}$.*

Proof (i) Proposition 11.3.1(i) and Lemma 11.3.1(iv) allow us to apply the Mountain Pass theorem to Φ on $\mathcal{M}$, yielding a critical point $\mathbf{u}^*$ of Φ. By the Mountain Pass theorem, it also follows that $\Phi(\mathbf{u}^*) > \max\{\Phi(\mathbf{u}_1), \Phi(\mathbf{u}_2)\}$.

(ii) By Lemma 11.3.1(iv), the $\inf_{\mathcal{M}} \Phi$ is achieved at some $\tilde{\mathbf{u}} \neq \mathbf{0}$. Moreover, if $\beta > \Lambda'$, then Proposition 11.3.1 (ii) implies that $\Phi(\mathbf{u}^*) < \min\{\Phi(\mathbf{u}_1), \Phi(\mathbf{u}_2)\}$. $\square$

Concerning ground states, the main result is the following.

Theorem 11.3.1 (Ambrosetti and Colorado [10]) *If $\beta > \Lambda'$, then (11.33) has a (positive) radial ground state $\tilde{\mathbf{u}}$.*

Proof Lemma 11.3.3(ii) yields a critical point $\tilde{\mathbf{u}} \in \mathcal{M}$ which is a non-trivial solution of (11.33). To complete the proof, we have to show that $\tilde{\mathbf{u}} > \mathbf{0}$ and is a ground state in the sense of Definition 11.3.1. To prove these facts, we argue as follows. Since $|\tilde{\mathbf{u}}| = (|\tilde{u}_1|, |\tilde{u}_2|)$ also belongs to $\mathcal{M}$ and $\Phi(|\tilde{\mathbf{u}}|) = \Phi(\tilde{\mathbf{u}}) = \min\{\Phi(\mathbf{u}) : \mathbf{u} \in \mathcal{M}\}$, we can assume that $\tilde{\mathbf{u}} \geq \mathbf{0}$ and $\tilde{\mathbf{u}} \neq \mathbf{0}$ because $\mathbf{u} \in \mathcal{M}$. If one of the components of $\tilde{\mathbf{u}}$, say $\tilde{u}_2$ is equal to 0, then $\tilde{u}_1$ satisfies $-\Delta \tilde{u}_1 + \lambda_1 \tilde{u}_1 = \mu_1 \tilde{u}_1^3$ and hence $\tilde{u}_1 = U_1$, namely $\tilde{\mathbf{u}} = \mathbf{u}_1$. This is a contradiction because, by Lemma 11.3.3(ii), $\Phi(\tilde{\mathbf{u}}) < \Phi(\mathbf{u}_1)$. So, both the components of $\tilde{\mathbf{u}}$ are such that $\tilde{u}_j \geq 0$, $\tilde{u}_j \neq 0$, $j = 1, 2$. Since $\tilde{u}_j$ satisfies

$$-\Delta \tilde{u}_j + \lambda_j \tilde{u}_j = \mu_j \tilde{u}_j^3 + \beta \tilde{u}_i^2 \tilde{u}_j, \quad i \neq j, \ j = 1, 2,$$

the maximum principle applied to each single equation implies that $\tilde{u}_j > 0$, $j = 1, 2$, that is, $\tilde{\mathbf{u}} > \mathbf{0}$.

It remains to prove that

$$\Phi(\tilde{\mathbf{u}}) = \min\{\Phi(\mathbf{v}) : \mathbf{v} \in \mathbb{E} \setminus \{\mathbf{0}\}, \ \Phi'(\mathbf{v}) = 0\}. \tag{11.50}$$

By contradiction, let $\tilde{\mathbf{v}} \in \mathbb{E}$ be a non-trivial critical point of Φ such that

$$\Phi(\tilde{\mathbf{v}}) < \Phi(\tilde{\mathbf{u}}) = \min\{\Phi(\mathbf{u}) : \mathbf{u} \in \mathcal{M}\}. \tag{11.51}$$

Setting $\mathbf{w} = |\tilde{\mathbf{v}}|$, there holds

$$\Phi(\mathbf{w}) = \Phi(\tilde{\mathbf{v}}), \qquad \Psi(\mathbf{w}) = \Psi(\tilde{\mathbf{v}}). \tag{11.52}$$

If $u \in H$, $u \geq 0$, then let u^* denote the Schwarz symmetric function associated to u, namely the radially symmetric, radially non-increasing function, equi-measurable with u. We will use its following properties (see Sect. 1.11, or [120], Corollary 2.33, and pp. 22, 23, equations (C), $(P1)$)

$$\int_{\mathbb{R}^n} |\nabla u^*|^2 \, dx \leq \int_{\mathbb{R}^n} |\nabla u|^2 \, dx \quad \forall u \in W^{1,2}(\mathbb{R}^n), \ u \geq 0,$$

$$\int_{\mathbb{R}^n} |u^*|^p \, dx = \int_{\mathbb{R}^n} |u|^p \, dx \quad \forall u \in L^p(\mathbb{R}^n), \ u \geq 0,$$

$$\int_{\mathbb{R}^n} u^* v^* \, dx \geq \int_{\mathbb{R}^n} uv \, dx \quad \forall u, v \in L^2(\mathbb{R}^n), \ u, v \geq 0.$$

If $\mathbf{w} = (w_1, w_2)$, then we set $w^* = (w_1^*, w_2^*)$. Using the preceding properties, we get $\|\mathbf{w}^*\|^2 \leq \|\mathbf{w}\|^2$ and $F(\mathbf{w}^*) + \beta G(\mathbf{w}^*) \geq F(\mathbf{w}) + \beta G(\mathbf{w})$. Thus $\Psi(\mathbf{w}^*) \leq \Psi(\mathbf{w})$. Using (11.52) and the fact that $\tilde{\mathbf{v}}$ is a critical point of Φ, we get $\Psi(\mathbf{w}) = \Psi(\tilde{\mathbf{v}}) = 0$, and there exists a unique $t \in (0, 1]$ such that $t\mathbf{w}^* \in \mathcal{M}$. Moreover,

$$\Phi(t\mathbf{w}^*) = \frac{1}{4} t^2 \|\mathbf{w}^*\|^2 \leq \frac{1}{4} \|\mathbf{w}\|^2 = \Phi(\mathbf{w}).$$

This, (11.52) and (11.51) yield

$$\Phi(t\mathbf{w}^*) \leq \Phi(\mathbf{w}) = \Phi(\tilde{\mathbf{v}}) < \Phi(\tilde{\mathbf{u}}) = \min\{\Phi(\mathbf{u}) : \mathbf{u} \in \mathcal{M}\},$$

which is a contradiction since $t\mathbf{w}^* \in \mathcal{M}$. This shows that (11.50) holds and completes the proof of Theorem 11.3.1. $\qquad\square$

Concerning the existence of positive bound states, the following result holds.

Theorem 11.3.2 (Ambrosetti and Colorado [10]) *If $\beta < \Lambda$, then (11.33) has a radial bound state $\mathbf{u}^*$ such that $\mathbf{u}^* \neq \mathbf{u}_j$, $j = 1, 2$. Furthermore, if $\beta \in (0, \Lambda)$, then $\mathbf{u}^* > \mathbf{0}$.*

Proof If $\beta < \Lambda$, then a straight application of Lemma 11.3.3(i) yields a non-trivial solution $\mathbf{u}^* \in \mathcal{M}$ of (11.33), which corresponds to a mountain-pass critical point of Φ on $\mathcal{M}$. Moreover, $\Phi(\mathbf{u}^*) > \max\{\Phi(\mathbf{u}_1), \Phi(\mathbf{u}_2)\}$ implies that $\mathbf{u}^* \neq \mathbf{u}_j$, $j = 1, 2$.

To show that $\mathbf{u}^* > \mathbf{0}$ provided $\beta \in (0, \Lambda)$, let us introduce the functional

$$\Phi^+(\mathbf{u}) = \frac{1}{2} \|\mathbf{u}\|^2 - F(\mathbf{u}^+) - \beta G(\mathbf{u}^+),$$

where $\mathbf{u}^+ = (u_1^+, u_2^+)$ and $u^+ = \max\{u, 0\}$. Consider the corresponding Nehari manifold

$$\mathcal{M}^+ = \{\mathbf{u} \in \mathbb{H} \setminus \{\mathbf{0}\} : (\nabla \Phi^+(\mathbf{u}) | \mathbf{u}) = 0\}.$$

Repeating with minor changes the arguments above, one readily shows that what was proved there, still holds with Φ and $\mathcal{M}$ substituted by Φ^+ and $\mathcal{M}^+$. In particular, Lemma 11.3.1 holds true for Φ^+ and $\mathcal{M}^+$. On the other hand, Proposition 11.3.1(i) cannot be proved as before, because Φ^+ is not C^2. To circumvent this difficulty, we argue as follows.

Consider an ε-neighborhood $V_\varepsilon \subset \mathcal{M}$ of $\mathbf{u}_1$. For each $\mathbf{u} \in V_\varepsilon$, there exists $T(\mathbf{u}) > 0$ such that $T(\mathbf{u})\mathbf{u} \in \mathcal{M}^+$. Actually $T(\mathbf{u})$ satisfies

$$\|\mathbf{u}\|^2 = 4T^2(\mathbf{u})\big[F(\mathbf{u}^+) + \beta G(\mathbf{u}^+)\big],$$

and since $\|\mathbf{u}\| = 4[F(\mathbf{u}) + \beta G(\mathbf{u})]$, we get

$$\big[F(\mathbf{u}) + \beta G(\mathbf{u})\big] = T^2(\mathbf{u})\big[F(\mathbf{u}^+) + \beta G(\mathbf{u}^+)\big]. \tag{11.53}$$

Let us point out that $F(\mathbf{u}^+) + \beta G(\mathbf{u}^+) \leq F(\mathbf{u}) + \beta G(\mathbf{u})$ and this implies that $T(\mathbf{u}) \geq 1$. Moreover, since $\lim_{\mathbf{u} \to \mathbf{u}_1}(F(\mathbf{u}^+) + \beta G(\mathbf{u}^+)) = F(\mathbf{u}_1) > 0$, it follows that there exist $\varepsilon > 0$ and $c > 0$ such that

$$F(\mathbf{u}^+) + \beta G(\mathbf{u}^+) \geq c \quad \forall \mathbf{u} \in V_\varepsilon.$$

This and (11.53) imply that the map $\mathbf{u} \to T(\mathbf{u})\mathbf{u}$ is a homeomorphism, locally near $\mathbf{u}_1$. In particular, there are ε-neighborhoods $V_\varepsilon \subset \mathcal{M}$, $W_\varepsilon \subset \mathcal{M}^+$ of $\mathbf{u}_1$ such that for all $\mathbf{v} \in W_\varepsilon$, there exists $\mathbf{u} \in V_\varepsilon$ such that $\mathbf{v} = T(\mathbf{u})\mathbf{u}$. Finally, from $\Phi^+(\mathbf{v}) = \frac{1}{4}\|\mathbf{v}\|^2$ (see (11.37)) and the fact that $T(\mathbf{u}) \geq 1$, we infer that

$$\Phi^+(\mathbf{v}) = \frac{1}{4}\|\mathbf{v}\|^2 = \frac{1}{4}T^2(\mathbf{u})\|\mathbf{u}\|^2 \geq \frac{1}{4}\|\mathbf{u}\|^2 = \Phi(\mathbf{u}).$$

According to Proposition 11.3.1, $\mathbf{u}_1$ is a local minimum of Φ on $\mathcal{M}$, and thus

$$\Phi^+(\mathbf{v}) \geq \Phi(\mathbf{u}) \geq \Phi(\mathbf{u}_1) = \Phi^+(\mathbf{u}_1) \quad \forall \mathbf{v} \in W_\varepsilon,$$

proving that $\mathbf{u}_1$ is a local strict minimum for Φ^+ on $\mathcal{M}^+$. A similar proof can be carried out for $\mathbf{u}_2$.

From the preceding arguments, it follows that Φ^+ has a mountain-pass critical point $\mathbf{u}^* \in \mathcal{M}^+$, which gives rise to a solution of

$$\begin{cases} -\Delta u_1 + \lambda_1 u_1 = \mu_1\big(u_1^+\big)^3 + \beta\big(u_2^+\big)^2 u_1^+, & u_1 \in W^{1,2}(\mathbb{R}^n), \\ -\Delta u_2 + \lambda_2 u_2 = \mu_2\big(u_2^+\big)^3 + \beta\big(u_1^+\big)^2 u_2^+, & u_2 \in W^{1,2}(\mathbb{R}^n). \end{cases} \tag{11.54}$$

In particular, one finds that $u_j^* \geq 0$. In addition, since $\mathbf{u}^*$ is a mountain-pass critical point, one has $\Phi^+(\mathbf{u}^*) > \max\{\Phi(\mathbf{u}_1), \Phi(\mathbf{u}_2)\}$. Let us also remark that $\mathbf{u}^* \in \mathcal{M}^+$ implies that $\mathbf{u}^* \neq \mathbf{0}$ and hence $u_2^* \equiv 0$ implies that $u_1^* \not\equiv 0$. Now we can argue as in the proof of Theorem 11.3.1. From $\Phi'(u_1^*, 0) = 0$, it follows that u_1^* is a non-trivial solution of

$$-\Delta u + \lambda_1 u = \mu_1 u_+^3, \quad u \in H.$$

Since $u_1^* \geq 0$ and $u_1^* \not\equiv 0$, it follows that $u_1^* = U_1$, namely $\mathbf{u}^* = (U_1, 0) = \mathbf{u}_1$. This is in contradiction to $\Phi^+(\mathbf{u}^*) > \Phi(\mathbf{u}_1)$, proving that $u_2^* \not\equiv 0$. A similar argument proves that $u_1^* \not\equiv 0$. Since both u_1^* and u_2^* are never 0, applying the maximum principle to each equation in (11.54), we get $u_1^* > 0$ and $u_2^* > 0$. $\square$

(2) Bartsch, Dancer and Wang [32] studied bifurcation structure

$$\begin{cases} -\Delta u + \lambda_1 u = \mu_1 u^3 + \beta v^2 u, \\ -\Delta v + \lambda_2 v = \mu_2 v^3 + \beta u^2 v, \\ u, v > 0 \quad \text{in } \Omega, \quad u, v \in H_0^1(\Omega) \end{cases} \tag{11.55}$$

in a possible unbounded domain $\Omega \subset \mathbb{R}^n$, $n \leq 3$.

As $\lambda_1 = \lambda_2$ (may assume $\lambda_1 = \lambda_2 = 1$), fixing $\mu_1, \mu_2 > 0$, $n = 1$, Ω can be bounded or unbounded; if $n = 2$ or $n = 3$, the domains Ω are bounded or radially symmetric (possibly unbounded). If $w \in H_0^1(\Omega)$ is a solution of

$$-\Delta w + w = w^3, \quad w > 0 \text{ in } \Omega$$

then a direct calculation shows that for $\beta \in (-\sqrt{\mu_1 \mu_2}, \mu_1) \cup (\mu_2, \infty)$ the pair

$$u_\beta = \left(\frac{\mu_2 - \beta}{\mu_1 \mu_2 - \beta^2} \right)^{1/2} w, \qquad v_\beta = \left(\frac{\mu_1 - \beta}{\mu_1 \mu_2 - \beta^2} \right)^{1/2} w$$

solves (11.55) (as $\lambda_1 = \lambda_2 = 1$).

If $\mu_1 = \mu_2 := \mu$ this simplifies to

$$u_\beta = v_\beta = \left(\frac{1}{|\mu + \beta|} \right)^{1/2} w$$

which is defined for $\beta \neq -\mu$. Thus if $0 < \mu_1 < \mu_2$ we have a "trivial" branch of (11.55).

$$\mathcal{T}_w := \left\{ (\beta, u_\beta, v_\beta) \in \mathbb{R} \times H_0^1(\Omega) \times H_0^1(\Omega) : \beta \in (-\sqrt{\mu_1 \mu_2}, \mu_1) \cup (\mu_2, \infty) \right\}.$$

They proved bifurcation of non-trivial solutions from this branch: There are infinitely many bifurcation points along this trivial branch; in case $n = 1$ or Ω radially symmetric, the bifurcation branches are global and unbounded to the left in the β-direction.

(3) Dancer, Wei and Weth [73] studied

$$\begin{cases} -\Delta u + \lambda_1 u = \mu_1 u^3 + \beta v^2 u, \\ -\Delta v + \lambda_2 v = \mu_2 v^3 + \beta u^2 v, \\ u, v > 0 \quad \text{in } \Omega, \quad u, v \in H_0^1(\Omega) \end{cases} \tag{11.56}$$

in a smooth bounded domain $\Omega \subset \mathbb{R}^n$, $n \leq 3$ with coupling parameter $\beta \in \mathbb{R}$. They show that the value $\beta = -\sqrt{\mu_1 \mu_2}$ is critical for the existence of a priori bounds for solution of (11.56). More precisely, they show that for $\beta > -\sqrt{\mu_1 \mu_2}$, solutions of (11.56) are a priori bounded. In contrast, when $\lambda_1 = \lambda_2$, $\mu_1 = \mu_2$, (11.56) admits an unbounded sequence of solutions if $\beta \leq -\sqrt{\mu_1\mu_2}$.

(4) Liu and Wang [141] considered

$$-\Delta u_j + \lambda_j u_j = \sum_{i=1}^{N} \beta_{ij} u_i^2 u_j \quad \text{in } \mathbb{R}^n, \qquad u_j(x) \to 0 \quad \text{as } |x| \to \infty,$$

$$j = 1, 2, \ldots, N. \tag{11.57}$$

λ_j and β_{jj} are positive constants for all j, $\beta_{ij} = \beta_{ji}$, $n = 2, 3$, $N \geq 2$.

Theorem 11.3.3 (Liu and Wang [141]) *Let $n = 2, 3$ and let λ_j and β_{jj} are fixed positive constants. Then for $k \in \mathbb{N}$, there exists $\beta_k > 0$ such that for $|\beta_{ij}| \leq \beta_k$, $i \neq j$, the system (11.57) has at least k pairs of non-trivial spherically symmetric solutions.*

Proof Using invariant sets, it is proved. See [141]. □

Let $N = 2$, rewrite $\mu_i = \beta_{ii}$, $\beta = \beta_{12}$

$$\begin{cases} -\Delta u + \lambda_1 u = \mu_1 u^3 + \beta v^2 u & \text{in } \mathbb{R}^n, \\ -\Delta v + \lambda_2 v = \mu_2 v^3 + \beta u^2 v & \text{in } \mathbb{R}^n, \\ u(x) \to 0, \quad v(x) \to 0, \quad \text{as } |x| \to \infty. \end{cases} \tag{11.58}$$

Theorem 11.3.4 (Liu and Wang [141]) *Let $n = 2, 3$ and for $i = 1, 2$ let λ_i and μ_i be fixed positive constants. Then for any $k \in \mathbb{N}$, there exists $\beta^k > 0$ such that for $\beta > \beta^k$ the system (11.58) has at least k pairs of non-trivial spherically symmetric solutions.*

Proof Using minimax procedure to distinguish solutions by analyzing their Morse indices, they proved. See [141]. □

11.3.2 The Limit State of Schrödinger Systems

In recent years, the following systems with $\kappa \to +\infty$:

$$-d_i \Delta u_i = f_i(u_i) - \kappa u_i \sum_{j \neq i} b_{ij} u_j^2 \quad \text{in } B_1(0), \tag{11.59}$$

and the parabolic analogue of (11.59):

$$\frac{\partial u_i}{\partial t} - d_i \Delta u_i = f_i(u_i) - \kappa u_i \sum_{j \neq i} b_{ij} u_j^2 \quad \text{in } B_1(0) \times (-1, 0) \tag{11.60}$$

has received a lot of attention. Here $B_1(0)$ is the open unit ball of $\mathbb{R}^n$, $b_{ij} > 0$ and $d_i > 0$ are equal constants, satisfying $b_{ij} = b_{ji}$, $1 \leq i, j \leq M$. $B_1(0)$ is the unit ball in $\mathbb{R}^n$ ($n = 1, 2, 3$). A typical model for $f_i(u_i)$ is $f_i(u) = a_i u - u^p$ with constants $a_i > 0$, $p > 1$.

Uniformly bounded solutions (we assume uniformly bounded in Theorem 11.3.8) of the above system (11.59) are critical points of the following functional (under suitable boundary conditions, e.g. $u_i \in H_0^1(B_1(0))$): $\forall u \in H_0^1(B_1(0)) \times H_0^1(B_1(0)) \times \cdots \times H_0^1(B_1(0))$:

$$J_\kappa(u) = \int_{B_1(0)} \frac{1}{2} \sum_i d_i |\nabla u_i|^2 + \frac{\kappa}{4} \sum_{i \neq j} b_{ij} u_i^2 u_j^2 - \sum_i F_i(u_i), \tag{11.61}$$

where $F_i(u) = \int_0^u f_i(t) \, dt$.

(1) Terracini and Verzini [178] seek radial solutions to the system of elliptic equations:

$$\begin{cases} -\Delta u_i + u_i = u_i^3 - \beta u_i \sum_{j \neq i} u_j^2, & i = 1, 2, \ldots, k \\ u_i \in H^1(\mathbb{R}^n), & u_i > 0, \end{cases} \tag{11.62}$$

with $n = 2, 3$, $k \geq 3$ and β (positive and) large, in connection with the changing-sign solutions of the scalar equation

$$-\Delta W + W = W^3, \quad W \in H^1(\mathbb{R}^n). \tag{11.63}$$

Bartsch and Willem [27] showed that for any $h \in \mathbb{N}$, (11.63) possesses radial solutions with exactly $h - 1$ changes of sign, that is, h nodal components ("bumps"), with a variational characterization.

Wei and Weth [191] have shown that $k = 2$, there are solutions (u_1, u_2) such that the difference $u_1 - u_2$ for large values of β, approaches some sign-changing solution W of (11.63).

Terracini and Verzini [178] showed as $k \geq 3$, β large, each component u_i is near the sum of some non-consecutive bumps of $|W|$.

(2) B. Noris, H. Tavares, S. Terracini, and G. Verzini [148] studied

$$\begin{cases} -\Delta u + u^3 + \beta v^2 u = \lambda u, \\ -\Delta v + v^3 + \beta u^2 v = \mu v, \\ u, v > 0 \quad \text{in } \Omega, \quad u, v \in H_0^1(\Omega), \end{cases} \tag{11.64}$$

with Ω a smooth bounded domain of $\mathbb{R}^n$, $n = 2, 3$ and the related scalar equation

$$-\Delta w + w^3 = \lambda w^+ - \mu w^-. \tag{11.65}$$

Open Problem (Question and conjecture of Terracini et al., see [148]) Is it true that every bounded family (u_β, v_β) of solutions of the system (11.64) converges, as $\beta \to +\infty$, up to a subsequence, to a pair (u_∞, v_∞), where $u_\infty - v_\infty$ solves the scalar equation (11.65)?

S. Terracini et al. conjectured that it is true.
Let

$$J_\beta(u, v) = \frac{1}{2}\left(\|u\|^2 + \|v\|^2\right) + \frac{1}{4}\int_\Omega \left(u^4 + v^4\right) dx + \frac{\beta}{2}\int_\Omega u^2 v^2 \qquad (11.66)$$

be constrained to the manifold

$$M = \left\{(u, v) \in H_0^1(\Omega) \times H_0^1(\Omega) : \int_\Omega u^2 \, dx = \int_\Omega v^2 \, dx = 1\right\}$$

so that λ and μ in (11.64) can be understood as Lagrange multipliers.
Define

$$J_\infty(u, v) = \sup_{\beta > 0} J_\beta(u, v) = \begin{cases} J_0(u, v), & \text{as } \int_\Omega u^2 v^2 \, dx = 0, \\ +\infty, & \text{otherwise.} \end{cases}$$

Theorem 11.3.5 (Noris, Tavares, Terracini and Verzini [148]) *Let $(u_\beta, v_\beta) \in M$, for $\beta \in (0, +\infty)$, be a minimizer of J_β constrained to M. Then, up to subsequences, $(u_\beta, v_\beta) \to (u_\infty, v_\infty)$, strongly in $H^1 \cap C^{0,\alpha}$, minimizer of J_∞ constrained to M. Moreover, $u_\infty - v_\infty$ solves (11.65).*

Recall *Lotka–Volterra competing system from population dynamics.*
There is another system, which has a similar form as (11.60):

$$\frac{\partial u_i}{\partial t} - d_i \Delta u_i = f_i(u_i) - \kappa u_i \sum_{j \neq i} b_{ij} u_j \quad \text{in } B_1(0) \times (-1, 0). \qquad (11.67)$$

This is the Volterra–Lotka competing model from population dynamics.

(This system has complex dynamics for multiple species, Smale [166]. *No variational structure.*)

We can prove (for example, see M. Conti, S. Terracini and G. Verzini [58]), as $\kappa \to +\infty$, uniformly bounded solutions of (11.59), (11.60) or (11.67) converge to a limiting configurations in some weak sense, $(u_1, u_2, \ldots, u_M)$. The limit satisfies a separation condition, that is, different components have separated supports:

$$u_i u_j \equiv 0, \quad \text{for } i \neq j.$$

For (11.67), we can get more: the limit $(u_1, u_2, \ldots, u_M)$ satisfies a remarkable system of differential inequalities:

$$\begin{cases} \dfrac{\partial u_i}{\partial t} - d_i \Delta u_i \leq f_i(u_i) \quad \text{in } B_1(0) \times (-1,0), \\[2mm] \left(\dfrac{\partial}{\partial t} - d_i \Delta\right) u_i - \sum_{j \neq i}\left(\dfrac{\partial}{\partial t} - d_j \Delta\right) u_j \geq f_i(u_i) - \sum_{j \neq i} f_j(u_j) \\[2mm] \quad \text{in } B_1(0) \times (-1,0), \\[2mm] u_i \geq 0 \quad \text{in } B_1(0) \times (-1,0), \\[2mm] u_i u_j = 0, \quad \text{for } i \neq j \text{ in } B_1(0) \times (-1,0). \end{cases} \tag{11.68}$$

Similar results also hold for the elliptic case. That is, the singular limit as $\kappa \to +\infty$ is

$$\begin{cases} -d_i \Delta u_i \leq f_i(u_i) & \text{in } B_1(0), \\[2mm] -d_i \Delta u_i + \sum_{j \neq i} d_j \Delta u_j \geq f_i(u_i) - \sum_{j \neq i} f_j(u_j) & \text{in } B_1(0), \\[2mm] u_i \geq 0 & \text{in } B_1(0), \\[2mm] u_i u_j = 0 \quad \text{for } i \neq j, & \text{in } B_1(0). \end{cases} \tag{11.69}$$

Starting with the work of Dancer and Zhang [72], there has arisen a lot of interest in the dynamics of (11.67), especially in the case of large κ (for example, see Dancer, Wang and Zhang [75]). The general principle in Dancer and Zhang [72] is, if we can prove that the singular limit of (11.67) has simple dynamics, then we can prove the dynamics of (11.67) with κ large is simple too.

It is natural to try to use similar ideas for (11.60). In order to achieve this, we will prove that the singular limit of (11.60) is (11.68), too. This conjecture (in the elliptic case) was first proposed in [148] and was studied in [177] by Hugo Tavares and Susanna Terracini.

The main purpose of our results [76] is to cover and to prove the conjecture of Terracini et al. This result is much harder to prove than the result for (11.67).

(3) Tavares–Terracini's related main results (see [177]).

Theorem 11.3.6 (Tavares and Terracini [177]) *Let $U = (u_1, \ldots, u_h) \in (H^1(\Omega)^h)$ be a vector of non-negative Lipschitz functions in $\Omega \subset \mathbb{R}^n, n \geq 2$, having mutually disjoint supports: $u_i \cdot u_j \equiv 0$ in Ω for $i \neq j$. Assume that $U \not\equiv 0$ and*

$$-\Delta u_i = f_i(x, u_i) \quad \text{whenever } u_i > 0, \ i = 1, \ldots, h,$$

where $f_i : \Omega \times \mathbb{R}^+ \to \mathbb{R}$ are C^1 functions such that $f_i(x, s) = O(s)$ when $s \to 0$, uniformly in x. Moreover, defining for every $x_0 \in \Omega$ and $r \in (0, \text{dist}(x_0, \partial\Omega))$ the energy

$$\tilde{E}(r) = \tilde{E}(x_0, U, r) = \frac{1}{r^{n-2}} \int_{B_r(x_0)} |\nabla U|^2,$$

assume that $\tilde{E}(x_0, U, \cdot)$ is an absolutely continuous function of r and that it satisfies the following differential equation:

$$\frac{d}{dr}\tilde{E}(x_0, U, r) = \frac{2}{r^{n-2}}\int_{\partial B_r(x_0)}(\partial_\nu U)^2\,d\sigma$$

$$+ \frac{2}{r^{n-2}}\int_{\partial B_r(x_0)}\sum_i f_i(x, u_i)\langle\nabla u_i, x - x_0\rangle.$$

Let us consider the nodal set $\Gamma_U = \{x \in \Omega : U(x) = 0\}$. Then we have $\mathcal{H}_{\dim}(\Gamma_U) \le n - 1$. Moreover there exists a set $\Sigma_U \subseteq \Gamma_U$, relatively open in Γ_U, such that

- *$\mathcal{H}_{\dim}(\Gamma_U \setminus \Sigma_U) \le n - 2$, and if $n = 2$ then actually $\Gamma_U \setminus \Sigma_U$ is a locally finite set;*
- *Σ_U is a collection of hypersurfaces of class $C^{1,\alpha}$ (for every $0 < \alpha < 1$). Furthermore for every $x_0 \in \Sigma_U$*

$$\lim_{x \to x_0^+}\left|\nabla U(x)\right| = \lim_{x \to x_0^-}\left|\nabla U(x)\right| \ne 0,$$

where the limits as $x \to x_0^{\pm}$ are taken from the opposite sides of the hypersurface.

Furthermore, if $n = 2$ then Σ_U consists of a locally finite collection of curves meeting with equal angles at singular points.

Notations For any vector function $U = (u_1, \ldots, u_h)$, define $\nabla U = (\nabla u_1, \ldots, \nabla u_h)$, $|\nabla U|^2 = |\nabla u_1|^2 + \cdots + |\nabla u_h|^2$, $(\partial_\nu U)^2 = (\partial u_1)^2 + \cdots + (\partial u_h)^2$ and $U^2 = u_1^2 + \cdots + u_h^2$. $F(x, U) = (f_1(x, u_1), \ldots, f_h(x, u_h))$. Denote by $\{U > 0\}$ the set $\{x \in \Omega : u_i(x) > 0 \text{ for some } i\}$.

Definition 11.3.2 (Tavares and Terracini [177]) Define the class $\mathcal{G}(\Omega)$ as the set of functions $U = (u_1, \ldots, u_h) \in (H^1(\Omega))^h$, whose components are all non-negative and Lipschitz continuous in the interior of Ω, and such that $u_i \cdot u_j \equiv 0$ in Ω for $i \ne j$. Moreover, $U \not\equiv 0$ and it solves a system of the type

$$-\Delta u_i = f_i(x, u_i) - \mu_i \quad \text{in } \mathcal{D}'(\Omega) = \left(C_c^\infty(\Omega)\right)', \; i = 1, \ldots, h, \tag{11.70}$$

where

(G_1) $f_i : \Omega \times \mathbb{R}^+ \to \mathbb{R}$ are C^1 functions such that $f_i(x, s) = O(s)$ when $s \to 0$, uniformly in x.

(G_2) $\mu_i \in \mathcal{M}(\Omega) = (C_0(\Omega))'$ are some non-negative Radon measures, each supported on the nodal set $\Gamma_U = \{x \in \Omega : U(x) = 0\}$, and moreover

(G_3) associated to system (11.70), define for every $x_0 \in \Omega$ and $r \in (0, \text{dist}(x_0, \partial\Omega))$ the energy

$$\tilde{E}(r) = \tilde{E}(x_0, U, r) = \frac{1}{r^{n-2}}\int_{B_r(x_0)}|\nabla U|^2,$$

assume that $\tilde{E}(x_0, U, \cdot)$ is an absolutely continuous function of r and that it satisfies the following differential equation:

$$\frac{d}{dr}\tilde{E}(x_0, U, r) = \frac{2}{r^{n-2}}\int_{\partial B_r(x_0)}(\partial_\nu U)^2 \, d\sigma$$
$$+ \frac{2}{r^{n-2}}\int_{\partial B_r(x_0)}\sum_i f_i(x, u_i)\langle \nabla u_i, x - x_0\rangle.$$

Next for nonlinear Schrödinger equations

$$\begin{cases} -\Delta u_i + \lambda_i u_i = \omega_i u_i^3 - \beta u_i \sum_{j \neq i} \beta_{i,j} u_j^2, & i = 1, \ldots, h, \\ u_i \in H_0^1(\Omega), \quad u_i > 0 \quad \text{in } \Omega, \end{cases} \tag{11.71}$$

with smooth bounded domain $\Omega \subset \mathbb{R}^n$, $n = 2, 3$, $\beta_{ij} = \beta_{ji} \neq 0$ which give a variational structure to the problem. For $\lambda_i, \omega_i \in \mathbb{R}$, $\beta \in (0, +\infty)$ large (the existence of solutions for β large is still an open problem for some choices of λ_i, ω_i), we have

Theorem 11.3.7 (Tavares and Terracini [177]) *Let U be a limit as $\beta \to +\infty$ of a family $\{U_\beta\}$ of L^∞ bounded solution of* (11.71). *Then $U \in \mathcal{G}(\Omega)$.*

(4) Caffarelli and Lin [43] proved corresponding results to Theorem 11.3.6 (Tavares–Terracini) for minimizers of the functional corresponding to (11.71).

There are many difficulties compared with the minimizing case. Hugo Tavares and Susanna Terracini have developed remarkable methods to attack this problem in the elliptic case, and they have proved a good deal of this result.

(5) We complete the roof of the conjecture in both the elliptic and parabolic case. We mainly consider the parabolic case. The elliptic case is almost the same (in fact, a little easier than the parabolic case in some places). We also solve the important case left open in Tavares–Terracini's result on the elliptic case (it is a gap in their proof). That is to exclude the possibility of the existence of multiplicity one points. Simply stated, it is impossible to have one component of U, say u_1, vanishing on a locally smooth hypersurface, where u_1 is strictly positive in a deleted neighborhood of this hypersurface. This seems fundamental to prove that (11.69) is the limit of (11.59).

Assumption

$$\int_{B_{1/2}(0)}|u(x, 0)|^2 \, dx > 0. \tag{11.72}$$

This assumption can be guaranteed, for example, if the solution u is in a fixed bounded smooth domain with zero Dirichlet boundary condition. (This can be proved by the backward uniqueness, using the classical method, the log-convexity of $\int_{B_{1/2}(0)}|u|^2 \, dx$.)

Our main result is

Theorem 11.3.8 (Dancer, Wang and Zhang [76]) *As* $\kappa \to +\infty$, *if a sequence of bounded solutions* $u_\kappa = (u_{1,\kappa}, u_{2,\kappa}, \ldots, u_{M,\kappa})$ *of* (11.60) *converges to* $u = (u_1, u_2, \ldots, u_M)$, *then u satisfies the system* (11.68).

Proof Please see [76]. $\qquad\qquad\qquad\qquad\qquad\qquad\qquad\qquad\qquad\qquad\qquad\qquad$ $\square$

Since we assume a uniform bound on u_κ, we can prove the uniform Hölder continuity of u_κ (see Sect. 11.3.3 below) with respect to the parabolic distance

$$d\big((x, t), (y, s)\big) := \max\big\{|t - s|, |x - y|^2\big\}^{\frac{1}{2}}.$$

Thus, without loss of generality, we can assume u_κ converges to u, uniformly on $Q_1(0)$. We denote $X = (x, t)$, parabolic ball $Q_r(X) = B_r(x) \times (t - r^2, t)$, parabolic cylinder $P_r(X) = B_r(x) \times (t - r^2, t + r^2)$ and $Q_r = Q_r(0)$, $B_r = B_r(0)$. We use u, v denoting the vector valued function, $(u_1, u_2, \ldots, u_M)$, $(v_1, v_2, \ldots, v_M)$ and so on. By saying a sequence of $\kappa \to +\infty$, we always means a subsequence of $\kappa_i \to +\infty$.

11.3.3 C^α *Estimate of the Solutions of Parabolic Systems*

We establish the uniform C^α bound of the solutions for the following two parabolic systems for $\kappa \in (0, +\infty)$:

$$\begin{cases} \dfrac{\partial u_i}{\partial t} - d_i \Delta u_i = f_i(u_i) - \kappa u_i \displaystyle\sum_{j \neq i} b_{ij} u_j, & \text{in } \Omega \times (0, +\infty), \\[2ex] u_i = \varphi_i, & \text{on } \partial\Omega \times (0, +\infty), \\[1ex] u_i = \phi_i, & \text{on } \Omega \times \{0\}, \end{cases} \tag{11.73}$$

and

$$\begin{cases} \dfrac{\partial u_i}{\partial t} - d_i \Delta u_i = f_i(u_i) - \kappa u_i \displaystyle\sum_{j \neq i} b_{ij} u_j^2, & \text{in } \Omega \times (0, +\infty), \\[2ex] u_i = \varphi_i, & \text{on } \partial\Omega \times (0, +\infty), \\[1ex] u_i = \phi_i, & \text{on } \Omega \times \{0\}. \end{cases} \tag{11.74}$$

Here $\Omega \subset \mathbb{R}^n$ ($n \geq 1$) is an open bounded domain with smooth boundary, $i, j = 1, 2, \ldots, K$, $K \geq 2$; $d_i > 0$ and $b_{ij} > 0$ are constants (in (11.74)), we also assume the symmetric condition $b_{ij} = b_{ji}$. φ_i are given non-negative Lipschitz continuous functions on $\partial\Omega \times (0, +\infty)$; and ϕ_i are given non-negative Lipschitz continuous functions on Ω, which satisfy $\phi_i(x) = \varphi_i(x, 0)$ for $x \in \partial\Omega$. They also satisfy the segregated property $\phi_i \phi_j = 0$ and $\varphi_i \varphi_j = 0$ for $i \neq j$. f_i are given Lipschitz functions, that is, $\exists C > 0$,

$$\big|f_i(u) - f_i(v)\big| \leq C|u - v|.$$

The uniform Hölder regularity in related problems have been studied by many authors.

In [58, 147], Susanna Terracini et al. proved the uniform Hölder regularity of solutions to the elliptic analogue of (11.73) and (11.74). Although they only state the result for dimension $n \leq 3$, it is essentially true in any dimension, as pointed out in their paper.

In [41], Caffarelli, Karakhanyan and F. Lin also proved these estimate for (11.73), both in the elliptic case and the parabolic case (see also [38]). However, their result is a local one, only concerning the interior regularity. We will prove a global result and the proof is different from the one in [41]. In fact, our method mainly follows the blow-up method, developed by Susanna Terracini and her coauthors in [58, 147]. This method is a blow-up analysis and needs us to prove some Liouville type theorems. This can be achieved by some monotonicity formulas of Alt–Caffarelli–Friedman type.

With minor assumptions on f_i (for example, if we take the classical logistic model $f_i(u) = a_i u - u^2$), for fixed κ, the existence of global solutions u_κ of both systems (11.73) and (11.74) can be guaranteed. Moreover, u_κ are non-negative and Lipschitz continuous on $\overline{\Omega} \times [0, +\infty)$ (but the Lipschitz constants may depend on κ). We also assume that $\exists C > 0$ independent of κ, such that $\sum_i u_{i,\kappa} \leq C$.

The main result is the following uniform regularity result:

Theorem 11.3.9 (Dancer, Wang and Zhang [74]) *For any $\alpha \in (0, 1)$, there exists a constant C_α independent of κ, such that if u_κ is a solution of (11.73) or (11.74), then*

$$\max_i \sup_{\Omega \times (0,+\infty)} \frac{|u_{i,\kappa}(x,t) - u_{i,\kappa}(y,s)|}{d^\alpha((x,t),(y,s))} \leq C_\alpha.$$

Proof The proof relies upon the blow-up technique and the monotonicity formula by Almgren and Alt, Caffarelli, and Friedman. Please see [74]. $\square$

Erratum to: Preliminaries

**Erratum to: Z. Zhang, *Variational, Topological, and
Partial Order Methods with Their Applications*,
Developments in Mathematics 29, pp. 1–33,
DOI 10.1007/978-3-642-30709-6_1,
© Springer-Verlag Berlin Heidelberg 2013**

The presentation of Definition 1.6.2 was incorrect. The correct version is given below.

Definition 1.6.2 Assume $L \in \mathcal{F}(X, Y)$, let

$$\mathrm{ind}(L) \triangleq \dim \mathrm{Ker}\, L - \dim \mathrm{Coker}\, L,$$

it is called the index of L.

The online version of the original chapter can be found at doi: 10.1007/978-3-642-30709-6_1.

Z. Zhang, *Variational, Topological, and Partial Order Methods with Their Applications*,   E1
Developments in Mathematics 29,
DOI 10.1007/978-3-642-30709-6_12, © Springer-Verlag Berlin Heidelberg 2013

References

1. R. Adimurthi, S.L. Yadava, An elementary proof of the uniqueness of positive radial solutions of a quasilinear Dirichlet problem. Arch. Ration. Mech. Anal. **127**, 219–229 (1994)
2. R.P. Agarwal, K. Perera, Z. Zhang, On some nonlocal eigenvalue problems. Discrete Contin. Dyn. Syst., Ser. S **5**(4), 707–714 (2012)
3. H.W. Alt, L.A. Caffarelli, A. Friedman, Variational problems with two phases and their free boundaries. Trans. Am. Math. Soc. **282**, 431–461 (1984)
4. C.O. Alves, F.J.S.A. Correa, On existence of solutions for a class of problem involving a nonlinear operator. Commun. Appl. Nonlinear Anal. **8**(2), 43–56 (2001)
5. C.O. Alves, D.G. de Figueiredo, Nonvariational elliptic systems. Discrete Contin. Dyn. Syst. **8**, 289–302 (2002)
6. C.O. Alves, F.J.S.A. Correa, T.F. Ma, Positive solutions for a quasilinear elliptic equation of Kirchhoff type. Comput. Math. Appl. **49**(1), 85–93 (2005)
7. H. Amann, Multiple positive fixed points of asymptotically linear maps. J. Funct. Anal. **17**, 174–213 (1974)
8. H. Amann, Fixed point equations and nonlinear eigenvalue problems in ordered Banach spaces. SIAM Rev. **18**, 620–709 (1976)
9. H. Amann, E. Zehnder, Nontrivial solutions for a class of nonresonance problems and applications to nonlinear differential equations. Ann. Sc. Norm. Super. Pisa **7**, 539–603 (1980)
10. A. Ambrosetti, E. Colorado, Standing waves of some coupled nonlinear Schrödinger equations. J. Lond. Math. Soc. **75**(1), 67–82 (2007)
11. A. Ambrosetti, P.H. Rabinowitz, Dual variational methods in critical point theory and applications. J. Funct. Anal. **14**, 349–381 (1973)
12. A. Ambrosetti, H. Brezis, G. Cerami, Combined effects of concave and convex nonlinearities in some elliptic problems. J. Funct. Anal. **122**, 519–543 (1994)
13. T. Ando, On fundamental properties of a Banach space with a cone. Pac. J. Math. **12**, 1163–1169 (1962)
14. D. Andrade, T.F. Ma, An operator equation suggested by a class of nonlinear stationary problems. Commun. Appl. Nonlinear Anal. **4**(4), 65–71 (1997)
15. G. Anello, A uniqueness result for a nonlocal equation of Kirchhoff type and some related open problem. J. Math. Anal. Appl. **373**(1), 248–251 (2011)
16. M. Arias, J. Campos, M. Cuesta, J.-P. Gossez, Asymmetric elliptic problems with indefinite weights. Ann. Inst. Henri Poincaré, Anal. Non Linéaire **19**, 581–616 (2002)
17. A. Arosio, S. Panizzi, On the well-posedness of the Kirchhoff string. Trans. Am. Math. Soc. **348**(1), 305–330 (1996)
18. C. Azizieh, Ph. Clement, A priori estimates and continuation methods for positive solutions of p-Laplace equations. J. Differ. Equ. **179**, 213–245 (2002)

Z. Zhang, *Variational, Topological, and Partial Order Methods with Their Applications*, Developments in Mathematics 29, DOI 10.1007/978-3-642-30709-6, © Springer-Verlag Berlin Heidelberg 2013

19. J.G. Azorero, I.P. Alonso, Existence and non-uniqueness for the p-Laplacian: Nonlinear eigenvalues. Comm. PDE **12**, 1389–1430 (1987)

20. J.G. Azorero, I.P. Alonso, Multiplicity of solutions for elliptic problems with critical exponent or with a nonsymmetric term. Trans. Am. Math. Soc. **323**, 877–895 (1991)

21. J.G. Azorero, J.J. Manfredi, I.P. Alonso, Sobolev versus Holder local minimizer and global multiplicity for some quasilinear elliptic equations. Commun. Contemp. Math. **2**, 385–404 (2000)

22. I.J. Bakelman, Variational problems and elliptic Monge–Ampère equations. J. Differ. Geom. **18**(4), 669–699 (1983)

23. C. Bandle, *Isoperimetric Inequalities and Applications* (Pitman, Boston, 1980)

24. T. Bartsch, M. Degiovanni, Nodal solutions of nonlinear elliptic Dirichlet problems on radial domains. Atti Accad. Naz. Lincei, Rend. Lincei, Mat. Appl. **17**, 69–85 (2006)

25. T. Bartsch, S. Li, Critical point theory for asymptotically quadratic functionals and applications to problems with resonance. Nonlinear Anal. **28**(3), 419–441 (1997)

26. T. Bartsch, T. Weth, A note on additional properties of sign changing solutions to superlinear elliptic equations. Topol. Methods Nonlinear Anal. **22**(1), 1–14 (2003)

27. T. Bartsch, M. Willem, Infinitely many radial solutions of a semilinear elliptic problem on $\mathbb{R}^N$. Arch. Ration. Mech. Anal. **124**(3), 261–276 (1993)

28. T. Bartsch, K.C. Chang, Z. Wang, On the Morse indices of sign-changing solutions of nonlinear elliptic problems. Math. Z. **233**, 655–677 (2000)

29. T. Bartsch, A. Pankov, Z.-Q. Wang, Nonlinear Schrödinger equations with steep potential well. Commun. Contemp. Math. **3**, 549–569 (2001)

30. T. Bartsch, T. Weth, M. Willem, Partial symmetry of least energy nodal solutions to some variational problems. J. Anal. Math. **96**, 1–18 (2005)

31. T. Bartsch, Z. Wang, Z. Zhang, On the Fučik point spectrum for Schrödinger operators on R^n. Fixed Point Theory Appl. **5**(2), 305–317 (2009)

32. T. Bartsch, E.N. Dancer, Z.-Q. Wang, A Liouville theorem, a-priori bounds, and bifurcating branches of positive solutions for a nonlinear elliptic system. Calc. Var. Partial Differ. Equ. **37**(3–4), 345–361 (2010)

33. S. Bernstein, Sur une classe d'équations fonctionnelles aux dérivées partielles. Izv. Akad. Nauk SSSR, Ser. Mat. **4**, 17–26 (1940)

34. B. Bollobas, *Linear Analysis: An Introduction Course* (Cambridge University Press, Cambridge, 1999)

35. H. Brezis, L. Nirenberg, Remarks on finding critical points. Commun. Pure Appl. Math. **64**, 939–963 (1991)

36. H. Brezis, L. Nirenberg, H^1 versus C^1 local minimizers. C. R. Math. Acad. Sci. Paris **317**, 465–472 (1993)

37. N.P. Cac, On nontrivial solutions of a Dirichlet problem whose jumping nonlinearity crosses a multiple eigenvalue. J. Differ. Equ. **80**, 379–404 (1989)

38. L.A. Cafarelli, J.M. Roquejorffre, Uniform Hölder estimates in a class of elliptic systems and applications to singular limits in models for diffusion flames. Arch. Ration. Mech. Anal. **183**(3), 457–487 (2007)

39. L. Caffarelli, B. Gidas, J. Spruck, Asymptotic symmetry and local behavior of semilinear elliptic equations with critical Sobolev growth. Commun. Pure Appl. Math. **42**, 217–297 (1989)

40. L.A. Caffarelli, L. Karp, H. Shahgholian, Regularity of a free boundary with applications to the Pompeiu problem. Ann. Math. **151**, 269–292 (2000)

41. L.A. Caffarelli, A.L. Karakhanyan, F. Lin, The geometry of solutions to a segregation problem for non-divergence systems. Fixed Point Theory Appl. **5**(2), 319–351 (2009)

42. L.A. Cafferelli, F. Lin, An optimal partition problems for eigenvalues. J. Sci. Comput. **31**(1/2), 5–18 (2007)

43. L.A. Cafferelli, F. Lin, Singularly perturbed elliptic systems and multi-valued harmonic functions with free boundaries. J. Am. Math. Soc. **21**(3), 847–862 (2008)

44. A. Castro, J. Cossio, J.M. Neuberger, A sign-changing solution for a superlinear Dirichlet problem. Rocky Mt. J. Math. **27**, 1041–1053 (1997)
45. A. Castro, J. Cossio, J.M. Neuberger, A minimax principle, index of the critical point, and existence of sign-changing solutions to elliptic boundary value problems. Electron. J. Differ. Equ. **2**, 1–18 (1998)
46. M.M. Cavalcanti, V.N. Domingos Cavalcanti, J.A. Soriano, Global existence and uniform decay rates for the Kirchhoff–Carrier equation with nonlinear dissipation. Adv. Differ. Equ. **6**(6), 701–730 (2001)
47. G. Cerami, An existence criterion for the critical points on unbounded manifolds. Rend. - Ist. Lomb., Accad. Sci. Lett. A **112**(2), 332–336 (1978)
48. K.C. Chang, Solutions of asymptotically linear operator equations via Morse theory. Commun. Pure Appl. Math. **34**, 693–712 (1981)
49. K.-C. Chang, *Infinite Dimensional Morse Theory and Multiple Solutions Problems* (Birkhäuser, Basel, 1993)
50. K.-C. Chang, *Methods in Nonlinear Analysis* (Springer, Berlin, 2005)
51. K.-C. Chang, S. Li, J. Liu, Remarks on multiple solutions for asymptotically linear elliptic boundary value problems. Topol. Methods Nonlinear Anal. **3**, 179–187 (1994)
52. X. Cheng, C. Zhong, Existence of positive solutions for a second-order ordinary differential system. J. Math. Anal. Appl. **312**, 14–23 (2005)
53. M. Chipot, B. Lovat, Some remarks on nonlocal elliptic and parabolic problems. Nonlinear Anal. **30**(7), 4619–4627 (1997)
54. M. Chipot, J.-F. Rodrigues, On a class of nonlocal nonlinear elliptic problems. Modél. Math. Anal. Numér. **26**(3), 447–467 (1992)
55. P. Clement, D.G. de Figueiredo, E. Mitidieri, Positive solutions of semilinear elliptic systems. Commun. Partial Differ. Equ. **17**, 923–940 (1992)
56. M. Conti, S. Terracini, G. Verzini, Neharis problem and competing species systems. Ann. Inst. Henri Poincaré, Anal. Non Linéaire **19**(6), 871–888 (2002)
57. M. Conti, S. Terracini, G. Verzini, An optimal partition problem related to nonlinear eigenvalues. J. Funct. Anal. **198**, 160–196 (2003)
58. M. Conti, S.S. Terracini, G. Verzini, A variational problem for the spatial segregation of reaction diffusion systems. Indiana Univ. Math. J. **54**(3), 779–815 (2005)
59. M. Conti, S.S. Terracini, G. Verzini, Asymptotic estimates for the spatial segregation of competitive systems. Adv. Math. **195**(2), 524–560 (2005)
60. M.G. Crandall, P.H. Rabinowitz, Bifurcation from simple eigenvalues. J. Funct. Anal. **8**, 321–340 (1971)
61. M. Cuesta, Eigenvalue problems for the p-Laplacian with indefinite weights. Electron. J. Differ. Equ. **2001**, 33 (2001)
62. M. Cuesta, D. de Figueiredo, J.P. Gossez, The beginning of the Fučik spectrum for the p-Laplacian. J. Differ. Equ. **159**, 212–238 (1999)
63. L. Damascelli, Comparison theorems for some quasilinear degenerate elliptic operators and applications to symmetry and monotonicity results. Ann. Inst. Henri Poincaré **15**, 493–516 (1998)
64. E.N. Dancer, On the Dirichlet problem for weakly nonlinear elliptic partial differential equations. Proc. R. Soc. Edinb., Sect. A **76**, 283–300 (1977)
65. E.N. Dancer, On the number of positive solutions of weakly nonlinear elliptic equations when a parameter is large. Proc. Lond. Math. Soc. **53**(3), 429–452 (1986)
66. E.N. Dancer, Multiple solutions of asymptotically homogeneous problems. Ann. Mat. Pura Appl. **152**, 63–78 (1988)
67. E.N. Dancer, Remarks on jumping nonlinearities, in *Topics in Nonlinear Analysis*, ed. by Escher, Simonett (Birkhäuser, Basel, 1999), pp. 101–116
68. E.N. Dancer, Y. Du, Competing species equations with diffusion, large interactions, and jumping nonlinearities. J. Differ. Equ. **114**, 434–475 (1994)
69. E.N. Dancer, Y. Du, Existence of changing sign solutions for some semilinear problems with jumping nonlinearities at zero. Proc. R. Soc. Edinb., Sect. A **124**, 1165–1176 (1994)

70. E.N. Dancer, Y. Du, On sign-changing solutions of certain semilinear elliptic problems. Appl. Anal. **56**, 193–206 (1995)

71. E.N. Dancer, Z. Zhang, Fucik spectrum, sign-changing and multiple solutions for semilinear elliptic boundary value problems with resonance at infinity. J. Math. Anal. Appl. **250**, 449–464 (2000)

72. E.N. Dancer, Z. Zhang, Dynamics of Lotka–Volterra competition systems with large interaction. J. Differ. Equ. **182**(2), 470–489 (2002)

73. E.N. Dancer, J. Wei, T. Weth, A priori bounds versus multiple existence of positive solutions for a nonlinear Schrödinger system. Ann. Inst. Henri Poincaré, Anal. Non Linéaire **27**(3), 953–969 (2010)

74. E.N. Dancer, K. Wang, Z. Zhang, Uniform Hölder estimate for singularly perturbed parabolic systems of Bose–Einstein condensates and competing species. J. Differ. Equ. **251**, 2737–2769 (2011)

75. E.N. Dancer, K. Wang, Z. Zhang, Dynamics of strongly competing systems with many species. Trans. Am. Math. Soc. **364**(2), 961–1005 (2012)

76. E.N. Dancer, K. Wang, Z. Zhang, The limit equation for the Gross–Pitaevskii equations and S. Terracini's conjecture. J. Funct. Anal. **262**, 1087–1131 (2012)

77. P. D'Ancona, S. Spagnolo, Global solvability for the degenerate Kirchhoff equation with real analytic data. Invent. Math. **108**(2), 247–262 (1992)

78. D.G. de Figueiredo, J.-P. Gossez, On the first curve of the Fučik spectrum of an elliptic operator. Differ. Integral Equ. **7**, 1285–1302 (1994)

79. D.G. de Figueiredo, O. Lopes, Solitary waves for some nonlinear Schrödinger systems. Ann. Inst. Henri Poincaré, Anal. Non Linéaire **25**(1), 149–161 (2008)

80. D.G. de Figueiredo, P.L. Lions, R.D. Nussbaum, A priori estimates and existence of positive solutions of semilinear elliptic equations. J. Math. Pures Appl. **61**, 41–63 (1982)

81. K. Deimling, *Nonlinear Functional Analysis* (Springer, Berlin, 1985)

82. M.A. del Pino, R. Manasevich, Global bifurcation from the eigenvalues of the p-Laplacian. J. Differ. Equ. **92**, 226–251 (1991)

83. M.A. del Pino, M. Elguest, R. Manasevich, A homotopic deformation along p of a Leray–Schauder degree result and existence for $(|u'|^{p-2}u')' + f(t, u) = 0, u(0) = u(T) = 0, p > 1$. J. Differ. Equ. **80**, 1–13 (1989)

84. P. Delanoe, Radially symmetric boundary value problems for real and complex elliptic Monge-Ampère equations. J. Differ. Equ. **58**(3), 318–344 (1985)

85. E. Di Benedetto, $C^{1+\alpha}$ local regularity of weak solutions of degenerate elliptic equations. Nonlinear Anal. **7**, 827–850 (1983)

86. G. Dinca, P. Jebelean, J. Mawhin, Variational and topological methods for Dirichlet problems with p-Laplacian. Port. Math. **58**(3), 339–377 (2001)

87. P. Drábek, S.B. Robinson, Resonance problems for the p-Laplacian. J. Funct. Anal. **169**, 189–200 (1999)

88. Y. Du, Fixed points for a class of noncompact operators and applications. Acta Math. Sin. **32**, 618–627 (1989) (In Chinese)

89. Y. Du, Fixed points of increasing operators in ordered Banach spaces and applications. Appl. Anal. **38**, 1–20 (1990)

90. Y.H. Du, Z.M. Guo, Liouville type results and eventual flatness of positive solutions for p-Laplacian equations. Adv. Differ. Equ. **7**, 1479–1512 (2002)

91. Y. Egorov, V. Kondratiev, *On Spectral Theory of Elliptic Operators* (Birkhäuser, Basel, 1996)

92. S. Fučik, Boundary value problems with jumping nonlinearities. Čas. Pěst. Math. Fys. **101**, 69–87 (1976)

93. S. Fučik, *Solvability of Nonlinear Equations and Boundary Value Problems* (Reidel, Dordrecht, 1980)

94. B. Gidas, J. Spruck, A priori bounds for positive solutions of nonlinear elliptic equations. Commun. Partial Differ. Equ. **6**, 883–901 (1981)

95. D. Gilbarg, N.S. Trudinger, *Elliptic Partial Differential Equations of Second Order* (Springer, Berlin, 2001)

96. M. Guedda, L. Veron, Local and global properties of solutions of quasilinear elliptic equations. J. Differ. Equ. **76**, 159–189 (1988)

97. M. Guedda, L. Veron, Bifurcation phenomena associated to the p-Laplace operator. Trans. Am. Math. Soc. **310**, 419–431 (1988)

98. M. Guedda, L. Veron, Quasilinear elliptic equations involving critical Sobolev exponents. Nonlinear Anal., Theory Methods Appl. **13**(8), 879–902 (1989)

99. D. Guo, Positive solutions of integral equations of polynomial type and its applications. Chin. Ann. Math., Ser. A **4**(5), 645–656 (1983) (In Chinese)

100. D. Guo, *Nonlinear Functional Analysis* (Shandong Science and Technology Press, Jinan, 1985) (In Chinese)

101. D. Guo, Fixed points of mixed monotone operators with application. Appl. Anal. **34**, 215–224 (1988)

102. D. Guo, Existence and uniqueness of positive fixed point for mixed monotone operators and applications. Appl. Anal. **46**, 91–100 (1992)

103. Z. Guo, Some existence and multiplicity results for a class of quasilinear elliptic eigenvalue problems. Nonlinear Anal. **18**, 957–971 (1992)

104. Z. Guo, Existence and uniqueness of positive radial solutions for a class of quasilinear elliptic equations. Appl. Anal. **47**, 173–190 (1992)

105. Z. Guo, On the number of positive solutions for quasilinear elliptic eigenvalue problems. Nonlinear Anal. **27**, 229–247 (1996)

106. D. Guo, *The Method of Partial Ordering in Nonlinear Analysis* (Shandong Science and Technology Press, Jinan, 2000) (in Chinese)

107. Z. Guo, Structure of large positive solutions of some semilinear elliptic problems where the nonlinearity changes sign. Topol. Methods Nonlinear Anal. **18**, 41–71 (2001)

108. Z. Guo, Uniqueness and flat core of positive solutions for quasilinear elliptic eigenvalue problems in general smooth domains. Math. Nachr. **243**, 43–74 (2002)

109. Z. Guo, Structure of nontrivial nonnegative solutions to singularly perturbed semilinear Dirichlet problems. Proc. R. Soc. Edinb., Sect. A **133**, 1–30 (2003)

110. D. Guo, V. Lakskmikantham, *Nonlinear Problems in Abstract Cones* (Academic Press, San Diego, 1988)

111. D. Guo, J. Sun, *Nonlinear Integral Equations* (Shandong Science and Technology Press, Jinan, 1987) (In Chinese)

112. Z. Guo, J.R.L. Webb, Uniqueness of positive solutions for quasilinear equations when a parameter is large. Proc. R. Soc. Edinb., Sect. A **124**, 189–198 (1994)

113. Z. Guo, J.R.L. Webb, Large and small solutions of a class of quasilinear elliptic eigenvalue problems. J. Differ. Equ. **180**, 1–50 (2002)

114. Z. Guo, Z.D. Yang, Some uniqueness results for a class of quasilinear elliptic eigenvalue problems. Acta Math. Sin. New Ser. **14**(2), 245–260 (1998)

115. Z. Guo, Z. Zhang, $W^{1,p}$ versus C^1 local minimizers and multiplicity results for quasilinear elliptic equations. J. Math. Anal. Appl. **286**(1), 32–50 (2003)

116. Q. Han, Nodal sets of harmonic functions. Pure Appl. Math. Q. **3**(3), 647–688 (2007)

117. Q. Han, F. Lin, *Elliptic Partial Differential Equations, Lecture Notes* (Am. Math. Soc., Providence, 2000)

118. N. Hirano, Multiple solutions for quasilinear elliptic equations. Nonlinear Anal. **7**, 625–638 (1990)

119. M.G. Huidobro, R. Manasevich, J. Serrin, M. Tang, C.S. Yarur, Ground states and free boundary value problems for the n-Laplacian in n dimensional space. J. Funct. Anal. **172**, 177–201 (2000)

120. B. Kawohl, *Rearrangements and Convexity of Level Sets in PDE*. Lecture Notes in Mathematics, vol. 1150 (Springer, Berlin, 1985)

121. G. Kirchhoff, *Mechanik* (Teubner, Leipzig, 1883)

122. M.A. Krasnosel'skii, *Positive Solutions of Operator Equations* (Noordhoft, Groningen, 1964)

123. M.K. Kwong, Uniqueness of positive solutions of $\Delta u - u + u^p = 0$ in $\mathbb{R}^n$. Arch. Ration. Mech. Anal. **105**, 243–266 (1989)

124. O.A. Ladyzhenskaya, N.N. Ural'tseva, *Linear and Quasilinear Elliptic Equations* (Academic Press, New York, 1968)

125. S. Li, J. Liu, Morse theory and asymptotically linear Hamiltonian systems. J. Differ. Equ. **78**, 53–73 (1989)

126. S. Li, M. Willem, Applications of local linking to critical point theory. J. Math. Anal. Appl. **189**(1), 6–32 (1995)

127. S. Li, Z. Zhang, Fucik spectrum and sign-changing and multiple solutions theorems for semilinear elliptic boundary value problems with jumping nonlinearities at zero and infinity. Sci. China **44**(7), 856–866 (2001)

128. G. Li, H. Zhou, Asymptotically linear Dirichlet problem for the p-Laplacian. Nonlinear Anal., Theory Methods Appl. **43**, 1043–1055 (2001)

129. K. Li, J. Liang, T. Xiao, New existence and uniqueness theorems of positive fixed points for mixed monotone operators with perturbation. J. Math. Anal. Appl. **328**, 753–766 (2007)

130. C. Li, S. Li, Z. Liu, J. Pan, On the Fučík spectrum. J. Differ. Equ. **244**, 2498–2528 (2008)

131. E.H. Lieb, M. Loss, *Analysis*. Graduate Studies in Mathematics, vol. 14 (Am. Math. Soc., Providence, 1997)

132. G.M. Lieberman, Boundary regularity for solutions of degenerate elliptic equations. Nonlinear Anal. **12**, 1203–1219 (1988)

133. P. Lindqvist, On the equation $\Delta_p u + \mathrm{div}(|\nabla u|^{p-2}\nabla u) = 0$. Proc. Am. Math. Soc. **109**, 157–164 (1990). Addendum in Proc. Am. Math. Soc. **116**, 583–584 (1992)

134. J.-L. Lions, On some questions in boundary value problems of mathematical physics, in *Contemporary Developments in Continuum Mechanics and Partial Differential Equations, Proc. Internat. Sympos.*, Inst. Mat., Univ. Fed. Rio de Janeiro, Rio de Janeiro, 1977. North-Holland Math. Stud., vol. 30 (North-Holland, Amsterdam, 1978), pp. 284–346

135. P.L. Lions, The concentration-compactness principle in the Calculus of Variations. The locally compact case, part 1. Ann. Inst. Henri Poincaré, Anal. Non Linéaire **1**(2), 109–145 (1984)

136. P.L. Lions, The concentration-compactness principle in the Calculus of Variations. The locally compact case, part 2. Ann. Inst. Henri Poincaré, Anal. Non Linéaire **1**(4), 223–283 (1984)

137. J. Liu, The Morse index of a saddle point. Syst. Sci. Math. Sci. **2**, 32–39 (1989)

138. Z. Liu, A topological property of the boundary of bounded open sets in $\mathbb{R}^2$. J. Shandong Univ. **29**(3), 299–303 (1994)

139. J. Liu, S. Li, An existence theorem for multiple critical points and its application. Kexue Tongbao **29**(17), 1025–1027 (1984) (In Chinese)

140. X. Liu, J. Sun, Computation of topological degree and applications to superlinear system of equations. J. Syst. Sci. Math. Sci. **16**(1), 51–59 (1996) (In Chinese)

141. Z. Liu, Z.-Q. Wang, Multiple bound states of nonlinear Schrödinger systems. Commun. Math. Phys. **282**(3), 721–731 (2008)

142. Z. Liu, F.A. van Heerden, Z.-Q. Wang, Nodal type bound states of Schröedinger equations via invariant set and minimax methods. J. Differ. Equ. **214**, 358–390 (2005)

143. Z. Liu, Z.-Q. Wang, T. Weth, Multiple solutions of nonlinear Schrödinger equations via flow invariance and Morse theory. Proc. R. Soc. Edinb., Sect. A **136**, 945–969 (2006)

144. T.F. Ma, J.E. Muñoz Rivera, Positive solutions for a nonlinear nonlocal elliptic transmission problem. Appl. Math. Lett. **16**(2), 243–248 (2003)

145. A. Mao, Z. Zhang, Sign-changing and multiple solutions of Kirchhoff type problems without the P.S. condition. Nonlinear Anal. **70**(3), 1275–1287 (2009)

146. Z. Nehari, Characteristic values associated with a class of nonlinear second order differential equations. Acta Math. **105**, 141–175 (1961)

147. B. Noris, H. Tavares, S. Terracini, G. Verzini, Uniform Hölder bounds for nonlinear Schrödinger systems with strong competition. Commun. Pure Appl. Math. **63**(3), 267–302 (2010)
148. B. Noris, H. Tavares, S. Terracini, G. Verzini, Convergence of minimax and continuation of critical points for singularly perturbed systems. J. Eur. Math Soc. **14**(3), 1245–1273 (2012)
149. R.S. Palais, Homotopy theory of infinite dimensional manifolds. Topology **5**, 1–16 (1966)
150. R.S. Palais, The principle of symmetric criticality. Commun. Math. Phys. **69**, 19–30 (1979)
151. K. Perera, Critical groups of critical points produced by local linking with applications. Abstr. Appl. Anal. **3**(3–4), 437–446 (1998)
152. K. Perera, M. Schechter, Double resonance problems with respect to the Fucik spectrum. Indiana Univ. Math. J. **52**(1), 1–17 (2003)
153. K. Perera, M. Schechter, Computation of critical groups in Fucik resonance problems. J. Math. Anal. Appl. **279**(1), 317–325 (2003)
154. K. Perera, Z. Zhang, Nontrivial solutions of Kirchhoff-type problems via the Yang index. J. Differ. Equ. **221**(1), 246–255 (2006)
155. S.I. Pohožaev, A certain class of quasilinear hyperbolic equations. Mat. Sb. **96**(138), 152–166 (1975)
156. P.H. Rabinowitz, Some global results for nonlinear eigenvalue problems. J. Funct. Anal. **7**, 487–513 (1971)
157. P.H. Rabinowitz, On bifurcation from infinity. J. Differ. Equ. **14**(4), 462–475 (1973)
158. P.H. Rabinowitz, A bifurcation theorem for potential operators. J. Funct. Anal. **25**(4), 412–424 (1977)
159. P.H. Rabinowitz, *Minimax Methods in Critical Point Theory with Applications to Differential Equations*. CBMS Regional Conference Series in Mathematics, vol. 65 (Am. Math. Soc., Providence, 1986). Published for the Conference Board of the Mathematical Sciences, Washington, DC
160. M. Schechter, The Fucik spectrum. Indiana Univ. Math. J. **43**(4), 1139–1157 (1994)
161. J. Serrin, H. Zou, Existence of positive solutions of the Lane–Emden system. Atti Semin. Mat. Fis. Univ. Modena **46**(suppl.), 369–380 (1998)
162. J. Shi, Exact multiplicity of solutions to superlinear and sublinear problems. Nonlinear Anal. **50**(5), 665–687 (2002)
163. J. Shi, J. Wang, Morse indices and exact multiplicity of solutions to semilinear elliptic problems. Proc. Am. Math. Soc. **127**(12), 3685–3695 (1999)
164. S. Smale, Morse theory and a nonlinear generalization of the Dirichlet problem. Ann. Math. **80**, 382–396 (1964)
165. S. Smale, An infinite dimensional version of Sard's Theorem. Am. J. Math. **87**(4), 861–866 (1965)
166. S. Smale, On the differential equations of species in competition. J. Math. Biol. **3**, 5–7 (1976)
167. M.A.S. Souto, A priori estimates and existence of positive solutions of nonlinear cooperative elliptic systems. Differ. Integral Equ. **8**, 1245–1258 (1995)
168. M. Struwe, *Variational Methods*, 3rd edn. (Springer, Berlin, 2000)
169. J. Sun, Non-zero solutions of superlinear Hammerstein integral equations and its applications. Chin. Ann. Math., Ser. A **7**(5), 528–535 (1986) (In Chinese)
170. J. Sun, Fixed point theorems for noncontinuous increasing operators and applications to nonlinear equations with discontinuous nonlinearities. Acta Math. Sin. **31**(1), 101–107 (1988) (In Chinese)
171. J. Sun, Fixed-point theorems for increasing operators and applications to nonlinear equations with discontinuous terms in Banach spaces. Acta Math. Sin. **34**(5), 665–674 (1991) (In Chinese)
172. J. Sun, Z. Liu, Calculus of variations and super- and sub-solution in reverse order. Acta Math. Sin., **37**(4), 512–514 (1994) (In Chinese)
173. J. Sun, Z. Zhao, Fixed point theorems of increasing operators and applications to nonlinear integro-differential equations with discontinuous terms. J. Math. Anal. Appl. **175**(1), 33–45 (1993)

174. A. Szulkin, Ljusternik–Schnirelman theory on C^1-manifolds. Ann. Inst. Henri Poincaré, Anal. Non Linéaire **5**, 119–139 (1988)

175. S. Takeuchi, Positive solutions of a degenerate elliptic equations with logistic reaction. Proc. Am. Math. Soc. **129**, 433–441 (2000)

176. Z. Tan, Z. Yao, The existence of multiple solutions of p-Laplacian elliptic equation. Acta Math. Sci. Ser. B **21**, 203–212 (2001)

177. H. Tavares, S. Terracini, Regularity of the nodal set of the segregated critical configurations under a weak reflection law. Calc. Var. Partial Differ. Equ. (2012). doi:10.1007/s00526-011-0458-z

178. S. Terracini, G. Verzini, Multipulse phases in k-mixtures of Bose–Einstein condensates. Arch. Ration. Mech. Anal. **194**(3), 717–741 (2009)

179. G. Tian, *Canonical Metrics in Kahler Geometry*. Lectures in Mathematics ETH Zurich (Birkhäuser, Basel, 2000)

180. P. Tolksdorf, On the Dirichlet problem for quasilinear equations in domains with conical boundary points. Commun. Partial Differ. Equ. **8**, 773–817 (1983)

181. P. Tolksdorf, Regularity for a more general class of quasilinear elliptic equations. J. Differ. Equ. **51**, 126–150 (1984)

182. K. Tso, On a real Monge–Ampère functional. Invent. Math. **101**, 425–448 (1990)

183. C.F. Vasconcellos, On a nonlinear stationary problem in unbounded domains. Rev. Mat. Univ. Complut. Madr. **5**(2–3), 309–318 (1992)

184. J.L. Vazquez, A strong maximum principle for some quasilinear elliptic equations. Appl. Math. Optim. **12**, 191–202 (1984)

185. C. Viterbo, Metric and isoperimetric problems in symplectic geometry. J. Am. Math. Soc. **13**(2), 411–431 (2000)

186. Z.Q. Wang, On a superlinear elliptic equation. Ann. Inst. Henri Poincaré, Anal. Non Linéaire **8**, 43–57 (1991)

187. X. Wang, A class of fully nonlinear elliptic equations and related functionals. Indiana Univ. Math. J. **43**(1), 25–54 (1994)

188. W. Wang, On a kind of eigenvalue problems of Monge–Ampère type. Chin. Ann. Math., Ser. A **28**(3), 347–358 (2007)

189. K. Wang, Z. Zhang, Some new results in competing systems with many species. Ann. Inst. Henri Poincaré, Anal. Non Linéaire **27**, 739–761 (2010)

190. S. Wang, M. Wu, Z. Jia, *Matrix Inequalities*, 2nd edn. (Science Press, Beijing, 2006) (In Chinese)

191. J.C. Wei, T. Weth, Radial solutions and phase separation in a system of two coupled Schrodinger equations. Arch. Ration. Mech. Anal. **190**, 83–106 (2008)

192. G.T. Whyburn, *Topological Analysis* (Princeton University Press, Princeton, 1958)

193. M. Willem, *Minimax Theorems*. Progress in Nonlinear Differential Equations and Their Applications, vol. 24 (Birkhäuser Boston, Boston, 1996)

194. C.-T. Yang, On theorems of Borsuk–Ulam, Kakutani–Yamabe–Yujobô and Dyson. II. Ann. Math. **62**(2), 271–283 (1955)

195. Z.D. Yang, Z.M. Guo, On the structure of positive solutions for quasilinear ordinary differential equations. Appl. Anal. **58**, 31–51 (1995)

196. Z. Zhang, New fixed point theorems of mixed monotone operators and applications. J. Math. Anal. Appl. **204**(1), 307–319 (1996)

197. Z. Zhang, Fixed point theorems of mixed monotone operators and its applications. Acta Math. Sin. **41**(6), 1121–1126 (1998) (In Chinese)

198. Z. Zhang, Fixed-point theorems for a class of noncompact decreasing operators and their applications. J. Syst. Sci. Math. Sci. **18**(4), 422–426 (1998) (In Chinese)

199. Z. Zhang, Some new results about abstract cones and operators. Nonlinear Anal., Theory Methods Appl. **37**, 449–455 (1999)

200. Z. Zhang, Existence of non-trivial solutions for superlinear systems of integral equations and its applications. Acta Math. Appl. Sin. **15**(2), 153–162 (1999)

201. Z. Zhang, On bifurcation, critical groups and exact multiplicity of solutions to semilinear elliptic boundary value problems. Nonlinear Anal. **58**(5–6), 535–546 (2004)
202. Z. Zhang, New results of nonlinear operator and applications, Ph.D. Thesis, Shandong University, China, 1997 (In Chinese)
203. Z. Zhang, X. Cheng, Existence of positive solutions for a semilinear elliptic system. Topol. Methods Nonlinear Anal. **37**, 103–116 (2011)
204. Z. Zhang, X. Li, Sign-changing solutions and multiple solutions theorems for semilinear elliptic boundary value problems with a reaction term nonzero at zero. J. Differ. Equ. **178**(2), 298–313 (2002)
205. Z. Zhang, S. Li, On sign-changing and multiple solutions of the p-Laplacian. J. Funct. Anal. **197**(2), 447–468 (2003)
206. Z. Zhang, K. Perera, Sign changing solutions of Kirchhoff type problems via invariant sets of descent flow. J. Math. Anal. Appl. **317**(2), 456–463 (2006)
207. Z. Zhang, K. Wang, Existence and non-existence of solutions for a class of Monge–Ampre equations. J. Differ. Equ. **246**(7), 2849–2875 (2009)
208. Z. Zhang, S. Li, S. Liu, W. Feng, On an asymptotically linear elliptic Dirichlet problem. Abstr. Appl. Anal. **7**(10), 509–516 (2002)
209. Z. Zhang, J. Chen, S. Li, Construction of pseudo-gradient vector field and sign-changing multiple solutions involving p-Laplacian. J. Differ. Equ. **201**, 287–303 (2004)
210. H.S. Zhou, Existence of asymptotically linear Dirichlet problem. Nonlinear Anal., Theory Methods Appl. **44**, 909–918 (2001)
211. W. Ziemer, *Weakly Differentiable Functions* (Springer, Berlin, 1989)
212. H. Zou, A priori estimates for a semilinear elliptic system without variational structure and their applications. Math. Ann. **323**, 713–735 (2002)
213. W. Zou, *Sign-Changing Critical Point Theory* (Springer, Berlin, 2008)

Index

Z. Zhang, *Variational, Topological, and Partial Order Methods with Their Applications*,
Developments in Mathematics 29,
DOI 10.1007/978-3-642-30709-6, © Springer-Verlag Berlin Heidelberg 2013